METAL FORMING
and
IMPACT MECHANICS

Pergamon Titles of Related Interest

ASHBY & BROWN	Perspectives in Creep Fracture
ASHBY & JONES	Engineering Materials
CARLSSON & OHLSON	Mechanical Behaviour of Materials (ICM 4)
COMINS & CLARK	Specialty Steels and Hard Materials
FROST & ASHBY	Deformation-Mechanism Maps
GIFKINS	Strength of Metals and Alloys (ICSMA 6)
HEARN	Mechanics of Materials, 2nd Edition
HOPKINS & SEWELL	Mechanics of Solids
HULL & BACON	Introduction to Dislocations, 3rd Edition
JOHNSON et al	Plane Strain Slip Line Fields for Metal Deformation Processes
McQUEEN et al	Strength of Metals and Alloys (ICSMA 7)
MILLER & SMITH	Mechanical Behaviour of Materials (ICM 3)
NIKU-LARI	Advances in Surface Treatments, Volumes 1 & 2
OSGOOD	Fatigue Design
SIH & FRANCOIS	Progress in Fracture Mechanics
VALLURI	Fracture 1984 (ICF 6)

Pergamon Related Journals (Specimen Copy Gladly Sent on Request)

Acta Metallurgica
Corrosion Science
Engineering Fracture Mechanics
Fatigue of Engineering Materials and Structures
International Journal of Impact Engineering
International Journal of Machine Tool Design and Research
International Journal of Mechanical Sciences
International Journal of Solids and Structures
Journal of the Mechanics and Physics of Solids
Materials Research Bulletin
Scripta Metallurgica

WILLIAM JOHNSON, FRS

METAL FORMING
and
IMPACT MECHANICS

WILLIAM JOHNSON COMMEMORATIVE VOLUME

Edited by

S. R. REID

University of Manchester Institute of Science and Technology, Manchester, UK

PERGAMON PRESS

OXFORD · NEW YORK · TORONTO · SYDNEY · FRANKFURT

U.K.	Pergamon Press Ltd., Headington Hill Hall, Oxford OX3 0BW, England
U.S.A.	Pergamon Press Inc., Maxwell House, Fairview Park, Elmsford, New York 10523, U.S.A.
CANADA	Pergamon Press Canada Ltd., Suite 104, 150 Consumers Road, Willowdale, Ontario M2J 1P9, Canada
AUSTRALIA	Pergamon Press (Aust.) Pty. Ltd., P.O. Box 544, Potts Point, N.S.W. 2011, Australia
FEDERAL REPUBLIC OF GERMANY	Pergamon Press GmbH, Hammerweg 6, D-6242 Kronberg-Taunus, Federal Republic of Germany

First edition 1985

British Library Cataloguing in Publication Data
Metal forming and impact mechanics : William
Johnson commemorative volume.
1. Metal-work
I. Reid, S. R.
671.3   TS205
ISBN 0–08–031679–4

Library of Congress Cataloging in Publication Data
Metal forming and impact mechanics
3. Metal-work.   2. Deformations (Mechanics)
I. Impact   I. Johnson, W. (William), 1922–
II. Reid, S. R.
TS213.M385   1985   671.3   85–12292
ISBN 0-08-031679-4

Printed in Great Britain by A. Wheaton & Co. Ltd., Exeter

EDITORIAL

This book has been compiled to honour Professor William Johnson. It grew out of a wish by a number of his friends, colleagues and former students to mark several significant events which have occured over the past few years. In 1982 Bill Johnson celebrated his sixtieth birthday, he was elected a Fellow of the Royal Society and he retired from the Chair of Mechanics in the Department of Engineering at Cambridge University. In 1983 he was elected to the Fellowship of Engineering and in 1985 he completes 25 years as Editor-in-Chief of the International Journal of Mechanical Sciences. In his Biographical Notes, Peter Mellow describes Bill's education and development as an engineer and as a scholar and indicates some of the many areas of research and education to which he has contributed.

Bill Johnson's research output has been phenomenal as evidenced by his list of publications which are included in this volume. What is also apparent from a careful study of this list are the names of a vast number of students and academic colleagues with whom he has interacted and sparked into productive life in research. Many of these have continued to be active and owe their start in research to Bill's guidance, encouragement and drive. Equally evident in this large array of publications is the wide range of his interests including many areas of fundamental and technological metal forming including the use of explosive, electromagnetic and direct impact forces; plastic collapse of structures; terminal ballistics; explosive fragmentation and cutting; dynamic structural plasticity; explosive welding; creep; vehicle crashworthiness; elastic impact mechanics; impact energy absorption; powder compaction and various aspects of biomedical engineering. Many of the publications which have resulted from his activities in these areas are characterised by a clarity of presentation and an exposure of the essence of the problem under discussion which have opened up whole fields of research into which others have been drawn.

No compendium of papers of the length of this volume could reflect all of Professor Johnson's interests. However, the eighteen technical chapters, split equally between the broad headings of metal forming and impact, give a flavour of some of the areas in

which Bill has been and continues to be active. The aim has been
to collect a set of papers which reflect current work in these
major areas of research and as such the volume should be of
interest to all those active in the field of solid mechanics. The
papers include both theoretical and experimental contributions and
as such reflect the essential interplay between these two
approaches in these areas of solid mechanics which are dominated
by non-linear effects.

Bill Johnson has an international reputation and this is reflected
in the countries of origin of the various authors who have
contributed to this volume. As Editor, I wish to thank all the
contributors for their assistance and patience and would like to
express particular thanks to Peter Mellor and Norman Jones for
their assistance in formulating the contents of the book and
during the preparation of the manuscripts for publication. Thanks
are also due to Pergamon Press for their enthusiasm in agreeing to
provide this very appropriate way of paying tribute to Bill
Johnson's contribution to research in solid mechanics.

This group of contributors is but a token of the large number of
friends, colleagues, former students and professional
acquaintances who would, I am sure, have been happy to make
similar contributions if space had allowed them to do so. On
their behalf, on behalf of the contributing authors and on my own
behalf, I hope that this volume will stimulate further work in
these important areas of mechanics. If it does then it will
provide a fitting tribute to Professor Johnson.

S. R. Reid
Manchester, 1985

CONTENTS

IMPACT MECHANICS

LIST OF CONTRIBUTORS

Dr W Abramowicz, Institute of Fundamental Technological Research, Warsaw, Poland

Dr J M Alexander, Department of Mechanical Engineering, University of Surrey, Guildford GU2 5XH, UK

Dr S K Ghosh, Department of Engineering, University of Aberdeen, Aberdeen AB9 1AS, UK

Professor W Goldsmith, Department of Mechanical Engineering, University of California, Berkeley, CA 94720, USA

Professor R Hill, Department of Applied Mathematics and Theoretical Physics, Silver Street, Cambridge CB3 9EW, UK

Professor N Jones, Department of Mechanical Engineering, University of Liverpool, Liverpool, UK

Professor S Kobayashi, Department of Mechanical Engineering, University of California, Berkeley, CA 94720, USA

Dr J A König, Institute of Fundamental Technological Research, Warsaw, Poland

Professor H Kudo, Department of Mechanical Engineering, Yokohama National University, Japan

Dr K Leers, Institut für Mechanik, Universität Hannover, FRG

Dr O Mahrenholtz, Arbeitsbereich Meerestechnik II/Strukturmechanik, TU Hamburg-Harburg, FRG

Dr Z Marciniak, Technical University of Warsaw, Poland

Mr J Pearson, Michelson Laboratories, Naval Weapons Center, China Lake, California, USA

Professor S R Reid, Department of Mechanical Engineering, UMIST, P O Box 88, Sackville Street, Manchester M60 1QD, UK

Contributors

Professor G W Rowe, Department of Mechanical Engineering,
University of Birmingham, Birmingham B15 2TT, UK

Dr C Q Ru, Department of Mechanics, Peking University, Beijing,
Peoples Republic of China

Professor R Sowerby, Department of Mechanical Engineering,
McMaster University, Hamilton, Ontario L8S 4L7, Canada

Dr W J Stronge, Department of Engineering, University of
Cambridge, Trumpington Street, Cambridge CB2 1PZ, UK

Professor P S Symonds, Division of Engineering, Brown University,
Providence, RI 02912, USA

Professor P S Theocaris, Section of Mechanics, Department of
Engineering Science, National Technical University of Athens,
5 Heroes of Polytechnion Avenue, Zographou, GR-157 73 Athens,
Greece

Dr F W Travis, Department of Mechanical Engineering, Sunderland
Polytechnic, UK

Professor R Wang, Department of Mechanics, Peking University,
Beijing, Peoples Republic of China

Dr Yong Taek Im, Department of Mechanical Engineering, University
of California, Berkeley, CA 94720, USA

WILLIAM JOHNSON

Biographical Note

Bill Johnson was born on 20 April 1922 and spent his boyhood in
Lower Openshaw, Manchester; a time he has recalled vividly in his
Presidential Address to the Manchester Technology Association. The
family home was opposite an engraving works and an ecclesiastical
mission, the latter sent from a church in Higher Openshaw to save
the working class from sin and drink. Next to "The Mission" was a
scrap iron yard and an Italian street organ manufacturer. This
latter must surely have been a dying industry at that time and, of
course, the Openshaw of the 1930's has long since been swept away.
Bill was a constant visitor to the wire and bar drawing plant of
Richard Johnson and Nephews, where his father spent 43 years of his
life, first as a labourer and later as a foreman. Also, with other
youngsters, he often watched steam hammers at work at Vaughan's and
it is tempting to think that these early impressions constantly drew
him back in later life to Manchester and to metal forming! Although
poverty and hunger among his childhood friends were not unknown to
him he remembers a happy childhood of playing endless football with
many evenings spent in library reading rooms, although he did not
know the meaning of set homework until, at the age of 12, he entered
the Central Grammar School.

Bill's progress at grammar school was conventional, while the world
outside erupted. His twin adolescent passions were mathematics and
politics, the latter much debated with his very intelligent father.
Trade union disputes, local strikes and labour party politics were
everyday affairs. This period was also full of occasions such as
Oswald Moseley's provocative weekend meetings at Belle Vue or in
Piccadilly (Manchester) and meetings at the "Co-op" to hear J. B. S.
Haldane talk about "Aid to Spain". Some sixth formers collectively
joined Victor Gollancz's "Left Book Club" - and read some of its
output. On the 3 September 1939 the School was evacuated to Black-
pool but returned to Manchester in February 1940 to await the end of
"The Phoney War".

In October 1940 Bill crossed the street from the School to enrol as
a student in mechanical engineering under Professor Wright Baker at
"Manchester Tech" where he was one of five first year honours students. The "Manchester Blitz" followed in December 1940 and he spent
much time fire-watching on the roof of the main building of the Tech.

or taking part in discussions in the basement shelters during air
raids. During this time he maintained his interest in politics and
his independence of mind - he was expelled from the Socialist Society
at the Tech. largely for refusing to be Marxist and for not joining
the Young Communist League. These experiences later made him sym-
pathetic to students who wished to stretch and flap their political
wings.

In 1942 C. P. Snow, in his role of Director of Personnel at the
Ministry of Labour, descended on Manchester University (and other
universities) to lecture science and engineering students about
their wartime destinies - as Light Aid Detachment commanders operating in the midst of, and after, tank battles. In the event few
were called and Bill seems to have been the only one from Manchester
destined for the Royal Electrical and Mechanical Engineers that year.
After training (primary, pre-OCTU, OCTU, Command Workshop Service in
Bury St Edmunds and a special Tank Repair and Recovery Course) he
was shipped out as a junior officer to Italy and deposited in an
industrial complex around Naples. As a member of a Control Purchas-
ing and Production Unit he had among other things to progress con-
tracts, including one that was carried out in the local gaol and
managed by a prisoner serving ten years for embezzlement. He later
worked in the same way in Rome and Milan and in Austria after the
War finished. There he admired the efficiency of his German workers
compared with his British or Italian ones but, on remarking to his
Colonel that after Italy, Austria was clean, ordered and hygienic
but very dull, he was returned to Italy and spent nine months in
charge of a Tank Transportation Platoon. In effect he became foreman of a group of 100 men who were really long distance heavy goods
drivers! He ended his army career as a Workshop Officer at Padua,
just 34 kilometres from Venice. Despite the death and destruction
around him his time in Italy turned out to be the equivalent of an
18th Century Grand Tour, where he was able to familiarise himself
with Opera in Rome, Naples, Venice and Milan and the architecture
of Assissi, Orvieto, Siena, Bologna and Florence. Cigarettes
throughout were the currency which opened museum and gallery doors.

On returning to England in 1947, and now married, he became an
Assistant Principal in the Administrative grade of the Home Civil
Service and toiled at Treasury Exchange Control Regulations and the
drafting of Purchase Tax Schedules. Younger in years than many of
his colleagues, but with more experience of life, he countered his
tedious work by becoming a part-time postgraduate student in the
History and Philosophy of Science at University College, London
after taking en route an External Degree in Mathematics.
Fascinated by the subject matter of the course, and no doubt
unsettled by the dichotomy of his life, he decided to return full-
time to the study of engineering science. His first move was to a
lectureship at Northampton Polytechnic (now City University) in
1950 - where the 20 hours of teaching per week left little time for
original work - and after 18 months to Sheffield University as a
lecturer, where he found his metier in the Department of Mechanical
Engineering under Professor H. W. Swift.

Swift at that time had two main lines of research, metal forming
and the lubrication of bearings. The metal forming research in the
department was particularly buoyant because of the presence of the
British Iron and Steel Research Association in Sheffield. Hugh Ford
was Head of the Mechanical Working Division of B I S R A from 1945-
47 and Rodney Hill the head of a new section in the Metal Flow Research Laboratory from 1948-50. Hill, and later his colleague, A.
P. Green, maintained close contact with the metal forming research

in Swift's department. In 1951 J. B. Hawkyard and myself were
appointed research students under Swift and paid from B I S R A
funds. My first contact with Bill Johnson occurred when I in-
advertently trod on the blueprint of a slip-line field which his
over-enthusiastic final year project student had extended until it
covered the floor of the laboratory. He must have remembered that
event when a few years later he "recruited" D. M. Woo and myself to
work on multiple hole extrusion using that slip-line field.

At Sheffield Bill quickly immersed himself in the application of
slip-line fields and in the experimental determination of yield loci
of metals. He also retained an interest in the history of science
as is seen by his presidential address to the local Engineering
Society in 1954 on "Mechanics before Newton" and a paper entitled,
"A Villainous Scientist", which discusses the life and times of 16th
Century Girolamo Cardan. A more onerous task, the first of many,
was the Secretaryship of the Yorkshire Branch of the Institution of
Mechanical Engineers which he held from 1953 to 1956.

In 1956 he moved to Professor Jack Diamond's department at Manchester
University as a Senior Lecturer with the thought of becoming involved
in nuclear engineering. In fact he became entranced with upper bound
solutions in plastic forming and in particular with the temperature
generated when metal is fast formed. For simple forging the theory
suggested the existence of visible forging crosses, generated by
block shearing processes. It was also possible to estimate the
magnitude of the temperature jumps inside the forged block. This
theoretical work was carried out with R. I. Tanner and published
in 1960. Some years later he visited Massey's forge in Openshaw
and was given a copy of H. F. Massey's lecture on "The flow of metal
during forging" which was printed in the Transactions of the
Manchester Association of Engineers, 1921-22. This is a remarkable
paper which describes in detail the use of wax as a model material
for studying the flow of steel during forging and there are charming
references to the work of Monsieur Tresca. But most exciting of all
Massey reported the appearance of "Heat Lines" when steel is forged
at a low temperature, say about 680°C. Massey tried but failed to
record the phenomenon on photograph and he had to be satisfied in
his paper with sketching the appearance of the lines. Bill returned
to Massey's a few days later with R. A. C. Slater and after much
trial and error they obtained successful colour photographs of the
"heat lines". The temperatures were also recorded using an optical
pyrometer and compared with the calculated temperature jumps. The
phenomenon has since been incorporated into a teaching film in
Sweden. Subsequently Frank Travis investigated the very high speed
blanking of plate and encountered similarly surfaces of highly
localised temperature generation, now called adiabatic "heat lines".
The subject still draws much attention having been found in processes
from bullet penetration to the machining of some particular metals
and enjoying the current title of Catastrophic Shearing.

In 1958 Bill bought a second hand copy of James Nasmyth's auto-
biography and was fascinated by a description of the invention of
the vee-anvil for forging crank shafts of steam ships (carried out
in the Manchester area). Nasmyth stated that the forgings were
sounder near the centre than when forged on flat anvils. Bill
applied slip-line field theory to these two situations and the
solutions which emerged coincided with the ideas of Nasmyth's
intuition. Later he received a photograph of an experiment by
Professor Tomlenov in Moscow which validated the result. In fact
the cracking zone was moved away from the shaft centre to a circular
zone some distance from the centre.

In 1960 Professor Wright Baker retired and Bill was appointed to the
vacant Chair, a post he was to hold for the next 15 eventful years.
The Manchester College of Science and Technology became the University
of Manchester Institute of Science and Technology which, inevitably
and thankfully, was shortened to UMIST. This was the age of expansion and, deciding on the direction the department would take, in
rapid succession Franz Koenigsburger was appointed to the Chair in
Machine Tool Engineering (then unique in the UK) and Rowland Benson
to a second Chair of Mechanical Engineering with special responsi-
bility for Thermodynamics and Fluid Mechanics. The department's
laboratories and activities at that time were in 11 separate
locations and it was not until 1974 that the locations were reduced
to three. Bill was responsible for reorganising the administration
of the Department into three Divisions and he acted as Chairman from
1960-69 and again from 1971-73. Under his guidance the undergraduate
course was modified and a wide range of options introduced into the
final year. New methods of assessing students were also introduced.
On the postgraduate side, he, with his colleagues, introduced new
Master's courses as well as expanding the research activities of
the Department into one of the largest postgraduate schools in the
country. Ron Kitching looked after day-to-day matters for Applied
Mechanics and John Parker for the whole of the undergraduate school.

January 1960 heralded the first issue of the International Journal
of Mechanical Sciences published by Pergamon Press, with W. Johnson
as its founder and Editor-in-Chief. For the first five years there
were two regional editors, W. Ruppert in Germany and H. Kudo in Japan.
Professor D. G. Christopherson wrote the following Foreword.

> "In introducing a new Journal in the field of Engineering
> Science, it may be useful to outline the purpose for which
> it has been established. For some years, research workers
> in the field of applied mechanics have felt that, although
> there are now several national journals of the highest
> academic quality available to them, there is no inter-
> national journal appearing regularly. Moreover, in some
> countries there are no adequate arrangements for the publi-
> cation of specialised experimental and theoretical studies
> in an Engineering field which might not be of sufficiently
> general interest to attract the attention of the pro-
> fessional Engineering Institutions. Accordingly, it is
> hoped that the new journal will fill a real need.
>
> The intention is to publish work of the highest quality in
> any field of science relevant to Mechanical Engineering;
> it is to be expected that in the first instance the majority
> of the papers will deal with questions of thermodynamics,
> fluid mechanics, dynamics, mechanics of materials, plas-
> ticity and elasticity, but other subjects will no doubt be
> added as the work proceeds. Papers may be accepted by
> authors from any country but the intention is that all will
> be published in English."

The journal has been highly successful and has certainly filled a
real need. 1983 saw the completion of the 25th volume each of about
800 pages per year. W. Johnson is still the Editor-in-Chief but
some of the burden was lifted from him when S. R. Reid became first
Associate Editor in 1978 and Executive Editor in 1983. Over the
years Mrs Heather Johnson has been very busy in the background
sorting and organising the typescripts and referees' reports and
generally giving support. An editor learns a lot about human nature
from reading referees' reports and he must have considerable under-

standing and forebearance to remain in the job for long. It is
interesting to note Bill's own comments on referees' reports,

> "Of course the spectrum or quality of assessments is enor-
> mous: at one extreme there is the referee who 'passes'
> everything without comment or criticism; at the other,
> there is the referee who will 'pass' nothing and offers
> minimum reasons for so doing. Fortunately - in my experi-
> ence - one in every two referees will do the task well;
> he will see the good, new features of a paper and con-
> structively criticise to improve a publication."

The Bulletin of Mechanical Engineering Education was founded in
1952 at Manchester College of Science and Technology. K. L.
Johnson, S. S. Gill, J. H. Lamble and J. Parker were early editors.
It was principally a forum for the discussion of new laboratory
experiments in mechanical engineering. Although it is widely read
it has had difficulty in making a profit. With W. Johnson as
"godfather" it has been published successively by Pergamon (1962-71),
the Institution of Mechanical Engineers (1973-81) and presently by
Ellis Horwood (1981-). Under the title of International Journal of
Mechanical Engineering Education it now sets out to promote better
technical teaching in all its aspects. The Editors, J. Parker,
C. M. Leech and S. S. Gill, with W. Johnson as Chairman of the
Editorial Advisory Committee, intend that the journal should continue
to be in the forefront of the debate on the education and training of
mechanical engineers.

In 1961 Bill received the degree of DSc from Manchester University
and in 1962 published two books which contained a large part of his
research work. The first was "Plasticity for Mechanical Engineers"
(written with P. B. Mellor, then a lecturer at Liverpool University),
which was translated officially into Japanese in 1964 and unof-
ficially, (as was learnt many years later) into Chinese. In
deference to the market in the USA a new expanded volume published
in 1973 was given the title, "Engineering Plasticity". This was
translated into Russian in 1979 for a print run of 10,000. An un-
official English edition has also been printed in Taiwan. The
Japanese edition sold extremely well in the 1960's and was widely
used in the education of students of engineering and metallurgy.
The book has recently been republished by Ellis Horwood with detailed
corrections.

The second book was "The Mechanics of Metal Extrusion" written with
H. Kudo, who was then with the Government Mechanical Engineering
Laboratory in Tokyo and is now a Professor in the Faculty of Engineering at Yokohama National University. This was the first book to
discuss the mechanics of the process in such detail and it introduced bounding techniques to a wide audience. This book was translated into Russian in 1965 and sold 25 percent more copies than the
English edition but unfortunately for the authors this was before
the reciprocal royalty rights had been agreed!

At this time Bill also fostered the translation and edited the
classical Russian engineering monograph, "Mechanisms for the Generation of Plane Curves" by I. I. Artobolevskii of the Academy of
Sciences of the USSR and "Stress Concentrations around Holes" by
G. N. Savin. The publisher was Pergamon in both cases. Later, in
1965, Bill was instrumental in recommending to Pergamon a new project which resulted in the Journal of Mechanisms edited by Frank
Erskine Crossley.

Once established at UMIST Bill pressed for the setting up of a
History of Science and Technology Department and his suggestion
was taken up by the Principal, Lord Bowden. Donald Cardwell, an
old friend from the History and Philosophy of Science course at
University College, was brought in to lead it and one consequence
was that mechanical engineering students at UMIST have had their
historical and cultural insights greatly expanded. Later Bill,
with Donald Cardwell and Professor Jack Diamond, urged the establishment of the North West Museum of Science and Technology and this
too came into being. Bill, then, has always been interested in
turning out well-rounded engineering graduates but at the same time
very conscious that British graduates have the shortest degree
course in the world. He has endeavoured to alert students to the
fact that engineering is multi-disciplinary and works against a
background of economics and sociology, that engineering problems
are always multi-dimensional and not ones simply of engineering
science.

One further important venture was the setting up of the Medical
Engineering Unit at UMIST in 1973. Bill's interest in this came
partly from the stimulus of medical members of his family (brother
and son) and partly from the design work Dr Storecki was encouraged
to do with the late Sir John Charnley. Professor Colin Adamson had
similar interests of an electrical engineering nature and therefore
they made a joint proposal to Lord Bowden that he champion a form
of cooperation between the Medical Faculty and the Faculty of Technology. This eventually resulted in the Medical Engineering Unit
of which Bill became the first Director from 1973 to 1975. An off-
shoot was the Manchester and Salford Medical Engineering club, which
still runs very successfully and of which he is now an Honorary
Member.

Despite his immersion in administration he nevertheless made time
for research and scholarship. In 1970 Edward Arnold published
"Plane-strain slip-line fields" by W. Johnson, R. Sowerby and J. B.
Haddow, a compendium of all known slip-line field solutions. This
was updated in 1982 by W. Johnson, R. Sowerby and R. D. Venter and
published by Pergamon under the title, "Plane-strain slip-line fields
for metal-deformation processes: a source book and bibliography".

The 1960's saw an intense interest in high rate forming. Bill
initiated a large programme of work on explosive forming of sheet
metal and tubes, magnetohydraulic forming, forming using electric
spark discharge, and water hammer forming. Associated with him in
this work were, among others, J. L. Duncan, R. Sowerby, S. T. S. Al-
Hassani, I. Donaldson and F. W. Travis. With R. A. C. Slater the
linear induction motor was developed as a drive for high speed
testing and forming. One conspicuous success was the expanding of
some very large tubes using high explosive. An "international team"
was lead by Frank Travis and included now Prof. Dr E. Doege from
Germany, Dr K. Kormi from Iran and an Irish technician. The work
resulted in a production process that is still used today. Bill's
interest in impact stresses extended to teaching and his "Impact
Strength of Materials" published in 1972 was written specifically
for engineering students to make them theoretically aware of the
elementary ideas surrounding stress waves in solids and high speed
metal deformation. The interest further developed into the constructive use of materials to absorb energy and resulted in the
text, "Crashworthiness of Vehicles" by W. Johnson and A. G. Mamalis
published by Mechanical Engineering Publications in 1978. Another
volume (2) on Engineering Plasticity, with A. G. Mamalis, was
edited by H. Lippman and derived from lectures given at C.I.S.M. in
Udine in 1979.

During the 1960's Bill's research became more and more concerned
with what the late Professor H. G. Hopkins called "dirty problems" -
those problems which could not be solved by elegant mathematics.
Among such problems must be counted ball-rolling, cold rolling of
ring gears, rotary forging, and peen-forming which were investigated
with N. R. Chitkara, J. B. Hawkyard and S. T. S. Al-Hassani respectively, among others. With such processes it is necessary to examine
the variables experimentally, sometimes using model materials, sometimes using full scale experiments.

His interests in explosive forming led to an invitation from the
British Council to lecture in Turkey and Greece. While in Turkey
he took the opportunity to see the cave dwellings and paintings of
the Greek anchorites who had been evicted by Kemal Ataturk in the
early 1920's. Circulating for two days - to Urgup, Uchisar and
Goreme in Cappadocia - in an ancient "taxi", he encountered two
underground "towns" at Derinkuyu both built about 1000 years ago.
Entry to both was as if into a sewer. One was seven storeys deep,
able to accommodate people and cattle and to sustain then for months
under siege. After lecturing in Izmir or Smyrna he flew to Athens
just in time to witness the overthrow of the ruling junta. However,
three days later he was able to deliver his lecture (with demon-
stration) on "The Constructive Uses of Explosives"! In a later, more
peaceful, visit to Greece in 1978 he was able to satisfy a long held
ambition to visit Mount Athos. This went back to his days in Florence
in 1945 when he had met Ralph Brewster, the author of the well known
early Penguin Book, "6000 Beards of Athos". It appears that Bill
was not yet ready for the full contemplative life in a solely male
society and it was with relief that he returned to the 20th Century.

In 1974 he became Chairman of the North Western Branch of the
Institution of Mechanical Engineers but 1975 brought a major change
when he left UMIST to become Professor of Mechanics at Cambridge
University. From the first he was deeply involved in starting the
new Production Engineering Tripos and acted as its director; it is
a four-year Dainton-style course in Manufacturing Science, with a
fourth year of 38 weeks spent in industry. Now that both Cambridge
and Oxford openly and officially embrace manufacturing at the undergraduate level the whole subject has, at last become academically
respectable.

The late 1970's found him in a more reflective mood; perhaps inspired
by the dreaming spires of Cambridge and perhaps more conscious than
previously of the passing of time. In 1978 he gave the opening
address at the 18th International Machine Tool Design and Research
Conference. In his address he reflected on the vacillation and
indecision of engineering higher education and research policy in
the UK which has been a conspicuous feature of the last quarter of a
century. He commented especially on the five yearly fashions at
SERC, noting that in that time we have been through nuclear engineering, gas turbines, machine tools, tribology and now it is the
turn of manufacturing technology. Surely, he commented, a steady
balance between the various branches of engineering is greatly to be
desired, but apparently hard to come by. He also commented on the
absence of representation of British industrial firms at the MTDR
Conferences and contrasted this with the situation in Germany, USA
and Japan where similar conferences attract large industrial support.

In Guest Editorials in the Journal of Mechanical Working Technology
he has again questioned the direction and quality of research work
today. He has noted the trend towards "one off" conferences, which
will in the future make it difficult to locate the papers discussed

there. Conferences, he believes, should be held regularly, at set
intervals, so that they establish a pattern and a reputation. One
trend he notes in conference papers is that references tend to go
back no more than ten years and that even then most references are
to works of the authors themselves! He fears that we are in danger
of becoming journalists, recording ephemeral events, rather than
researchers knowledgeable of the past contributions of other people
and therefore better placed to have a clearer view of the future.

A major activity of any good teacher is that of putting together a
course of lectures - structuring or setting in order material which
is initially relatively formless or uncoordinated; and in engineering
this must be placed against a background of relevant experience.
This is an arduous, time-consuimg activity, which is often un-
appreciated. Bill has been pre-eminent in this field, as can be
seen by the text books he has written. In his research work he has
sought to understand and explain the variables of complex processes.
Wherever possible he has endeavoured to simplify and clarify the
mechanics of the process. He has also inspired, persuaded or coerced
his friends and colleagues to take part in many such successful
ventures!

The honorary degree of Doctor of Technology was conferred on Bill
by the University of Bradford in 1976. When awarding honorary
degrees to people of distinction the University also likes to
recognise some association of the person with either the City or
the University, and in this case the University particularly welcomed
Bill for the support he had given to the University over many years
especially to the Department of Mechanical Engineering.

In 1982 he was elected a Fellow of the Royal Society which gave much
pleasure to his engineering colleagues. That same year he was
elected a Foreign Fellow of the National Academy of Greece and
presented a lecture on "Aspects of Metal Plasticity" to the Academy
in Athens. Later in 1982 he was elected a Fellow of University
College, London, where over 30 years before he had been a postgraduate student on the History and Philosophy of Science Course.

In 1983 he retired from his Chair and is now an emeritus professor
of the University of Cambridge. Lest it be thought that he had also
retired from his life's work he simultaneously brought out a new
journal, the "International Journal of Impact Engineering" published
by Pergamon Press, of which he is Editor-in-Chief. Professor N.
Jones is Editor and Professor S. R. Reid the Associate Editor. The
first issue dealt with Impact Crashworthiness and printed some of
the papers presented at the First International Symposium on Structural Crashworthiness held at the University of Liverpool. This is
a subject which is now one of Bill's major interests.

Such a busy eventful life would not have been possible without the
tremendous support of Bill's wife, Heather. It is clear from acknowledgements in some of Bill's books that the children (now grown
up) - Philip, Christopher, Helen, Jeremy and Sarah - were also
brought in from time to time to do some indexing here or typing
there. They form a remarkably united family. Bill and Heather have
their comfortable eyrie above Chapel-en-le-Frith in Derbyshire but
are as likely to be found in Singapore, Korea, Greece or the USA.
The record of W. Johnson FRS does not end here.

                              P. B. Mellor
                              Department of Mechanical Engineering,
                              University of Bradford.

LIST OF PUBLICATIONS

BY WILLIAM JOHNSON

BOOKS

(with P B Mellor) <u>Plasticity for Mechanical Engineers</u>.
Van Nostrand, 1962 (Also published in Japanese and Chinese).

(with H Kudo) <u>Mechanics of Metal Extrusion</u>. Manchester University
Press, 1962 (Also published in Russian).

(with J B Haddow and R Sowerby) <u>Plane Strain Slipline Fields</u>.
Ed Arnold, 1970.

<u>Impact Strength of Materials</u>. Ed Arnold, 1972.

(with P B Mellor) Engineering Plasticity. Van Nostrand-Reinhold,
1973 (Also published in Russian).

(with A G Mamalis) <u>Engineering Plasticity: Metal Forming
Processes</u>. Lectures delivered at CISM, Udine, Italy, 1976.

(with A G Mamalis) <u>The Crashworthiness of Vehicles</u>. MEP, 1978.

(with R Sowerby and R D Venter) <u>A Source Book of Plane Strain Slip
Line Fields for Metal Deformation Processes</u>. Pergamon, 1982.

Translation Editor from Russian of:

   Savin G N <u>Stress Concentration Around Holes</u>. Pergamon,
   1961.

   Artobolevskii I I <u>Mechanisms for the Generation of Plane
   Curves</u>. Pergamon, 1964.

PAPERS

Extrusion through wedge-shaped dies, <u>J. Mech. Phys. Solids</u> 3, Pt.I
218-23; Pt.II, 224-30, 1955.

The effect of curvature on the centre of shear, J. R. Aero. Soc. 59, 562-65, 1955.

Approximate expressions for bow girders of circular cross-section, Engineer 145-46, July 1955.

Research into some metal forming and shaping operations, J. Inst. Met. 165-79, 1956.   Sheet Met. Ind. 41-50, 121-27, 1957

The twist due to bending moment in cantilevers curved in plan, J. R. Aero. Soc. 60, 227-81, 1956.

Extrusion through square dies of large reduction, J. Mech. Phys. Solids 4, 191-98, 1956.

The compression of circular rings, J. R. Aero. Soc. 60, 484-87, 1956.

Experiments in plane strain extrusion, J. Mech. Phys. Solids 4, 269-82, 1956.

The pressure for the cold extrusion of lubricated rod through square dies of moderate reduction at slow speeds, J. Inst. Met. 85, 403-8, 1957.

The plane strain extrusion of solids, short slugs, J. Mech. Phys. Solids 5, 202-14, 1957.

Partial sideways extrusion from a smooth container, J. Mech. Phys. Solids 5, 193-201, 1957.

(with P B Mellor) The centre of shear for a material having a non-linear stress strain curve, Appl. Sci. Res., A 467-77, 1957.

(with D M Woo) The pressure for indenting material resting on a rough foundation, J. Appl. Mech. 25, 64-7, 1958.

(with L C Dodeja) The multiple hole extrusion of sheets of equal thickness, J. Mech. Phys. Solids 5, 267-80, 1957.

(with L C Dodeja) The cold extrusion of circular rods through square multiple hole dies, J. Mech. Phys. Solids 5, 281-95, 1957.

The cutting of round wire with knife-edge and flat-edge tools, Appl. Sci. Res., A 7, 65-87, 1957.

(with B W Senior) The plastic bending of heavily curved beams, J. R. Aero. Soc. 61, 824-30, 1957.

(with P B Mellor) Approximate deflections in cantilevers curved in plan, J. R. Aero. Soc. 62, 64-6, 1958.

(with P B Mellor and D M Woo) Extrusion through single-hole staggered and unequal multi-hole dies, J. Mech. Phys. Solids 6, 203-22, 1958.

Indentation and forging and Nasymth's Anvil, Engineer 25, 348-50, 1958.

Overestimates of load for some two-dimensinal forging operations, Proc. 3rd US Cong. Applied Mechanics, pp. 571-79.   ASME, 1958.

(with M C Derrington) The onset of yield in thick spherical shells subject to internal pressure and a temperature gradient, Appl. Sci. Res., A 7, 408-20, 1959.

Experiments in the cold extrusion of non-circular cross-section, J. Mech. Phys. Solids 7, 37-44, 1958.

Estimation of upper bound loads for some extrusion and coining operations, Proc. Inst. Mech. Eng. 173, 61-72, 1957.

Upper bound loads for extrusion through circular dies, Appl. Sci. Res., A 7, 437-48, 1959.

An elementary consideration of some extrusion defects, Appl. Sci. Res., A 8, 52-60, 1959.

Cavity formation and enfolding defects in plane strain extrusion using a shaped punch, Appl. Sci. Res., A 8, 228-38.

(with R I Tanner) Temperature distribution in some fast metalworking operations, Int. J. Mech. Sci. 1, 28-44, 1960.

(with H Lippmann) Temperature development based on technological analysis: fast rolling as an example, Appl. Sci. Res., A 345-56, 1960.

(with H Kudo) Plane strain compression between rough inclined plates, Appl. Sci. Res., A 9, 206-12, 1960.

(with M B Derrington) The defect of Mohr's circle for 3-dimensional stress states, BMEE 16-22, July 1960.

(with H Kudo) The cutting of metal strips between partly rough-knife tools, Int. J. Mech. Sci. 2, 224-30, 1961.

(with H Lippmann and O Mahrenholtz) Thin heavy elastic strips at large deflexions, Int. J. Mech. Sci., 2, 294-310, 1961.

(with H Kudo) The compression of rigid-perfectly plastic materials between rough parallel dies of unequal width, Int. J. Mech. Sci. 1, 336-41, 1960.

(with H Kudo) The use of upper-bound solutions for the determination of temperature distribution in fast hot rolling and axi-symmetric extrusion processes, Int. J. Mech. Sci. 1, 175-91, 1960.

(with M Vasudevan) On tri-metal thermostats, J. R. Aero. Soc. 507-9, July 1961.

(with J B Haddow) Experiments in the piercing of soft metals, Int. J. Mach. Tool Des. Res. 2, 1-18, 1962.

(with I E McShane) A note on calculations concerning the plastic compression of thin material between smooth plates under conditions of plane strain, Appl. Sci. Res., A 9, 269-96, 1960.

(with J B Haddow) Indenting with pyramids, Int. J. Mech. Sci. (I) Theory 3, 229-38, 1961; (II) Experimental 4, 1-13, 1962.

An analogy between upper-bound solutions for plane strain metalworking and minimum weight two-dimensional frames, Int. J. Mech. Sci. 3, 239-47, 1961.

(with F W Travis) Experiments in the dynamic deformation of clamped circular sheets of various metals subject to an underwater explosive charge, Sheet Met. Ind. 17, April 1962.

(with J B Haddow) Bounds for the load to compress plastically square disc between rough dies, Appl. Sci. Res., A, 10, 476-77, 1962

(with O Mahrenholtz) Zur Theorie des Bimetall-Streitens, Z. Angew. Math. Mech. 41, 99-101, 1961. On bi-metal thermostats Int. J. Mech. Sci. 4, 35-52, 1962.

(with P B Mellor) Elastic-plastic behaviour of thick-walled spheres of non-hardening material subject to a steady-state radial temperature gradient, Int. J. Mech. Sci. 4, 147-58, 1962.

Some slip-line fields for swaging or expanding indenting, extruding and machining for tools with curved dies, Int. J. Mech. Sci. 4, 323-47, 1962.

(with F U Mahtab) The mechanics of cutting strip with knife-edge and flat-face dies, Int. J. Mach. Tool Des. Res. 2, 335-37, 1962.

The high rate forming of metals, Manchester Assoc. Eng. 1159(6) 1962-3. Met. Treat. Drop Forg. 227-86, 305-14, 349-56, 1963. BMEE 2, 111-48.

(with R A C Slater and E C Laithwaite) An appraisal of the linear induction motor concept for high energy rate metal forming, Sheet Met. Ind. 237-42, April 1963.

(with R A C Slater and E C Laithwaite) An experimental investigation relating to the accelerated motion of various translators in the air gap of a linear induction motor, Int. J. Mach. Tool Des. Res. 3, 111-35, 1963.

Slip-line field and discontinuous velocity solutions for some metal forming opertions Pts.I, II, Heft 3, 180-88, Pt.III Heft 4, 275-78 CIRP-Ann. 10, 1963.

(with M Vasudevan) On multi-metal thermostats, Appl. Sci. Res., B 420-30, 1963.

A note of calculating the collapse load for knee frames, Appl. Sci. Res., A 11, 318-20, 1961.

Extrusion, forging, machining and indenting minimum weight frames and high rate sheet metal forming. Proc. Int. Production Engineering Research Conf., pp.342-55. ASME, 1963.

(with M J Hillier) Some slip line fields for indenting the rotating dies, Int. J. Mech. Sci. 203-12, 1963.

(with M J Hillier) Plane strain extension through partly rough dies, Int. J. Mech. Sci. 5 191-201, 1963.

Dynamic indention using rigid slow speed conical indenters, J. Basic Eng. 1964.

(with I S Donaldson) Sheet metal forming using a water hammer type technique, 3rd Conf. Machine Tool Design and Research, pp.385-97, 1962.

(with F W Travis) The explosive forming of cones. 3rd Conf. Machine Tool Design and Research, pp.341-64, 1962.

(with J L Duncan) The free forming of sheet aluminium using and electic spark discharge method, 3rd Conf. Machine Tool Design and Research, pp.389-401, 1962.

A review of high rate metal forming, Institute of Metallurgists, Iliffe, pp.79-111, 1964.

(with H L Duncan, K Korml, R Sowerby and F W Travis) Some contributions to high rate metal forming - I: Some simple calculations concerning the deflection of peripherally clamped thin circular blanks subject to underwater explosive forming; II: A rapid method of determining stress-strain curves for sheet metal using the bulge test; III: The explosive drawing of hemispherical efficiency, 4th Conf. Machine Tool Design and Research, pp.257-317. Pergamon, 1963.

(with A H El-Behery and J H Lamble) The measurement of container wall pressure and friction coefficient axisymmetric extrusion, 4th Conf. Machine Tool Design and Research, pp.319-35. Pergamon 1963.

(with F Drabble) The development of the zones of yielding in thick-walled spherical shells of non-hardening material subject to a steady state radial temperature gradient and an internal or external pressure, Conf. Creep and Thermal Loading, p.15. Institute of Mechanical Engineers, 1964.

(with A Vafladakis and I S Donaldson) Graphodynamics applied to weight-piston-liquid impact, Int. J. Mech. Sci. 6, 153-60, 1964.

(with K Korm and F W Travis) An investigation into the explosive deep drawing of circular blanks using the plug-cushion technique, Int. J. Mech. Sci. 6, 287-301, 1964.

(with R A C Slater and E R Laithwaite) An experimental impact-extrusion machine by a linear induction motor, Proc. Inst. Mech. Eng. 179, 15-36, 1964-5.

An approximate treatment of metal deformation in rolling, rolling contact and rotary forming, Int. J. Prod. Res. 3, 51-64, 1964.

(with F U Mahtab and J B Haddow) The indentation of a semi-infinite block by a wedge of comparable hardness: I Theoretical, Int. J. Mech. Sci. 6, 329-36, 1964.

(with H Kudo) Plane strain deep indentation, 5th Conf. Machine Tool Design and Research, pp.441-47. Pergamon, 1965.

(with G L Baraya) Flat bar forging, 5th Conf. Machine Tool Design and Research, pp.449-69, Pergamon, 1965.

(with N R Chitkara) Some results for rolling with circular and polygonal rolls, 5th Conf. Machine Tool Design and Research pp.391-410. Pergamon, 1965.

(with J H Lamble and A M El-Behery) Container wall pressure in the slow speed plane strain direct extrusion of pure lead, 5th Conf. Machine Tool Design and Research, pp.429-40. Pergamon, 1965.

(with R Bell and J L Duncan) The evaluation of a prototype machine for automatically recording the true stress-strain curve for sheet metal using the hydrostatic bulge test, 5th Conf. Machine Tool Design and Research, pp.411-29. Pergamon, 1965.

(with R A C Slater) Some considerations relevant to the design of a high rate of energy transfer machine employing a linear induction motor, 5th Conf. Machine Tool Design and Research, 267-93. Pergamon, 1965.

(with F W Travis) The explosive drawing of square and flat-bottomed circular cups and bubble pulsation phenomena, 5th Conf. Machine Tool Design and Research, pp.293-328. Pergamon, 1965.

Encylopaedic Dictionary of Physics, 3, 277-78 577-79, 584; 4, 322-25, 709-10; 5, 250-51, 320-21; 6, 340-41; 7, 396-97. Pergamon, 1965.

(with F U Mahtab and A Williams) Experiments concerning geometric similarity in indentation, Int. J. Mech. Sci. 7, 389-98, 1965.

(with J H Lamble and M Safdar Ali) Low cycle high strain bending fatigue tests on ductile metals, Int. J. Mech. Sci. 7, 1-14, 1965.

(with G L Baraya and R A C Slater) On heat lines or lines of thermal discontinuity, Int. J. Mech. Sci. 6, 409-14, 1964.

(with F U Mahtab) Upper bounds for restricted edge machining pp.447-62; Some slip line fields and the method of upper bounds for milling, turning and boring, pp.463-86, 6th Conf. Machine Tool Design and Research. Pergamon, 1965. Also in Proc. 4th Machining Conf. Liptovsky Mikulas, Czechoslovakia, 1965.

(with R A C Slater) Accelerated motion of various translators in the air gap of a linear induction motor, Int. J. Mach. Tool Des. Res. 120-35, 1964.

(with N R Chitkara) Some experimental results concerning spread in the rolling of lead, J. Basic Eng. Paper No.65-WA/Met 11, 1966.

Further rotational figurations: Straight starting slip lines, 6th Conf. Machine Tool Design and Research, Paper III. Pergamon, 1966.

(with R Sowerby) Experiments on clamped circular blanks subject to an underwater explosion, Proc. Inst. Mech. Eng. 179, 197-221, 1964-5.

(with I S Donaldson and A P Vafiadakis) Deep drawing and free forming using a water-hammer technique, Proc. Inst. Mech. Eng. 179, 222-23, 1964-5.

(with J L Duncan) Comparison of the behaviour of different sheet materials formed by the underwater spark discharge method, Proc. Inst. Mech. Eng. 179, 234-239, 1964-5

(with E Doege and F W Travis) The explosive expansion of unrestrained tubes, Proc. Inst. Mech. Eng. 179, 240-56, 1964-5.

(with R A C Slater) A comparison of the energy required for slow speed and dynamic blanking using an improved linear motor, Proc. Inst. Mech. Eng. 179, 275-63, 1964-5.

(with A Poynton, H Singh and F W Travis) Experiments in the underwater explosive stretch forming of clamped circular blanks, Int. J. Mech. Sci. 8, 237-70, 1966.

(with J L Duncan) The use of the biaxial extensometer, Sheet Met. Ind. 271-75, April 1965.

(with F W Travis) Explosive hydrodynamic extrusion of metals, Engineer 219, 546-49, 1965.

(with G L Baraya and R A C Slater) The dynamic compression of circular cylinders of super-pure aluminium at elevated temperatures, Int. J. Mech. Sci. 7, 621-45, 1965.

(with F U Mahtab and R A C Slater) Dynamic indention of copper and aluminium alloy and mild steel with conical projectiles and dynamic tip flattening of conical projectiles at ambient temperature, Int. J. Mech. Sci. 7, 685-719, 1965.

(with F U Mahtab and R A C Slater) Dynamic indentation of copper and an aluminium alloy with a conical projectile at elevated temperatures, Proc. Inst. Mech. Eng. 180, 285-94, 1965-66.

(with F W Travis) Peeling, hot forming of tungsten and steel discs and the use of a vacuum chest with explosives, 6th Conf. Machine Tool Design and Research, pp.765-84. Pergamon, 1966.

(with F W Travis) Explosive fracturing, 6th Conf. Machine Tool Design and Research, pp.741-64. Pergamon, 1966.

(with J H Lamble and H Abdel-aziz) Low endurance fatigue of mild steel in torsion, Proc. 2nd Conf. Dimensioning and Strength Calculations, pp.313-37. Hungarian Academy of Sciences, 1965.

(with K Baines and J L Duncan) Electromagnetic metal forming, Proc. Inst. Mech. Eng. 180, 93-110, 1965-66.

(with F Drabble and R N Haward) Adiabatic models for deformation and cavitation in polystyrene, Br. J. Appl. Phys. 179, 241-52, 1966.

(with F U Mahtab and A Williams) Further experiments concerning geometrically similar indentions, Int. J. Mech. Sci. 8, 49-59, 1966.

The plane strain extrusion of initially curved sheet, Int. J. Mech. Sci. 8, 163-70, 1966.

(with P D Soden) Hot discharge characteristics of confined rubber cylinders, Int. J. Mech. Sci. 8, 213-25, 1966.

(with D G Dalrymple) A study of thin tube forming using non-uniform explosive charges, Int. J. Mech. Sci. 8, 353-81, 1966.

(with R A C Slater) Further experiments in quasi-static and dynamic blanking of circular discs from various materials, Paper 5 Applied Mechanics Convention 1965-66, Proc. Inst. Mech. Eng. 180, Pt. 31, 1966.

(with G Needham) An experimental study of asymmetrical rolling, Paper 11, Applied Mechanics Convention 1965-66, Proc. Inst. Mech. Eng. 180, Pt. 31, 1966.

(with H A Abdel-Aziz and J H Lamble) Low endurance fatigue studies in torsion for steel, and aluminium alloy, Paper 13, Applied Mechanics Convention 1965-66, <u>Proc. Inst. Mech. Eng.</u> 180, Pt. 31, 1966.

(with F W Travis) High speed blanking of copper, Paper 16, Applied Mechanics Convention 1965-66, <u>Proc. Inst. Mech. Eng.</u> 180, Pt. 31, 1966.

(with D G Dalrymple) An exploration of explosive swaging in the fastening of thin brass tubes, Paper 25, Applied Mechanics Convention 1965-66, <u>Proc. Inst. Mech. Eng.</u> 180, Pt. 31, 1966.

(with J L Duncan and A Ovreset) Some experiments and theory for plane strain side-extrusion, Paper 21, Applied Mechanics Convention 1965-66, <u>Proc. Inst. Mech. Eng.</u> 180, Pt. 31, 1966.

(with G Needham) Further experiments in asymmetrical rolling, <u>Int. J. Mech. Sci.</u> 8, 443-55, 1966.

A review of some experiments in, and efficiency considerations of, high rate energy forming, Int. Symp. HERF Prague 1966 and 20th Annual Met Conf. <u>J. Aust. Inst. Met.</u> May 1967.

(with M B Bassett) The bending of plate using a three-roll pyramid type plate bending machine, <u>J. Strain Anal.</u> 5, 398-414, 1966.

(with F W Travis) Further experiments in explosive fracturing, <u>7th Conf. Machine Tool Design and Research</u>, 1966.

(with J L Duncan A Ovreset) Effect of tool geometry in extrusion pressure in side-extrusion, <u>CIRP</u> 14, 89-95, 1966.

(with R A C Slater and A S Yu) The quasi-static compression of non-circular prismatic blocks between very rough platens using the friction hill concept, <u>Int. J. Mech. Sci.</u> 8, 731-38, 1966.

(with H A Abdel-Aziz and J H Lamble) Low-endurance fatigue of an aluminium alloy and a stainless steel in plane bending at ambient and elevated temperatures, <u>Int. J. Mech. Sci.</u> 8, 717-30, 1966.

Explosives: New uses, <u>Advance</u>, 60-69, 1966. Also <u>Student Technol.</u> 9-12 April 1967.

Cutting with tools having a rounded edge: Some theoretical considerations, <u>CIRP</u>, 14, 315-19, 1967.

(with J L Duncan and E Goni) Measurement of normal plastic anisotropy in sheet metal, <u>J. Aust. Inst. Met.</u> 127-38, May 1967

(with R A C Slater) The effects of temperature, speed and strain-rate on the force and energy required in blanking, <u>Int. J. Mech. Sci.</u> 9, 271-305, 1967.

(with J B Hawkyard) An analysis of the changes in geometry of a short hollow cylinder during axial compression, <u>Int. J. Mech. Sci.</u> 9, 163-82, 1967.

(with R Sowerby) On the collapse of some simple structures, <u>Int. J. Mech. Sci.</u> 9, 433-71, 1967.

Introductory address, <u>Applied Mechanics Conf.</u> Institute of Mechanical Engineers, 1966.

(with S Y Aku and R A C Slater) The use of plasticine to simulate the dynamic compression of prismatic blocks of hot metal, Int. J. Mech. Sci. 9, 495-527, 1967.

(with F W Travis) A tool for hydrodynamic extrusion, Proc. High Pressure Engineering Conf., pp.9-17. Institute of Mechanical Engineers, 1967.

(with F W Travis) Explosive hydrodynamic extrusion: Pressures generated and other phenomena, Proc. High Pressure Engineering Conf., pp.1-9. Institute of Mechanical Engineers, 1967.

Resume and critique of papers in Pt. 3, Sect. A on new methods of forming, CIRP-ASTME Int. Conf. Manufacturing Technology, pp.765-74, 1967.

(with S T S Al Hassani and J L Duncan) Analysis of the electro-magnetic metal forming process, CIRP-ASTME Int. Conf. Manufacturing Technology, pp.858-82, 1967.

(with F W Travis) High speed expansion and bore finishing of austenitic stainless steel tubes, CIRP-ASTME Int. Conf. Manufacturing Technology, pp.895-905, 1967.

(with A Poynton and F W Travis) Metal forming using explosive gas mixtures: The free forming of circular blanks, CIRP-ASTME Int. Conf. Manufacturing Technology pp.801-15, 1967.

(with R A C Slater) A survey of the slow and fast blanking of metals at ambient and high temperatures, CIRP-ASTME Int. Conf. Manufacturing Technology, pp.825-51, 1967.

(with R Sowerby) An analysis of a rigid-plastic thick-walled cylinder using slip line field theory, BMEE, 201-19, 1967.

(with S T S Al-Hassani and J L Duncan) The influence of the electrical and geometrical parameters in magnetic forming, 8th Conf. Int. Machine Tool Design and Research, 1967.

(with F W Travis) Explosive welding of stellite to stainless steel, 8th Int. Machine Tool Design and Research Conf. 1967.

(with J L Duncan) The ultimate strength of rectangular diaphragms, Int. J. Mech. Sci. 9, 681-96, 1967.

(with D Horrocks) On anticlastic curvature with special reference to plastic bending: A literature survey and some experimental investigations, Int. J. Mech. Sci. 9, 835-61, 1967.

(with G Needham) Experiments on ring rolling, Int. J. Mech. Sci. 9, 95-113, 1968.

(with J L Duncan) The ultimate strength of rectangular anisotropic diaphragms, Int. J. Mech. Sci. 10, 143-55, 1968.

(with J L Duncan) Plastic deformation and failure of thin square diaphragms, Int. J. Mech. Sci. 10, 157-68, 1968.

(with R A C Slater an S Y Aku) Experiments in the fast upsetting of short pure lead cylinders and a tentative analysis, Int. J. Mech. Sci. 10, 169-86, 1968.

(with R Sowerby and S K Samanta) Plane strain drawing and extrusion of a rigid-perfectly plastic material through concave dies, Int. J. Mech. Sci. 10, 231-38, 1968.

(with F W Travis) High speed blanking of steel, Conf. Engineering Plasticity, 1968.

(with R Sowerby) The diametral compression of circular rings Int. J. Mech. Sci., 10, 369-83, 1968.

(with R Sowerby) On the yielding of an initially curved rigid-plastic wide plate subjected to pure bending, BMEE 6, 307-16, 1967.

(with D G Dalrymple) Explosive cladding of thin steel tubes with Copper: An exploratory study, Int. J. Mach. Tool Des. Res. 7, 257-67, 1967.

(with W A Poynton and F W Travis) The free radial expansion of thin cylindrical brass tubes using explosive gas mixtures, Int. J. Mech. Sci. 10, 385-401, 1968.

(with I Macleod and G Neednam) An experimental investigation into the process of ring or metal tyre rolling, Int. J. Mech. Sci. 10, 455-68, 1968.

(with G Needham) Plastic hinges in ring indentation in relation to ring rolling, Int. J. Mech. Sci. 10, 487-90, 1968.

(with F W Travis and S Y Loh) High speed cratering in wax and plasticine, Int. J. Mech. Sci. 10 593-605, 1968.

(with J Carrell and F W Travis) High speed impact of plasticine projectiles with laminated plasticine targets, Int. J. Mech. Sci. 10, 677-80, 1968.

(with S T S Al-Hassani and J L Duncan) The effect of scale in electromagnetic forming when using geometrically similar coils, Proc. 1st Int. Conf. Center for High Energy Rate Forming, Vol.2, p.5.3.1, 1968.

(with J L Duncan) The use of a large tool in the forming of metals by explosive gasses, 9th Int. Conf. Machine Tool Design and Research, 1968.

(with J L Duncan) Approximate analyses of loads in axi-symmetric deep drawing, 9th Int. Conf. Machine Tool Design and Research, 1968.

(with R A C Slater and S Y Aku) Fast upsetting of short circular cylinders of plain medium carbon (0.55%C) steel at room temperature, 9th Int. Conf. Machine Tool Design and Research, 1968.

(with F W Travis) High speed riveting of mild steel, 9th Int. Conf. Machine Tool Design and Research, 1968.

(with R M Caddell and G Needham) Yield strength variations in ring-rolled aluminium, Int. J. Mech. Sci. 10, 749-56, 1968.

(with R M Caddell and J L Duncan) Fracture and limit strains in annealed and cold rolled brass and aluminium sheet, Proc. 3rd Conf. Dimensioning and Strength Calculations, pp.385-404. Hungarian Academy of Sciences, 1968.

(with J B Hawkyard and D Eaton) The mean dynamic yield strength of copper and low carbon steel at elevated temperatures from measurements of the 'mushrooming' of flat-ended projectiles, Int. J. Mech. Sci. 10, 929-48, 1968.

(with S T S Al-Hassani and H G Hopkins) A note on the fragmentation of tubular bombs, Int. J. Mech. Sci. 11, 545-49, 1969.

(with R Sowerby) The mechanics of wire drawing, Pts. I and II, Wire Ind., pp.137-44 and 249-56, Feb and March 1969.

The explosive forming of tube, 9th Commonwealth Mining and Metallurgical Congr., Paper No. 17, 1969.

(with S T S Al-Hassani) The dynamics of the fragmentation process for spherical shells containing explosives, Int. J. Mech. Sci. 11, 811-23, 1969.

Upper bounds to the load for the transverse bending of flat rigid-perfectly plastic plates. Pt. 1: Using hodographs - a new and simple approach. Pt. 2: An analogy - slip line fields for analysing the bending and torsion of plates, Int. J. Mech. Sci. 11, 913-38, 1969.

(with J B Hawkyard and D Horrocks) An investigation into a novel bubble casting process for forming thin-walled aluminium containers, Proc. Inst. Mech. Eng. 184, 615-32, 1969.

(with R A C Slater, N K Barooah and E Appleton) The rotary forging concept and initial work with an experimental machine, Proc. Inst. Mech. Eng. 1, 577-92, 1969.

(with S T S Al-Hassani and J L Duncan) The magnetohydraulic forming of tube: Experiment and theory, Int. J. Mech. Sci. 12, 371-92, 1970.

Latest trends in metal forming processes. Proc. Com. Tecn. Conf. Materials Processing and Manufacturing, University of Massachusetts, March 1969.

(with J L Duncan and S T S Al-Hassani) Techniques for designing electromagnetic forming coils, 2nd Int. Conf. Center for High Energy Rate Forming, 1969.

(with J L Duncan) Plasticity and superplasticity: Exploitation through engineering development, Inst. Metall. Series 2(3), 1969.

(with M I Yousif and J L Duncan) Plastic deformation and failure of thin elliptical diaphragms, Int. J. Mech. Sci. 12, 959, 1971.

(with A J O Zaid and J B Hawkyard) Experiments in plate cutting by shaped high explosive charges, J. Mech. Eng. Sci. 13, 1971.

(with J Lynch and J B Hawkyard) Laboratory scale experiments into cavity and crater formation by high explosive charges, J. Mech. Eng. Sci. 12, 339, 1970.

(with R Sowerby) Use of slip line field theory for the plastic design of pressure vessels, Experimental Stress Analysis Conf., p.74. Institute of Mechanical Engineers, 1970.

(with M C de Malherbe and R Venter) On geometrical similarity in compound circular cylinders and spherical shells under internal pressure, CIRP, 1970.

(with R Sowerby) Analysing flange drawing in non-circular cups using slip line fields, CIRP, 1970.

(with R Sowerby and R M Caddell) Redundant deformation factors and maximum reduction in plane strain and axisymmetric wire drawing. I: A theoretical investigation, Ann. CIRP, 17, 311-16, 1971.

(with R A C Slater and S Y Aku) Strain rate and temperature effects during the fast upsetting of short circular cylinders of 0.55% plain carbon steel at elevated temperatures, Ann. CIRP, 19, 513-27, 1971.

(with S T S Al-Hassani) Dynamic deformation and fragmentation of strain-hardening, strain rate sensitive shells containing high explosive, 11th Conf. Machine Tool Design and Research, 1970.

(with S T S Al-Hassani) A magnetohydraulically activated system for high strain rate testing and forming of thin cylindrical tubes, 11th Conf. Machine Tool Design and Research, 1970.

(with R A C Slater) The dynamic blanking, forging, indenting and upsetting of hot metal, Proc. IUTAM Symp. Thermo-inelasticity, pp.120-55. Springer, 1970.

(with R Sowerby) Upper bound techniques applied to plane strain extrusion minimum weight two-dimensional frames and rotationally symmetric flat plates, BMEE, 8, 269-84, 1969.

(with P D Soden and S T S Al-Hassani) Analysis of a simple heart-aorta analogue, Int. J. Mech. Sci., 13, 615, 1970.

(with P D Soden and E R Trueman) A study in jet propulsion: An analysis of the motion of squids, J. Exptl. Biol. 56, 155-65, 1972.

(with J B Hawkyard) Newer hot forming processes, Conf. Competitive Methods of Forming, J. Iron Steel Inst. 35-49, 197.

(with S T S Al-Hassani) The magnetomotive loading of cantilevers, beams and frames, Int. J. Mech. Sci. 12, 711, 1970.

(with R Venter ad M C de Malherbe) The limit strains of inhomogeneous sheet metal in biaxial tension, Int. J. Mech. Sci. 13, 299-309, 1971.

(with R G Hall and S T S Al-Hassani) The impulsive loading of cantilevers, Int. J. Mech. Sci. 13, 415-30, 1971.

(with S R Reid) Amplitude of interface waves in explosive welding, Nat., Phys. Sci. 231, 205, 1971.

(with J B Hawkyard) Recent developments in metal forming, Prod. Eng. 239, 1971.

The mechanics of metal working plasticity, Appl. Mech. Rev. Sept. 1971.

(with N R Chitkara, S R Reid and I F Collins) The displacement field and its significance for certain minimum weight two-dimensional frames using the analogy with perfectly plastic flow in metal working, Int. J. Mech. Sci. 13, 547, 1971.

(with S T S Al-Hassani) Stress wave fracturing of a bar, 12th Int. Conf. Machine Tool Design and Research, Paper 112. Macmillan, 1971.

(with M Nasim and S T S Al-Hassani) Stress wave propagation and fracture of thin curved bars, Int. J. Mech. Sci. 13, 599-603, 1971.

(with I H Wilson and J L Duncan) Biaxial creep testing, J. Mech. Sci. 13, 397-403, 1971.

(with R Venter and M C de Malherbe) The plane strain indentation of anisotropic aluminium using a frictionless, flat rectangular punch, J. Mech. Eng. Sci. 13, 416-28, 1971.

(with T Y M Al-Naib and J L Duncan) Superplastic forming techniques and strain distribution in a zinc-aluminium alloy, J. Inst. Met. 100, 45-50, 1972.

(with S T S Al-Hassani and M Nasim) Fracture of triangular plates due to contact explosive pressure, J. Mech. Eng. Sci. 14, 173-83, 1972.

Review of metal working plasticity: Quasi-static conventional forming processes, pp.89-99. Hydrostatic forming and super-plasticity, pp.128-29. High speed forming, pp.147-51, Metall. Met. Form. 1972.

Plane strain compression of pre-shaped material between wedge-shaped dies, Int. J. Mech. Sci. 14, 151-64, 1972.

(with W T Lowe and S T S Al-Hassani) Impact behaviour of small scale model motor coaches, Proc. Inst. Mech. Eng. Auto Division 186, 36-72, 1972. Also: J. Automotive Eng. 19, 1972.

(with J B Hawkyard and E Appleton) An experimental wide ring rolling mill of novel design, Conf. Machine Tool Design and Research, Paper 85, 1972.

(with J B Hawkyard and I O Uton) Simple analysis of the non-symmetric dynamic expansion of cylindrical cavities, Int. J. Mech. Sci., 14, 603-13, 1972.

(with M C de Malherbe and R Venter) Upper bounds to the load for the plane strain indentation of anisotropic metals, J. Mech. Eng. Sci. 14, 297, 1972.

(with G M Vickers) The development of an impression and the threshold velocity for erosian damage in alpha-brass and perspex due to repeated water jet impact, Int. J. Mech. Sci. 14, 765-79, 1972.

(with S T S Al-Hassani and W T Lowe) Characteristics of inversion tubes under axial loading, J. Mech. Eng. Sci. 14, 370-81, 1972.

(with S J Hashmi and S T S Al-Hassani) Dynamic plastic deformation of rings under impulsive load, Int. J. Mech. Sci. 14, 823-41, 1972.

(with S J Hashmi and S T S Al-Hassani) Large deflexion elastic-plastic response of certain structures to impulsive load: Numerical solutions and experimental results, Int. J. Mech. Sci. 14, 843-60, 1972.

(with A S Ranshi and N R Chitkara) Limit loads for the plastic bending in plane strain of cantilevers containing rectangular holes under end shear, Int. J. Mech. Sci. 15, 15-35, 1973.

(with M C de Malherbe and R Venter) Some slip line field results for the plane strain extrusion of anisotropic materials through frictionless wedge-shaped dies, Int. J. Mech. Sci. 15, 109-11, 1973.

(with S R Reid and R E Trembaszowski-Ryder) The impact, rebound and flight of a well inflated pellicle as exemplified in association football, Manchester Assoc. Eng. 1972-3, No.5, 1973.

(with N R Chitkara) Corrugated plate formed by side extrusion with two coaxial rams moving at different speeds, Int. J. Mech. Sci. 15, 199-210, 1973.

(with A S Ranshi and N R Chitkara) Plane strain plastic yielding due to bending of end-loaded cantilevers containing circular, triangular or diamond-shaped holes, Int. J. Mech. Sci. 15, 329-43, 1973.

An elementary analysis of an energy absorbing device: The rolling torus load limiter, Int. J. Mech. Sci. 15, 357-66, 1973.

(with S T S Al-Hassani and G W Vickers) The dynamic loading of variable thickness cantilevers using a magnetomotive impulse, Arch. Budowy Masz. 20(1), 1973.

(with N R Chitkara, A H Ibrahim and A K Dasgupta) Hole flanging and punching of circular plates with conically headed cylindrical punches, J. Strain Anal. 8(3), 223-41, 1973.

(with G W Vickers) Some results in the erosion of prestressed materials due to water jet impact, J. Mech. Eng. Sci. 15(4), 295-301, 1973.

(with G W Vickers) Transient stress distribution caused by water jet impact, J. Mech. Eng. Sci. 15, 302-10, 1973.

(with J B Hawkyard, J Kirkland and E Appleton) Analyses for roll force and torque in ring rolling with some supporting experiments, Int. J. Mech. Sci. 15, 873-94, 1973.

Some contrasting static and dynamic damage or deformation patterns in bodies of simple geometry, Presidential Inaugural Address, Manchester Assoc. Eng. Oct. 1973.

(with S R Reid) The rolling torus: Some elastic-plastic considerations, Int. J. Mech. Sci. 16, 45-62, 1974.

(with P Dewhurst and I F Collins) A class of slip line field solutions for the hot rolling of strip, J. Mech. Eng. Sci. 439-47, 1973.

(with R Sowerby) Prediction of earing in cups drawn from anisotropic sheet using slip line field theory, J. Strain Anal. 9(2), 102-8, 1974.

(with A S Ranshi and N R Chitkara) Plane stress yielding of cantilevers in bending due to combined shear and axial load, J. Strain Analysis 9(2), 67-77, 1974.

(with P Dewhurst and I F Collins) A theoretical and experimental investigation into asymmetrical not rolling, Int. J. Mech. Sci. 16, 389, 1974.

(with S T S Al Hassani and J L Duncan) On the parameters of the magnetic forming process, J. Mech. Eng. Sci. 16, 1, 1974.

(with E Lovell and S T S Al-Hassani) Fracture in solid spheres and circular disks due to a 'point' explosive impulse on the surface, Int. J. Mecn. Sci. 16, 193-99, 1974.

A note on the radial vibrations of an isotropic spherical shell, Int. J. Mech. Sci. 201-7, 1974.

Metal forming research at the University of Manchester Institute of Science and Technology, Metall. Met. Form. 44-46 Feb. 1984.

(witn P Dewhurst and J B Hawkyard) A theoretical and experimental investigation of dynamic cylindrical expansions in metal, J. Mech. Pnys. Solids 22, 267-83, 1974.

Elastic wave transmission and reflection in long uniform bars due to the end impact of a rigid mass, 15th Conf. Machine Tool Design and Research, pp.267-78, 1975.

(with N R Chitkara and P A Bex) Characteristic features in the hole flanging and piercing of thin and thick circular plates using conical and ogival punches, 15th Conf. Machine Tool Design and Research, pp.695-702, 1975. (with N R Cnitkara) Hole flanging and piercing of circular plates, Sheet Met. Ind. 635-40, 1974.

(with N R Chitkara and J R S Uttley) Ball rolling: A literature survey and some experimental results, 15th Conf. Machine Tool Design and Researcn, pp.497-506, 1975.

(with J B Hawkyard and F R Navaratne) Cold rolling of gears, 15th Conf. Machine Tool Design and Research, pp.507-14, 1975.

(with P D Soden and S T S Al-Hassani) The crumpling of polyvinylchloride tubes under static and dynamic axial load, Conf. High Strain Rate, Inst. Phys. Series No.21, 327-38, 1974.

(with A S Ranshi and N R Cnitkara) Plane stress-plastic collapse loads for tapered cantilevers and haunched beams, Int. J. Mech. Sci. 16, 867-85, 1974.

(with S A L Salem and S T S Al-Hassani) Aspects of the mechanics of driving nails into wood, Int. J. Mech. Sci. 17, 211-25, 1975.

(with R S Sowerby) A review of texture and anisotropy in relation to metal forming, Mater. Sci. Eng. 20(2), 101-11, 1975.

(with S R Reid) Ricocnet of spneres off water, J. Mech. Eng. Sci. 17, 71-81, 1975.

High energy rate forming of metals, Mater. Sci. Club Bulletin No.41, 2-17 March 1975.

(with J Skorecki and S R Reid) The Gadd severity index and measurements of acceleration when heading an association football, 2nd Int. Research Conf. Biokinetics of Impacts, 1975.

(with N R Chitkara and M Ueda) Incremental forging of balls from
cylindrical specimens and of rollers from rectangular solids, 16th
Conf. Machine Tool Design and Research, pp.431-46.

(with G G W Clemas and S T S Al-Hassani) An investigation into the
hole-flanging of Zn-Al superplastic sheet, 16th Conf. Machine Tool
Design and Research, pp. 479-86, 1975.

(with J B Hawkyard) Recent developments in ring rolling
applications and rotary forging, 5th Int. Cong. Cold Forging,
pp.209-23, 1975. Also: Metall. Met. Form. 4-11, Jan 1976.

Notes on methods of studying the flow of material in metal working
operations. Associazone Italiana di Analisi delle Solecitazioni,
General Lecture to 3rd National Congress (Bologna), October 1975.

(with A G Mamalis and J B Hawkyard) Cavity formation in rolling
profiled rings, Int. J. Mech. Sci. 17, 669, 1975.

(with L B Aksenov and N R Cnitkara) Pressure and deformation in
the plane strain pressing of circular section bar to form turbine
blades, Int. J. Mech. Sci. 17, 681, 1975.

(with R Kitching and R Houlston) A theoretical and experimental
study of hemispherical shells subjected to axial loads between
flat plates, Int. J. Mech. Sci. 17, 693, 1975.

(with G G W Clemas and S T S Al-Hassani) The bulging of a
superplastic sheet from a square die, Int. J. Mech. Sci. 17, 711,
1975.

(with M J Vojdani and J B Hawkyard) Transmission of strong
pressure pulses through curved and angled water-filled pipes in
connection with electro-hydraulic forming, Conf. Electrical
Methods of Machining, Forming and Coating. Institute of
Electrical Engineers, 1975.

(with J L Duncan and J Miller) Reducing thin-walled tube by
electro-hydraulic and other processes, Conf. Electrical Methods of
Machining, Forming and Coating. Institute of Electrical
Engineers, 1975.

Simple linear impact, Int. J. Mech. Eng. Educ. 2, 167-81, 1976.

(with A G Mamalis and J B Hawkyard) A ring rolling literature
review and some recent experimental results, Metall. Met. Form.
132-40, 1976.

(with A S Ranshi and N R Cnitkara) Collapse loads for thin
cantilevers with rectangular holes along the centre line,
J. Strain Anal. 11, 84-96, 1976.

(with A G Mamalis and J B Hawkyard) Spread and flow patterns in
ring rolling , Int. J. Mech. Sci. 18, 11, 1976.

(with S Clyens and S T S Al-Hassani) The compaction of powder
metallurgy bars using high voltage electrical discharges, Int. J.
Mech. Sci. 18, 37, 1976. Also (with S Clyens and S T S Al-Hassani)
Compaction of metal powders using high voltage electrical
discharges and rotary swaging, Metall. Met. Form. 382-85, 1976.

(with F Malko and C Leech) Damage in plates due to surface rings
of explosive, Int. J. Mech. Sci. 18, 33, 1976.

(with A G Mamalis and H Hunt) Small spherical lead shot forming from the liquid, using a shot tower, Metall. Met. Form. 68-73, March 1976.

(with S T S Al-Hassani and R B Lloyd) Aspects of pole vaulting mechanics, Proc. Inst. Mech. Eng. 189, 53-75, 1975.

(with S R Reid and L B Singh) Experimental study of the rolling torus load limiter, Int. J. Mech. Sci. 17, 603-15, 1975.

(with N R Chitkara and N Das) A class of slip-line field solutions for unsymmetrical extrusion and some experimental results, 17th Conf. Machine Tool Design and Research, pp.435-44, 1976.

(with S A Meguid and S T S Al-Hassani) Some factors in the shot-peening and peen-forming processes, 17th Conf. Machine Tool Design and Research, pp.653-59, 1976.

(with S A Meguid and S T S Al-Hassani) Multiple impact erosion of ductile metals by spherical particles, 17th Conf. Machine Tool Design and Research, pp.661-68, 1976.

Recent Japanese Work on Peen Forming (Translated by Y Mihara and Edited by W Johnson), pp.647-51, 1976.

(with A G Mamalis) A survey of some physical defects arising in metal working processes, 17th Conf. Machine Tool Design and Research, pp.607-21, 1976.

(with A G Mamalis and J B Hawkyard) On the pressure distribution between stock and rolls in ring rolling, J. Mech. Eng. Sci. 18, 184-95, 1976. (with A G Mamalis and J B Hawkyard) Neue Untersuchungen beim Ringwalzen, Arch. Eisenhuttenwes. 47, 613-18, 1976.

(with A G Mamalis and J B Hawkyard) Pressure distribution, roll force and torque in cold ring rolling, J. Mech. Eng. Sci. 18, 196-209, 1976. Untersuchungen uber die Brietung und den Werkstoff-fluss beim Ringwalzen, Arch. Eisenhuttenwes., 47, 619-22, 1976.

(with S A Soliman and S R Reid) The effect of spherical projectile speed in ricochet off water and sand, Int. J. Mech. Sci. 18, 279-84, 1976.

(with S G Thomas and S R Reid) Large deformations of thin-walled circular tubes under transverse loading - I: An experimental survey of the bending of simply supported tubes under a central load, Int. J. Mech. Sci. 18, 325-33, 1975.

(with N R Chitkara and H V Minh) Deformation modes and lip fracture during hole flanging of circular plates of anisotropic materials, J. Ind., Trans. ASME, Paper 75-WA/Prod-5, 1976.

(with R Hudson) Elementary rock climbing mechanics, Int. J. Mech. Eng. Educ. 4, 357-68, 1976.

(with A S Ranshi and N R Chitkara) Plastic yielding of I-beams under shear, and shear and axial loading, Int. J. Mech. Sci. 18, 375-87, 1976.

(with A R Watson, S R Reid and S G Thomas) Large deformations of thin-walled circular tubes under transverse loading II, Int. J. Mech. Sci. 18, 387-99, 1976.

(with I F Collins and S A Meguid) The co-indentation of a layer of two flat plane or spherical-headed rigid punches, Int. J. Mech. Sci. 19, 1-10, 1977.

(with A G Mamalis) Gegenuberstellung Statischer und Dynamischer Schadens-order Deformationserscheinungen, Fortschritts Ber. VDI, Z., 5(32), 72, 1977.

(with A R Watson and S R Reid) Large deformations of thin-walled circular tubes under transverse loading III, Int. J. Mech. Sci. 18, 501, 1976.

(with S Clyens and D J Williams) The compaction of some powdered foodstuffs, Int. J. Mech. Sci. 18, 449, 1976.

(with J F Silva-Gomes and S T S Al-Hassani) A note on times to fracture in solid perspex spheres due to point explosive loading, Int. J. Mech. Sci. 18, 541, 1976.

Longitudinal elastic stress waves due to impact, Engineering Feb, 106-8, 1977.

(with A G Mamalis and J Lewis) Forming of super-plastic zinc-aluminium sheet into a re-entrant die, Met. Technol. March, 160-66, 1977.

(with A G Mamalis) The fracture in some explosively end-loaded bars of plaster of paris and perspex containing transverse holes of changes in section, Int. J. Mech. Sci. 19, 169-76, 1977.

(with A G Mamalis) Aspects of mechanics in some sports and games, Fortschritt-Ber. VDI-Z. 17(4), 1977.

(with N R Chitkara) Some theoretical results of plane strain compression of a square compressed between flats and as a diamond, Proc. 1st All India MTDR Conf. pp.73-77, 1967.

(with A G Mamalis and Y Minara) The compression loading of solid spheres of plaster of paris, Int. J. Mech. Sci. 19, 373-77, 1977.

(with S R Reid and T Y Reddy) The compression of crossed layers of thin tubes, Int. J. Mech. Sci. 19, 423-37, 1977.

(with Y Minara) Crop loss: Front and back end deformation during slab and bloom rolling, Metall. Met. Form. 332-39, Aug. 1977.

Opening Address, 18th Int. Conf. Machine Tool Design and Research. Macmillan, 1977. Also: Int. J. Mech. Eng. Educ.

(with S Clyens) The dynamic compaction of powdered material, Mater. Sci. Eng. 30, 121-39, 1977.

(with A G Mamalis) Fracture development of solid perspex spheres with short cylindrical projections (bosses) due to point explosive loading, Int. J. Mech. Sci. 19, 309-14, 1977.

(with G H Daneshi) The ricochet of spherical projectiles off sand, Int. J. Mech. Sci. 19, 498, 1977.

(with A G Mamalis) Force polygons to determine upper bounds and force distribution in plane strain metal forming processes, 18th Int. Conf. Machine Tool Design and Research, pp.11-25, Macmillan, 1977.

(with A G Mamalis and H H Marczinski) A current state review of the warm working of metals, 18th Int. Conf. Machine Tool Design and Research, pp.173-82, Macmillan, 1977.

(with J B Hawkyard and C K S Gurnani) Pressure distribution measurements in rotary forging, Int. J. Mech. Eng. Sci. 19, 135-42, 1977.

(with P D Soden and S T S Al-Hassani) Inextensional collapse of thin-walled tubes under axial compression, J. Strain Anal. 12, 317-30, 1977.

(with G H Daneshi) The ricochet of dumb-bell shaped projectiles, Int. J. Mech. Sci. 19, 555-63, 1977.

(with G H Daneshi) Forces developed during the ricochet of projectiles of spherical and other shapes, Int. J. Mech. Sci. 19, 661-72, 1977.

(with S Clyens) A shear cell for the determination of the dynamic shear strength and flow behavior of particulate materials, Int. J. Mech. Sci. 19, 745-52, 1977.

(with A G Mamalis) Some force plane diagrams for plane strain slip-line fields, Int. J. Mech. Sci. 20, 47-56, 1978.

(with R D Venter and R L Hewitt) A matrix operator technique for slip-line field construction, Proc. 6th North American Metal-Working Research Conf., 1978.

On metal deformation processes, 21st Sagamore Conf. Plenum, 1978.

(with D J Williams and S Clyens) The electro-magnetic forming of powdered materials, Proc. 6th North American Metal-Working Research Conf., 1978.

(with S R Reid) Metallic energy dissipating systems, Appl. Mech. Rev. 277-88, March 1977.

(with G H Daneshi) The trajectory of a projectile when fired parallel and near to the free surface of a plastic solid, Int. J. Mech. Sci. 20, 255-64, 1978.

(with A G Mamalis) Fracture and void development in hemi-spherical end perspex rods due to stress waves from detonators, Proc. IUTAM Conf. High Velocity Deformation of Solids, pp.228-46. Springer, 1977.

(with G H Daneshi) Results for the single ricochet of spherical-ended projectiles off sand and clay at up to 400 m/sec, Proc. IUTAM Conf. High Velocity Deformation of Solids, pp.317-43. Springer, 1977.

(with J F Silva Gomes and S T S Al-Hassani) The plastic extension of a chain of rings due to an axial impact load, Int. J. Mech. Sci. 20, 529-38, 1978.

(with R Sowerby) Metal forming analysis and technology, Applications of Numerical Methods to Forming Processes, pp.1-14. ASME, 1978.

(with R D Venter and R L Hewitt) Applications of the matrix
inversion techniques to extrusion and drawing problems,
Applications of Numerical Methods to Forming Processes, pp.143-54.
ASME, 1978.

(with A G Mamalis) The perforation of circular plates with
four-sided pyramidally-headed square-section punches, Int. J.
Mech. Sci. 20, 849-66, 1978.

(with A G Mamalis) Plasticity and Metal Forming, p.85. Hellenic
Steel Pub., 1978.

(with A G Mamalis) Aspects of the Plasticity Mechanics of Some
Sheet Metal Forming Processes, p.52. Hellenic Steel Pub., 1978.

(with A G Mamalis) Ring-rolling processes, J. Prod. Eng., Indian
Soc. Mech. Eng. 2(3), 145-62, 1978.

(with M C de Malherbe, A G Mamalis and H K Tonshoff) Ultra-high
pressure equipment and techniques mainly for synthesising diamond
and cubic boron nitride, VDI Z. Forschrittsber. 2(37), 101, 1979.

(with S A L Salem and S T S Al-Hassani) Measurements of surface
pressure distribution during jet impact by a pressure pin
technique, 5th Conf. Erosion by Liquid and Solid Impact, 1979.

(with D J Williams and S Clyens) The production of a tungsten
copper pseudo-alloy by mechanical alloying using an unusual small
volume attritor and liquid phase sintering, 20th Int. Conf.
Machine Tool Design and Research, pp.103-10, 1979.

(with D J Williams) The production of porous ferrous components by
extension of the high voltage discharge forming process, 20th Int.
Conf. Machine Tool Design and Research, pp.607-12, 1979.

Applications: Processes involving high rates of strain. Inst.
Phys. Conf. Series No. 47, 337-59.

National Symp. Large Deformations: The State of Research,
pp.1-36. South Asian Pub Pvt, 1982.

(with S K Ghosh) The Quasi-static and dynamic perforation of thin
aluminium plates, Aluminium 56(2), 142-46, 1980.

(with S K Ghosh, A G Mamalis, T Y Reddy and S R Reid) The
quasi-static piercing of cylindrical tubes or shells, Int. J.
Mech. Sci. 22, 9-20, 1980.

(with A G Mamalis) Ingenious alternatives to the press in metal
forming, Weld. Met. Fabr. 375-83, 1979.

(with D J Williams) The production of porous ferrous components by
extensions of the high voltage discharge forming process,
Metallurgia 761-66, 1979.

(with N K Gupta and A G Mamalis) Sand spillage from an open-topped
rectangular container when in axial collison, Int. J. Vehicle Des.
1, 61-73, 1979.

(with A G Mamalis) Rolling of rings, Int. Met. Rev. 242, 137-48,
1979.

(with A C Walton) Airshipwreck: An analysis of the safety of rigid airships in terms of impact mechanics and fire, 5th Symp. Engineering Applications of Mechanics, pp.237-41, 1980.

(with D J Williams) Heat generation in the high voltage discharge forming of sponge iron powders, 21st Int. Conf. Machine Tool Design and Research. Macmillan, 1980.

(with N R Chitkara) An investigation into the possibility of roll peen-forming: Some experimental results, 21st Int. Conf. Machine Tool Design and Research, pp.75-84. Macmillan, 1980.

(with A G Mamalis and S K Ghosh) On the peen-forming of metals, 21st Int. Conf. Machine Tool Design and Research, pp.85-93, Macmillan, 1980.

(with S R Reid and T Y Reddy) Pipe-whip restraint systems, Chart. Mech. Eng. June, 55-60, 1980.

(with D Y Yang, J S Ryoo and J C Chol) The analysis of roll torque in profile ring-rolling of L-sections, 21st Int. Conf. Machine Tool Design and Research, pp.69-74, Macmillan, 1980..

(with S K Ghosh and S R Reid) Piercing and hole flanging of sheet metals: A survey, Proc. 11th Biennial IDDRG Congr., Metz, France, 1980, Mimonine Scientific Review Metallurgie, pp.585-606.

(with S R Reid and S K Ghosh) Piercing of cylindrical tubes, ASME Conf. Pressure Vessels and Piping, Paper No 80-C2/PVP-150, 1980. Also: J. Press. Vessel Technol., Trans ASME 103, 255-60, 1981.

(with D J Williams) A note on the formation of fulgurites, Geol. Mag. 3, 293-96, 1980.

(with A N Singh) Research into the elastic springback experienced with circular blanks formed by bending with a punch and die, Metallurgia 47, 275-80, 1980.

(with S K Ghosh) Demolition and dismantling mainly of building structures: An overview for use by lecturers, Int. J. Mech. Eng. Educ. 8(4) 111-26, 1980.

(with D J Williams and S Clyens) The production of tungsten-iron-nickel alloys by a mechanical alloying technique, Powder Metall. (2), 92-94, 1980.

A short review of metal forming mechanics and some processes of current research interest, 3rd Int. Conf. Mechanical Behaviour of Materials, Vol. 1, pp.167-225. Pergamon, 1980.

(with T Y Reddy and S R Reid) Model road-tank metal vessels subject to internal explosion of static penetration, J. Strain Anal. 15(4) 225-33, 1980.

(with S K Ghosh) Some physical defects arising in composite material fabrication, J. Mater. Sci. 16, 285-301, 1981.

(with T X Yu) Approximate calculations on the large plastic deformation of helical springs, and their possible application in a vehicle arresting system, J. Strain Anal. 16, 111-21, 1981.

(with A G Mamalis, A Kandell and M C de Malherbe) Defects in cold-hydrostatically extruded aluminium, iron and nickel base powder compacts, J. Met. Working Technol. 4, 327-40, 1981.

(with T X Yu) The angle of fold and the plastic work done in the folding of developable flat sheets of metal, J. Mech. Eng. Sci. 22, 233-41, 1980.

(with S T S Al-Hassani and S A L Salem) The catering of stationary and slowly moving targets by a high speed water jet, J. Ballistics 4(4), 909-33, 1980.

(with S G Ghosh, S R Reid and T X Yu) On thin rings and short tubes subjected to centrally opposed concentrated loads, Int. J. Mech. Sci. 23, 183-94, 1981.

(with N K Das Talukder) On the arrangements of rolls in cross-roll straighteners, Int. J. Mech. Sci. 23, 213-20, 1981.

Reflections engendered by introducing the papers on metal forming at the 21st International Machine Tool Design and Research Conference, J. Met. Working Technol. 5, 1-3, 1981.

(with T X Yu) The press-brake bending of rigid/linear work-hardening plates, Int. J. Mech. Sci. 23, 307-18, 1981.

(with T X Yu) Estimating the curvature of bars after cross-roll straightening, 23rd Int. Conf. Tool Design and Research, 1981.

(with S K Ghosh, S R Reid and M Golshekan) Pipe whip: Tubular cantilever beams subjected to localised constant force pulses, Conf. Structural Mechanics in Reactor Technology, Paper F8/3, 1981.

(with S K Ghosh and A G Mamalis) Ball-drop plate bending: An experimental study of some of its process variables, 1st Int. Conf. Shot Peening, pp.573-79. Pergamon, 1981.

(with A G Mamalis and K Isobe) Closed forging of rectangular billets with diamond-shaped dies, Ann. CIRP 30, 1981.

(with Y L Bai) An approximate estimate of energy absorption in plugging, II Meeting on Explosive Working of Materials, 1981.

(with A C Walton) Fires in public service vehicles in the United Kingdom, Int. J. Vehicle Des. 2, 322-34, 1981.

(with Y L Bai) The effects of projectile speed and medium resistance in ricochet off sand, J. Mech. Eng. Sci. 23(1), 69-75, 1981.

(with S R Reid and A J Edmunds) Bending of long steel and aluminium rods during impact with a rigid target, J. Mech. Eng. Sci. 23(2), 85-92, 1981.

(with T X Yu) Springback after the biaxial elastic-plastic pure bending of a rectangular plate - I, Int. J. Mech. Sci. 23, 619-30, 1981.

(with T X Yu) On the range of applicability of results for the springback of an elastic/perfectly plastic rectangular plate after subjecting it to biaxial pure bending - II, Int. J. Mech. Sci. 23, 631-37, 1981.

(with Y L Bai) An estimate of the effect of back spin on the critical angle in tne unyawed ricochet of a cylinder, J. Mech. Eng. Sci. 23, 201-5, 1981.

(with Y L Bai and B Dodd) On tangential velocity discontinuities being coincident with stress discontinuities, Int. J. Mech. Sci. 23, 323-28, 1981.

(with T X Yu) On springback after tne pure bending of beams and plates of elastic work-hardening material - III, Int. J. Mech. Sci. 23, 687-95, 1981.

(with T X Yu) On the range of applicability of results for tne springback of an elastic-work hardening rectangular plate after subjecting it to biaxial pure bending - IV, Int. J. Mech. Sci. 23, 697-701, 1981.

(with T X Yu) The location of the plastic hinge in a quadrantal circular curved beam struck in its plane by a jet at its tip, ASME Advances in Aerospace Structures Conf., pp.175-80, 1981.

(with T X Yu) The influence of axial force on the elastic-plastic bending and springback of a beam, J. Met. Working Technol. 6, 5-21, 1982.

Tne mechanics of some industrial pressing, rolling and forging processes. In: Hopkins H G and Sewell M J, Mechanics of Solids, pp.303-55. Pergamon, 1981. Also: republished, in part, in Seminar on Manufacturing Technology, UNESCO-CIRP Conf., pp.132-66, 1982.

(witn Y L Bai) Plugging: Physical understanding and energy absorption, Met. Technol. 9, 182-90, 1982.

(with Y L Bai, S K Ghosn and A K Sengupta) Tne fracture of brittle circular discs under impact load, Indian J. Prod. Eng. 4, 1-12, 1981.

(with A K Sengupta, S K Ghosh and S R Reid) Mechanics of high speed impact at normal incidence between plasticine long rods and plates, J. Mech. Phys. Solids 29, 413-45, 1981.

(with T X Yu) The buckling of annular plates in relation to the deep drawing process, Int. J. Mech. Sci. 24, 175-88, 1982.

(with A K Sengupta, C J Wrigglesworth, S K Ghosh and S R Reid) The normal impact of square section long rods on rigid and soft targets, J. Mech. Eng. Sci. 24, 31-35, 1982.

(with A G Mamalis and S R Reid) Aspects of car design and human injury. In: Ghista D N (ed.) Human Body Dynamics: Impact, Occupational and Athletic Aspects, pp.164-80. Oxford University Press, 1982.

(with S R Reid and A G Mamalis) Human body mechanics in some sports and games. In: Ghista D N (ed.) Human Body Dynamics: Impact, Occupational and Athletic Aspects, pp.524-42. Oxford University Press, 1982.

(with A K Sengupta and S K Ghosn) High velocity oblique impact and ricochet mainly of long rod projectiles: An overview, Int. J. Mech. Sci. 24, 425-36, 1982.

(with A K Sengupta and S K Ghosh) Plasticine modelled high
velocity oblique impact and ricochet of long rods, Int. J. Mech.
Sci. 24, 437-55, 1982.

(with D J Williams) Neck formation and growth in the high voltage
discharge forming of metal powders, Powder Metallurgy, 25, 85-89,
1982.

(with Y L Bai and S K Ghosh) A computerised long time 1-d elastic
stress wave analysis of a large impact system of oil and metal
columns of varying geometry with some results. Numerical Methods
in Industrial Forming Processes, pp.267-78. Pineridge Press, 1982.

(with P E Rees) The plastic collapse of short hollow cylinders or
rings under diametral load, National Symp. Large Deformations,
pp.88-97. IIT, Delhi, South Asian Pub Pvt Ltd, 1982.

(with S K Ghosh) Large plastic deformation in thin wide
rectangular strips loaded inertially and the notion of damage
number, National Symp. Large Deformations, pp.98-107. IIT,
Delhi, South Asian Pub Pvt Ltd, 1982.

(with A N Singh) Springback after cylindrically bending metal
strips National Symp. Large Deformations, pp.236-50. IIT, Delhi,
South Aisan Pub Pvt Ltd, 1982.

Large deformations in structures and people caused by impact and
blast, National Symp. Large Deformations, pp.323-43. IIT, Delhi,
South Asian Pub Pvt Ltd, 1982. Also, J. Met. Working Technol. 10,
145-64, 1984.

(with D J Williams) The radiusing of brass components by short
term dry self-tumbling, J. Met. Working Technol. 355-66, 7(4),
1983.

(with S R Reid, A K Sengupta and S K Ghosh) Modelling with
plasticine the low speed impact of long rods against inclined
rigid targets, Int. J. Impact Eng. 1, 73-84, 1983.

(with R Sowerby and R D Venter) Theory of the plastic deformation
of metals. In: Unksov E P (ed.) Slip Line Fields, pp.121-211.
Mashinostroyeniye, Moskva, 1983.

(witn A G Atkins and G W Rowe) Snear strains and strain-rates in
kinematically admissible velocity fields, Int. J. Mech. Eng. Educ.
10(4), 265-78, 1982.

(with T X Yu) The plastica: The large elastic-plastic deflection
of a strut, J. Non-Linear Mech. 17, 195-210, 1983.

(with T X Yu) The large elastic plastic deflection with springback
of circular plates subjected to circumferential moments, J. Appl.
Mech. 49, 507-15, 1982.

(with C D Austin and B J Walters) Bursting and bulging of carbon
fibre composite discs, J. Composites 383-88, October, 1982.

Aspects of damage to buildings from uncased explosives. In:
Dvorak C J and R T Shield (eds.) Mechanics of Materials Behavior,
pp.175-90. Elsevier, 1984.

Structural damage in airship and rolling stock collision,
Structural Crashworthiness, pp.417-39. Butterworths, 1983.

Aspects of metal plasticity, Akad. Athenon, 57, 391-401, 1982.

Industries and academics, Br. Impact Bull. 2(1), 3, 1982.

(with T X Yu and W J Stronge) Long stroke energy dissipation in splitting tubes, IJMS, 25, 637-48, 1983.

(with Y L Bai, G M Low and S K Ghosh) Asympototic behaviour in slumping from a cylindrical tank, J. Fluids Eng. 106, 279-84, 1984.

(with T X Yu) Experiments in the cylindrical bending of strips, Met. Technol. 10, 439-47, 1983. Also: (with T X Yu and W J Stronge) Conf. Theoretical Plasticity, 1984.

Guest Editorial, J. Mech. Working Tecn. 8, 1-11, 1983.

(with T X Yu) The large elastic deflection of thin strip in V-die bending and the onset of plasticity, Pt. C (JMES), Proc. Inst. Mech. Eng. 1-7, 1980.

(with A C Walton) Protection of car occupants in frontal impacts with heavy lorries: Frontal structures, Int. J. Impact Eng. 1(2), 111-24, 1983.

(with A C Walton) An experimental investigation of the energy dissipation of a number of car bumpers under quasi-static lateral loads, Int. J. Impact Eng. 1(3), 301-8, 1983.

(with Y L Bai and S K Ghosh) Gross crack initiation and propagation in brittle thin solid and annealed discs subjected to impact loading, Materials Division, J. Mater. Eng. Tech. 106, 167-72, 1984.

(with D J Williams and B J Walters) Crack patterns in cylinders explosively loaded at an hemispherical end, Int. J. Fracture, 23, 271-79, 1983.

(with A G Mamalis) The quasi-static crumpling of thin-walled circular frusta under axial compression, Int. J. Mech. Sci. 25, 713-32, 1983.

(with N K Gupta) The compression of crossed layers of wooden bars, Int. J. Mech. Sci. 25, 697-712, 1983.

(with D Y Yang and J Ryoo) Ring rolling: The inclusion of pressure roll speed for estimating torque by using a velocity superposition method, 24th Conf. Machine Tool Design and Research, pp.19-24. Macmillan, 1983.

Common defects in metal formed products, Plasticity Today: Modelling, Methods and Applications. Elsevier, 1983.

(with T X Yu) The cylindrical bending of strips and the biaxial stamping of rectangular plates into curved dies, Berichte, Nov 1983, pp.38, Reihe 38. Also: (with T X Yu and W J Stronge) Indian J. Prod. Eng. (In Press).

(with T X Yu) A theoretical analysis of the bending into cylindrical dies of metal strips, Proc. Inst. Mech. Eng. 198C, 99-108, 1984.

(with T X Yu and W J Stronge) Stamping of rectangular plates into doubly-curved dies, Proc. Inst. Mech. Eng. 198C, 109-25, 1984.

(with T X Yu and W J Stronge) Stamping and springback of circular plates deformed in hemispherical dies, IJMS 26, 131-48, 1984.

Presidential Address to Manchester Technology Association, Roots and Branches, pp.7-14, Nov 1983.

Vehicle impact and fire hazards, Structural Impact and Crashworthiness Conf., Vol. 1, pp.75-113. Elsevier, 1984.

(with T X Yu and W J Stronge) Energy dissipation by splitting and curling tubes, Structural Impact and Crashwortniness Conf., Vol. 2, pp.576-87. Elsevier, 1984.

(with S K Gnosh and S R Reid) The large deflection impulsive loading of clamped circular membranes, Structural Impact and Crashworthiness Conf., Vol. 2, pp.471-81. Elsevier, 1984.

(with J S Ryoo and D Y Yang) Lower upper bound analysis of the ring rolling process by using a force polygon diagram and dual velocity field, Int. Conf. Theoretical Plasticity, 1984.

(with A G Mamalis and G L Veigelahn) The crumpling of steel thin-walled tubes and frusta under axial compression at elevated strain-rates: Some experimental results, IJMS, 27, (In Press).

(with T X Yu and E Appleton) Buckling in rotary forging and ring rolling, J. Eng. Mater. Technol. (In Press).

(with D Y Yang and J S Ryoo) A theoretical and experimental investigation into roll torque and pressing load in tne plane strain ring rolling process, J. Ind. (In Press).

(with T X Yu and P S Symonds) A quadrantal circular beam subjected to radial impact in its own plane and at its tip by a rigid mass, Proc. R. Soc. (In Press).

(with T Alp and S T S Al Hassani) Electrical discharge compaction of powder: Mechanics and material structure, J. Eng. Mater. Technol. (In Press).

(witn J S Ryoo and D Y Yang) The influence of process parameters on rolling torque and load in the ring rolling process, J. Mech. Working Technol. (In Press).

(with T X Yu and Y L Hua) The plastic ninge position in a circular cantilever when struck normal to its plane by a constant jet at its tip, Int. J. Impact Eng. (In Press).

Fire and damage after vehicular impact, 1st Int. Symp. Fire Safety Science, 1985.

Historical aspects of fire, after impact, in vehicles of war, 1st Int. Symp. Fire Safety Science, 1985.

(with V Isobe and A G Mamalis) Deformation of billets in plane strain forging with Vee-shaped dies, Proc. 34th Japan Joint Conf. Telch Plasticity, pp.535-38, 1983.

(with A G Mamalis and K Isobe) Close die forging of rectangular billets with diamond-shaped dies, Ann. CIRP 30, 147-52, 1981.

(with N K Gupta) Rate effects in lateral compression for an energy absorbing unit of crossed layers of wooden bars, J. Impact Eng. 1985 (In Press).

METAL FORMING
MECHANICS

ON THE KINEMATICS OF STEADY PLANE FLOWS IN ELASTOPLASTIC MEDIA

R. HILL

*Department of Applied Mathematics and Theoretical Physics,
Silver Street, Cambridge CB3 9EW, UK*

ABSTRACT

Arbitrary plane flows in steady-state forming are envisaged here.
The usual assumption of rigid/plastic response is relaxed by incor-
porating an elastic constituent of incremental deformation, proportional to the Jaumann flux of deviatoric stress. The medium remains incompressible, non-hardening, and incrementally isotropic.
The differential equations for the components of velocity are still
hyperbolic, but the characteristics no longer coincide with the sliplines as in classical theory. However, the respective nets intersect
at a constant angle, dependent only on the ratio of yield stress to
elastic modulus. Unexpectedly simple relations along the characteristics are found in terms of an optimal decomposition of the velocity vector. These results suggest that steady flows in elastoplastic
media are amenable to computing techniques broadly similar to those
applied to rigid/plastic media. Nevertheless the separation of
characteristics and sliplines does pose unfamiliar questions of procedure in the solution of boundary-value problems. Their resolution
is expected to throw light on difficulties experienced with the
classical theory.

1. HISTORICAL BACKGROUND

The steady flows envisaged here are quasi-static, free from body
force, and plane. The medium is assumed to be elastoplastic (but
incompressible), non-hardening, and incrementally isotropic. Otherwise its general response is subject only to such minimal restrictions
as will ensure that the maximum shearing stress at yield in the planes
of flow has an invariant value, k. In brief, the particular response
differs from a classical rigid/plastic medium in just one respect:
incremental elastic shearing is governed by a finite modulus, μ.
When defined objectively via the Jaumann flux of Cauchy stress, μ
is supposed unaffected by deformation and pressure.

The earliest rigorous treatment of a non-trivial problem of steady
flow in a rigid/plastic medium is to be found in a classified report
dated 1945 (published by H.M.S.O. in 1952). In that connexion a
personal reminiscence may be of interest. During the war I was

3

engaged on armaments research, and one assignment was to estimate
the stresses in a malleable driving-band that aligned a high-velocity
shell within a prototype gun-barrel. As an idealization I elected
to study a related type of deformation in plane extrusion. Since the
computation of slipline nets was then an arduous undertaking, I set-
tled for a square die, a 50% reduction, and 90° centred fans. Eventually I noticed that this configuration could accommodate the redundant kinematic data (at entry, exit, and on the dead-metal interface).
It would hence deliver the contemplated extrusion; with hindsight
this perception could seem unremarkable. However, in the face of
received doctrine (which paid no heed to the kinematic imperatives),
it amounted to a radical shift in standpoint and cleared the way for
rigorous analyses of forming processes in general (Hill, 1947, 1950).

Subsequent mathematical developments have established the subject
(along with non-steady flows) as a distinctive branch of solid mechanics. Its special features are that the field equations are hyperbolic, with an elegant canonical structure, while the boundary-value
problems are unusually realistic, often with highly specific and
intricate geometries. In the early years, exact solutions were
quickly forthcoming where the slipline nets could be validated by
merely qualitative reference to the kinematic data, without calculations of hodographs. On the other hand, notionally simple problems
were soon encountered where the data and geometry were not so oblig-
ing, and a slipline net and hodograph had to be constructed in tandem.
Likely complexities in this second category were starkly revealed
by Green (1951), in an iterative solution of a capital problem (albeit
for initial motion). Computer capacity temporarily restrained further
progress on this front, save for an heroic assault on hot rolling
by Alexander (1955). Meanwhile solutions in the first category were
accumulated steadily, more especially in asymmetric forging and extrusion (Johnson and Kudo, 1962). A select third category, illustrated by strip drawing with convexly curved dies, was remarked by
Sokolovsky (1962); here the hodograph could be validated in advance
by merely qualitative reference to a schematic net, whose own specific
geometry was determinable uniquely thereafter from the die profile.
On another tack Richmond and Devenpeck (1962) exhibited a particular
sigmoidal profile that eliminated redundant work in frictionless
drawing. The general possibility of such ideal flows, with a variety
of profiles, was established by Hill (1966). Finally, the most recent
phase of development was initiated by Collins (1968), who devised a
special matrix algebra for slipline nets. This is well suited to
digital computers and has led to new insights in ever more complex
configurations (Collins, 1982; Johnson, Sowerby and Venter, 1983).

Interest in rigid/plastic theory has been sustained over this long
period not just by the attractions of distinctive mathematics, but
more by repeated demonstrations that its predictions are useful to
engineers and metallurgists. Nonetheless, some limitations have
been apparent from the outset, arising from the degeneracy of the
constitutive model. It is not so much that this is short on realism
(which to an extent can be allowed for), but rather that the mathematical consequences of the model itself can be uncertain. Most
conspicuously, without extraneous considerations, the stresses are
indeterminate in those parts of a medium that are currently rigid.
It is notoriously difficult even to construct *any* self-equilibrated,
sub-yield, field there which is compatible with a proposed net in
the working zone. Some such state is demanded by the uniqueness
theorem for slipline nets (Hill, 1951), which is relevant to steady
flow when the overall boundary is a datum. The dilemma is not yet
resolved, but it has been somewhat ameliorated by a criterion to
test for overstressing at critical localities (Hill, 1954). Even

when a degree of overstressing is demonstrable, especially where an
actual medium would be relatively harder (as at exit), a proposed net
may still be acceptable pragmatically. Somewhat in this spirit,
though at a further remove, engineers have often resorted to flows
that are admissible kinematically but not statically, in the knowledge
that these can deliver close estimates, always from above, to the work
of frictionless forming (Johnson and Kudo, 1962; Johnson and Mellor,
1973; Johnson and Mamalis, 1977). The relevant extremum principle,
in the precise context of rigid/plastic media, was given by Hill (1951)
along with an initial application to steady flow (drawing).

Effects of constitutive degeneracy are felt also in the working zone
itself. The absence of hardening often leads to theoretical patterns
of flow with angular streamlines. However, this feature of the model
is acceptable as an idealization of what is seen in zones of intense
shearing in forming processes with pre-strained materials. Potentially more troublesome for the theory in some ostensibly well-set
problems is an apparent non-uniqueness of *deformation*. This phenomenon
has been studied fairly thoroughly for non-steady processes, when
treated incrementally so that the medium boundary and slipline net
are known at each instant. It has been shown (Hill, 1956) that any
branching deformations allowed by the kinematic data are inhibited
by a sufficient rate of hardening; this can be arbitrarily small
when changes in geometry are negligible, and also on occasion when
they are not. In steady flow, on the other hand, an apparent non-
uniqueness is encountered mainly when the overall configuration is
not known in advance, as in machining (Dewhurst, 1978; Petryk, 1979).
In this respect the role of flow-induced hardening is little understood since the field equations are peculiarly intractable (Collins,
1982).

Lack of elasticity in the working zone counts as degeneracy, just as
it does elsewhere, but the mathematical consequences are not known.
General studies of plane deformations in elastoplastic media
have concentrated hitherto on non-steady processes, as for example
bifurcations from homogeneous states after finite strain (Hill and
Hutchinson, 1975; Hill, 1979). Steady plane flows, on the other
hand, seem not to have been treated analytically in systematic
fashion.

A start in that direction is made here, in order to gauge the more
immediate effects of degeneracy. With the constitutive model as
outlined, the slipline net is still classical since there is no flow-
induced hardening. Further, as will be shown, the field equations
for the components of velocity remain hyperbolic when $k < \mu$. But the
characteristic curves are not the sliplines, nor are they orthogonal.
Fortunately, the two nets stand in simple relation, their angle of
intersection being constant, namely $\frac{1}{2}\sin^{-1}(k/\mu)$. Any standard method
will furnish integrable relations along these characteristics, in
generalization of Geiringer's along sliplines when μ is infinite.
What is not automatic, however, is the optimal version of these relations, which depends on well-chosen kinematic variables. These
proved elusive, in fact, but the canonical structure is remarkably
concise.

        2.   CONSTITUTIVE MODEL

In a typical plane of flow let (x,y) be any cartesian coordinates in
a frame of reference relative to which the motion is steady. Denote
the corresponding components of Cauchy stress by σ_x, σ_y, τ_{xy}. In the

working zone it is assumed that

$$\frac{1}{4}(\sigma_x - \sigma_y)^2 + \tau_{xy}^2 = k^2 \qquad (2.1)$$

where k is a material constant. Denote the tensor components of Eulerian strain-rate by ε_x, ε_y, γ_{xy}. Then

$$\varepsilon_x + \varepsilon_y = 0, \qquad (2.2)$$

assuming both elastic and plastic incompressibility.

The response to incremental shearing is taken to be

$$\left.\begin{matrix} \frac{1}{2}(\varepsilon_x - \varepsilon_y) \\ \\ \gamma_{xy} \end{matrix}\right\} = \frac{1}{2\mu}\left(\frac{D}{Dt} + \lambda\right)\left\{\begin{matrix} \frac{1}{2}(\sigma_x - \sigma_y) \\ \\ \tau_{xy} \end{matrix}\right. \qquad (2.3)$$

where μ is an elastic modulus, λ is arbitrarily positive or zero, and D/Dt is the Jaumann rate of change for a material element. It is formed from tensor components on rotating axes which coincide momentarily with (x,y) but spin with the element. The response may be described as incrementally isotropic, since the constitutive relations are the same for components on any other coordinates attached to the background frame of reference. In particular, the λ terms represent a plastic strain-rate which is always coaxial with the Cauchy stress but is unspecified in magnitude. The classical rigid/plastic model is recovered when $\mu \to \infty$, keeping λ/μ finite.

Within the constraint of incremental isotropy, there is still some theoretical freedom of choice as to the elastic contribution to the total strain-rate. It could be defined by association with an objective flux other than the Jaumann rate, for instance $\delta/\delta t$ where

$$\left(\frac{D}{Dt} - \frac{\delta}{\delta t}\right)\left\{\begin{matrix} \frac{1}{2}(\sigma_x - \sigma_y) \\ \\ \tau_{xy} \end{matrix}\right. = m(\sigma_x + \sigma_y)\left\{\begin{matrix} \frac{1}{2}(\varepsilon_x - \varepsilon_y) \\ \\ \gamma_{xy} \end{matrix}\right.$$

and m is typically ± 1 but may have any value. This family of fluxes is standard, and in particular was considered by Hill and Hutchinson (1975) in regard to elastoplastic models with hardening and orthotropy. When (2.3) is expressed in terms of $\delta/\delta t$, the algebraic structure remains the same and λ is invariant in value; but μ is replaced by $\mu - m\sigma$ where

$$\sigma = \frac{1}{2}(\sigma_x + \sigma_y) \qquad (2.4)$$

is the hydrostatic tension. Since σ is a function of position in a steady flow, only one modulus in the family $\mu - m\sigma$ can be stipulated to be a *material constant*. It will be assumed that the favoured modulus is μ itself ($m = 0$), in order to simplify the field equations in the working zone. Since σ/μ is usually small for actual materials, the point at issue is more academic than practical. Constancy of μ is an acceptable approximation also for purely elastic deformation, whose order of magnitude is k/μ during loading at entry or unloading at exit.

Since the left side of (2.1) is a tensor invariant, its Jaumann flux vanishes and hence

$$\frac{1}{2}(\sigma_x - \sigma_y) \frac{D}{Dt} \frac{1}{2}(\sigma_x - \sigma_y) + \tau_{xy} \frac{D}{Dt} \tau_{xy} = 0 \qquad (2.5)$$

in the working zone. The rate of working per unit volume is

$$\sigma_x \epsilon_x + \sigma_y \epsilon_y + 2\tau_{xy}\gamma_{xy} = \frac{1}{2}(\sigma_x - \sigma_y)(\epsilon_x - \epsilon_y) + 2\tau_{xy}\gamma_{xy}$$

$$- 2k^2\lambda \geq 0 \qquad (2.6)$$

by (2.2) and (2.3) successively. The elastic contribution has disappeared by virtue of the orthogonality (2.5) between the stress deviator and its objective flux during plastic flow. This orthogonality further implies that *the elastic contribution to the strainrate tensor is coaxial with the sliplines*, whereas the plastic contribution is coaxial with the stress.

3. CARTESIAN FIELD EQUATIONS

Let (u, v) be the (x, y) components of velocity on arbitrary axes in the same frame of reference as before. Then

$$\epsilon_x = \frac{\partial u}{\partial x}, \quad \epsilon_y = \frac{\partial v}{\partial y}, \quad \gamma_{xy} = \frac{1}{2}(\frac{\partial u}{\partial y} + \frac{\partial v}{\partial x}), \qquad (3.1)$$

while the material spin is

$$\omega = \frac{1}{2}(\frac{\partial v}{\partial x} - \frac{\partial u}{\partial y}) \qquad (3.2)$$

anticlockwise relative to the axes.

By a standard result the required Jaumann rates in a steady flow are

$$\frac{D}{Dt}\begin{Bmatrix} \frac{1}{2}(\sigma_x - \sigma_y) \\ \tau_{xy} \end{Bmatrix} = (u\frac{\partial}{\partial x} + v\frac{\partial}{\partial y})\begin{Bmatrix} \frac{1}{2}(\sigma_x - \sigma_y) \\ \tau_{xy} \end{Bmatrix} + 2\omega\begin{Bmatrix} \tau_{xy} \\ -\frac{1}{2}(\sigma_x - \sigma_y) \end{Bmatrix}, \qquad (3.3)$$

where the operator $(u\partial/\partial x + v\partial/\partial y)$ takes care of material convection through an inhomogeneous field of stress. In the working zone these rates reduce to

$$\frac{D}{Dt}\begin{Bmatrix} \frac{1}{2}(\sigma_x - \sigma_y) \\ \tau_{xy} \end{Bmatrix} = 2k(u\frac{\partial\theta}{\partial x} + v\frac{\partial\theta}{\partial y} - \omega)\begin{Bmatrix} -\sin 2\theta \\ \cos 2\theta \end{Bmatrix}, \qquad (3.4)$$

having regard to

$$\frac{1}{2}(\sigma_x - \sigma_y) = k\cos 2\theta, \quad \tau_{xy} = k\sin 2\theta \qquad (3.5)$$

where θ is reckoned anticlockwise from the x-axis to the major principal stress, $\sigma + k$. From (3.4) we recognize again that the elastic contribution to the strain-rate tensor is a pure shear relative to the axes of stress. Also, the bracketed quantity is the rate at which these axes rotate relative to a material element as it traverses the field. It is understandable that this quantity should be the sole factor governing changes in the objective stress felt by an element, since the deviator here has fixed principal values k, $-k$.

By substituting (3.4) in (2.3) and eliminating λ, we finally obtain

$$\cos2\theta \ (\frac{\partial u}{\partial y} + \frac{\partial v}{\partial x}) + \sin2\theta \ (\frac{\partial v}{\partial y} - \frac{\partial u}{\partial x}) = \frac{2k}{\mu}(u\frac{\partial\theta}{\partial x} + v\frac{\partial\theta}{\partial y} - \omega). \quad (3.6)$$

Here ω is expressible in derivatives by (3.2), while $\theta(x,y)$ is known from the slipline net. Together with the incompressibility relation

$$\frac{\partial u}{\partial x} + \frac{\partial v}{\partial y} = 0, \qquad\qquad\qquad\qquad\qquad (3.7)$$

these are the basic first-order equations for the (u,v) components.

Any characteristic can be located by testing the field equations for possible jumps in the derivatives across a curve along which (u,v) are presumed known. Let the tangent to the curve be inclined to the x-axis at an angle ψ anticlockwise. Having regard to (3.7) and the constraints

$$\cos\psi \ [\frac{\partial u}{\partial x}] + \sin\psi \ [\frac{\partial u}{\partial y}] = 0 = \cos\psi \ [\frac{\partial v}{\partial x}] + \sin\psi \ [\frac{\partial v}{\partial y}]$$

imposed by the data, any jumps $[...]$ are subject to

$$\frac{[\frac{\partial u}{\partial x}]}{\sin\psi\cos\psi} = \frac{-[\frac{\partial u}{\partial y}]}{\cos^2\psi} = \frac{[\frac{\partial v}{\partial x}]}{\sin^2\psi} = \frac{-[\frac{\partial v}{\partial y}]}{\sin\psi\cos\psi} \ .$$

The convective terms in (3.6) are continuous across the curve, so there remains

$$\cos2\theta \ [\frac{\partial u}{\partial y} + \frac{\partial v}{\partial x}] + \sin 2\theta \ [\frac{\partial v}{\partial y} - \frac{\partial u}{\partial x}] = \frac{k}{\mu} \ [\frac{\partial u}{\partial y} - \frac{\partial v}{\partial x}].$$

Whence, if jumps are to be possible, ψ must satisfy

$$\cos2(\psi-\theta) = \frac{k}{\mu}, \qquad\qquad\qquad\qquad\qquad (3.8)$$

Therefore, when $0 \leq k/\mu < 1$, there are two characteristic directions at every point, such that

$$\left.\begin{array}{ll} \psi = \theta \pm \eta, & \eta = \frac{1}{4}\pi - \varepsilon, \\[2mm] \varepsilon = \frac{1}{2}\sin^{-1}(k/\mu), & 0 \leq \varepsilon < \frac{1}{4}\pi. \end{array}\right\} \qquad (3.9)$$

Since μ is a material constant, as explained, angles ε and η do not vary with position. The local disposition of the characteristics, sliplines, and major principal stress is shown schematically in Fig. 1. As k/μ notionally increases from zero, the characteristics move away from the sliplines towards the major principal axis, which always bisects the acute angle between them. When $k = \mu$ the field equations are parabolic, and when $k > \mu$ they are elliptic; these cases have no relevance for metal forming.

The space derivative along a characteristic will be denoted by

$$\frac{\partial}{\partial s} = \cos\psi \ \frac{\partial}{\partial x} + \sin\psi \ \frac{\partial}{\partial y}. \qquad\qquad\qquad (3.10)$$

Since jumps are possible, the field equations do not determine the (x,y) derivatives of (u,v) uniquely in terms of $\partial u/\partial s$ and $\partial v/\partial s$. It follows that a linear combination of the basic pair can be found which involves no derivatives other than $\partial u/\partial s$ and $\partial v/\partial s$, and hence serves merely as a constraint on the joint values of u and v along the particular curve. Such characteristic relations may be generated systematically by any one of several methods; the most economical in effort is due to Hill (1979) and caters for any n linear differential equations of first or second order in n dependent variables. The present pair is so simple, however, that the critical combination can be divined without recourse to formalism.

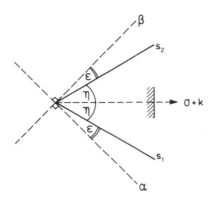

Fig. 1.   Relative local directions of sliplines
(α,β), characteristics (s_1, s_2), and major
principal stress, in terms of the angles
$\varepsilon = \tfrac{1}{2}\sin^{-1}(k/\mu)$ and $\eta = \tfrac{1}{2}\cos^{-1}(k/\mu)$ when
$0 < k < \mu$.

Let (3.7) be multiplied by $\sin 2(\psi-\theta)$ and added to the left side of (3.6). Replace $2k\omega/\mu$ by

$$\cos 2(\psi-\theta)\left(\frac{\partial v}{\partial x} - \frac{\partial u}{\partial y}\right)$$

in accordance with (3.2) and (3.8). The outcome, which is immediate, is

$$\sin(\psi-2\theta)\,\frac{\partial u}{\partial s} + \cos(\psi-2\theta)\,\frac{\partial v}{\partial s} = \frac{k}{\mu}\left(u\frac{\partial\theta}{\partial x} + v\frac{\partial\theta}{\partial y}\right),\qquad(3.11)$$

where ψ can take either value given by (3.9). For a rigid/plastic medium, where $k/\mu = 0$ and $\psi-\theta = \pm\tfrac{1}{4}\pi$, both relations reduce to

$$\cos\psi\,\frac{\partial u}{\partial s} + \sin\psi\,\frac{\partial v}{\partial s} = 0,$$

which is a statement of zero strain-rate in each ψ direction. Rearrangement as

$$\frac{\partial}{\partial s}\,(u\cos\psi + v\sin\psi) = (v\cos\psi - u\sin\psi)\,\frac{\partial\psi}{\partial s}$$

displays the slipline components of velocity, as in the relations of Geiringer (1937).

4. CANONICAL FIELD EQUATIONS

In order to reduce the characteristic relations to canonical form, the cartesian components (u,v) have to be replaced by appropriate kinematic variables. Since the characteristics are not orthogonal, the optimal choice is not apparent in advance, nor is there any helpful precedent or guideline. Several decompositions of velocity might well be considered as candidates: for instance, perpendicular components on the sliplines, or on the axes of stress, or on the streamline and its normal, or on either characteristic and its normal, or vector (contravariant) components on the two characteristics, or projected (covariant) components on the two characteristics. All these and other possibilities were reviewed, the last not least since it was effective for incipient flow of a rigid/plastic medium under plane stress (Hill, 1950, p.306). The best choice emerged only by degrees, as the inner structure of the field equations revealed it-self.

Let the velocity vector be decomposed by the parallelogram rule into components (p,q) on the characteristic directions $\theta-\eta$, $\theta+\eta$ respectively. The curves themselves are labelled s_1, s_2 (Fig. 2), the associated space derivatives being $\partial/\partial s_1$, $\partial/\partial s_2$. Further, the derivatives normal to these directions in the senses shown are $\partial/\partial n_1$, $\partial/\partial n_2$. Then

$$\left.\begin{aligned}
\frac{\partial}{\partial s_1} &= \cos\eta\,\frac{\partial}{\partial x_1} - \sin\eta\,\frac{\partial}{\partial x_2}, & \frac{\partial}{\partial s_2} &= \cos\eta\,\frac{\partial}{\partial x_1} + \sin\eta\,\frac{\partial}{\partial x_2},\\[2mm]
\frac{\partial}{\partial n_1} &= \sin\eta\,\frac{\partial}{\partial x_1} + \cos\eta\,\frac{\partial}{\partial x_2}, & \frac{\partial}{\partial n_2} &= \sin\eta\,\frac{\partial}{\partial x_1} - \cos\eta\,\frac{\partial}{\partial x_2}
\end{aligned}\right\}(4.1)$$

where (x_1, x_2) are cartesian axes aligned with the principal stresses. Immediate corollaries are

$$\left.\begin{aligned}
\sin 2\eta\,\frac{\partial}{\partial s_1} &= \cos 2\eta\,\frac{\partial}{\partial n_1} + \frac{\partial}{\partial n_2}, & \sin 2\eta\,\frac{\partial}{\partial s_2} &= \frac{\partial}{\partial n_1} + \cos 2\eta\,\frac{\partial}{\partial n_2},\\[2mm]
\sin 2\eta\,\frac{\partial}{\partial n_1} &= \frac{\partial}{\partial s_2} - \cos 2\eta\,\frac{\partial}{\partial s_1}, & \sin 2\eta\,\frac{\partial}{\partial n_2} &= \frac{\partial}{\partial s_1} - \cos 2\eta\,\frac{\partial}{\partial s_2}.
\end{aligned}\right\}$$

$$(4.2)$$

It is convenient to note at this stage the equivalent formula for the convection operator:

$$u\frac{\partial}{\partial x} + v\frac{\partial}{\partial y} \equiv p\frac{\partial}{\partial s_1} + q\frac{\partial}{\partial s_2} \;. \tag{4.3}$$

Here, in effect, the scalar product of the vectors (u,v) and $(\partial/\partial x, \partial/\partial y)$ is evaluated, as in tensor algebra, from the contravariant representation (p,q) of one and the covariant representation $(\partial/\partial s_1, \partial/\partial s_2)$ of the other.

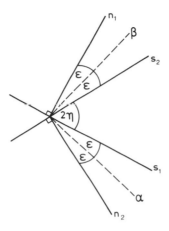

Fig. 2.   Local directions of characteristics (s_1,
s_2) and their normals (n_1, n_2), showing
the conventional senses assumed in the
canonical relations.

We could now proceed directly to transform (3.11). In the interests
of symmetry and structure it is preferred to begin afresh from field
equations based on the stress axes (x_1, x_2). Let (ε_{11}, ε_{22}, ε_{12}) be
the associated tensor components of strain-rate. The *intrinsic* com-
ponents of velocity along the trajectories of principal stress are
($p+q$)cosη, ($q-p$)sinη. From these the gradients of *cartesian* components of velocity are generated by a standard technique, leading to

$$\varepsilon_{11} + \varepsilon_{22} = \frac{\partial p}{\partial s_1} + \frac{\partial q}{\partial s_2} + p\frac{\partial \theta}{\partial n_1} - q\frac{\partial \theta}{\partial n_2}, \tag{4.4}$$

$$\varepsilon_{11} - \varepsilon_{22} = \frac{\partial p}{\partial s_2} + \frac{\partial q}{\partial s_1} + p\frac{\partial \theta}{\partial n_2} - q\frac{\partial \theta}{\partial n_1}, \tag{4.5}$$

$$2\varepsilon_{12} = \frac{\partial q}{\partial n_1} - \frac{\partial p}{\partial n_2} + p\frac{\partial \theta}{\partial s_2} + q\frac{\partial \theta}{\partial s_1}, \tag{4.6}$$

$$2\omega = \frac{\partial q}{\partial n_2} - \frac{\partial p}{\partial n_1} + p\frac{\partial \theta}{\partial s_1} + q\frac{\partial \theta}{\partial s_2}, \tag{4.7}$$

in view of (4.1). Since η is constant, changes in the directions of
characteristics and their normals have been transcribed as changes
in θ, the orientation of the major stress.

By choosing (x,y) in (3.6) and (3.7) as (x_1, x_2), we obtain

$$\varepsilon_{11} + \varepsilon_{22} = 0, \qquad \varepsilon_{12} = \frac{k}{\mu}(p\frac{\partial \theta}{\partial s_1} + q\frac{\partial \theta}{\partial s_2} - \omega), \tag{4.8}$$

having regard to (4.3). Now introduce (4.4), (4.6), (4.7) and replace k/μ by cos2η as authorized by (3.8). The outcome is

$$\frac{\partial p}{\partial s_1} + \frac{\partial q}{\partial s_2} + p\frac{\partial \theta}{\partial n_1} - q\frac{\partial \theta}{\partial n_2} = 0, \tag{4.9}$$

$$-\frac{\partial p}{\partial s_1} + \frac{\partial q}{\partial s_2} + p\frac{\partial \theta}{\partial n_1} + q\frac{\partial \theta}{\partial n_2} = 0, \tag{4.10}$$

on dropping the factor sin2η after use of (4.2). The resemblance
between this pair of equations is quite astonishing, considering
their very different origins. By addition and subtraction we have
finally

$$\frac{\partial p}{\partial s_1} = q\frac{\partial \theta}{\partial n_2}, \quad \frac{\partial q}{\partial s_2} = -p\frac{\partial \theta}{\partial n_1}. \tag{4.11}$$

These express the variations of p and q along the respective characteristics, since the gradients of θ are known from the slipline net.
The accolade *canonical* seems well merited by the relative simplicity
in the circumstances prevailing. It is reiterated that (p,q) are
the components of velocity on the characteristics by the parallelogram rule of decomposition. The same differential relations are
satisfied by $(p,q)\sin 2\eta$, which are the projections of the velocity
vector on the normals n_2, n_1 respectively. Another variant, not so
attractive, is obtained by using (4.2) to eliminate $\partial\theta/\partial n_1$, $\partial\theta/\partial n_2$
in favour of $\partial\theta/\partial s_1$, $\partial\theta/\partial s_2$.

So far the special geometry of the slipline net has played no part.
It is invoked now through an elementary formula, which has not been
traced in the literature. In the present notation this is

$$\frac{\partial \sigma}{\partial s_1} - 2k\frac{\partial \theta}{\partial n_2} = 0 = \frac{\partial \sigma}{\partial s_2} + 2k\frac{\partial \theta}{\partial n_1} \tag{4.12}$$

where s_1, n_2 and s_2, n_1 are pairs of directions equally inclined to
the α and β sliplines respectively (Fig. 2). Equally,

$$\frac{\partial \sigma}{\partial n_2} - 2k\frac{\partial \theta}{\partial s_1} = 0 = \frac{\partial \sigma}{\partial n_1} + 2k\frac{\partial \theta}{\partial s_2} \tag{4.13}$$

since the directions in a pair are interchangeable. Any one equation
of these four suffices as a full statement, if η is considered to
range over the interval $-\frac{1}{2}\pi$, $\frac{1}{2}\pi$. From that standpoint the formula
includes Hencky's canonical relations along the sliplines, and also
relations intrinsic to the trajectories of stress:

$$\frac{\partial \sigma}{\partial x_1} + 2k\frac{\partial \theta}{\partial x_2} = 0 = \frac{\partial \sigma}{\partial x_2} + 2k\frac{\partial \theta}{\partial x_1}, \tag{4.14}$$

where axes (x_1, x_2) are coaxial with the principal values $\sigma+k$, $\sigma-k$.
Conversely, these relations or Hencky's imply (4.12) and (4.13). All
might be regarded as subsumed in Geiringer's verbal statement (1937,
p. 33), to the effect that the vectors representing grad($\sigma/2k$) and
gradθ are mirror images in the α-direction; the explicit decompositions were not mentioned.

In the present context we need (4.12) specifically with regard to
the characteristics and their normals. Its substitution in (4.11)
produces

$$\frac{\partial p}{\partial s_1} = \frac{q}{2k}\frac{\partial \sigma}{\partial s_1}, \qquad \frac{\partial q}{\partial s_2} = \frac{p}{2k}\frac{\partial \sigma}{\partial s_2}, \qquad (4.15)$$

or their differential equivalents along the characteristics. This canonical variant expresses increments of (p,q) directly in terms of the field of hydrostatic tension, and hence involves the net geometry at one remove.

The same elementary formula can be applied to other kinematical expressions that involve gradients of θ. Take, for example, the bracketed quantity that governs the fluxes (3.4) and the reduced constitutive relations (4.8). This is

$$p\frac{\partial \theta}{\partial s_1} + q\frac{\partial \theta}{\partial s_2} - \omega = \tfrac{1}{2}(\frac{\partial p}{\partial n_1} - \frac{\partial q}{\partial n_2} + p\frac{\partial \theta}{\partial s_1} + q\frac{\partial \theta}{\partial s_2})$$

$$= \tfrac{1}{2}(\frac{\partial p}{\partial n_1} - \frac{q}{2k}\frac{\partial \sigma}{\partial n_1}) - \tfrac{1}{2}(\frac{\partial q}{\partial n_2} - \frac{p}{2k}\frac{\partial \sigma}{\partial n_2}) \qquad (4.16)$$

from (4.7) and (4.13) in turn. Note the resemblance to the expressions in the preceding canonical variant; the point of this remark will appear later.

5. SPECIAL FLOWS

Suppose that a plane flow is specified by its speed $w(x,y)$ and direction $\theta(x,y)$, so that $(u,v) = w(\cos\theta, \sin\theta)$. Let (x_1, x_2) be local axes tangential and normal to a streamline, the sense of x_1 being that of w. Then

$$\varepsilon_{11} + \varepsilon_{22} = \frac{\partial w}{\partial x_1} + w\frac{\partial \theta}{\partial x_2}, \qquad 2\varepsilon_{12} = \frac{\partial w}{\partial x_2} + w\frac{\partial \theta}{\partial x_1} \qquad (5.1)$$

from first principles, or by specializing intrinsic expressions for the rates of dilation and shear in a general flow (Hill, 1967). Now compare these formulae with the equations of balance for a plastic state, in the intrinsic form (4.14), and identify the two sets of local axes. It is evident that a flow along the trajectories of $\sigma+k$, with varying speed such that

$$w = w_0 \exp(\sigma/2k), \qquad (5.2)$$

is non-dilatant while its strain-rate is coaxial with the stress. Further, if the medium is considered to be rigid/plastic and isotropic, this flow is constitutively possible when directed along *converging* streamlines (so that $\lambda > 0$). Similarly, by putting $-k$ for k in (4.14), it is evident that a flow along the trajectories of $\sigma-k$, with varying speed such that

$$w = w_0 \exp(-\sigma/2k), \qquad (5.3)$$

is also constitutively possible when directed along *diverging* streamlines. A general account of such flows and their ideal properties in steady-state forming was given by Hill (1966, 1967), along with a method for constructing the requisite die profiles at will.

We examine the possibility of similar flows in elastoplastic media. Continuing with the same line of approach, it is noted that

$$u\frac{\partial\theta}{\partial x} + v\frac{\partial\theta}{\partial y} = w\frac{\partial\theta}{\partial x_1}, \qquad 2\omega = w\frac{\partial\theta}{\partial x_1} - \frac{\partial w}{\partial x_2} ,$$

and hence from (5.1) that

$$u\frac{\partial\theta}{\partial x} + v\frac{\partial\theta}{\partial y} - \omega = \varepsilon_{12} \tag{5.4}$$

on axes intrinsic to a streamline. In (4.8) the coordinates are intrinsic to the trajectories of stress. Therefore, when the two sets of local axes are identified, the shearing component vanishes as before. It follows that ideal flows exist regardless of the value of k/μ. This conclusion can be reached alternatively from the canonical relations. By inspection it is apparent from (4.15) that

$$p = q \propto \exp(\sigma/2k) \text{ and } p = -q \propto \exp(-\sigma/2k) \tag{5.5}$$

are possible solutions, with resultants $2p\cos\eta$ and $2p\sin\eta$ respectively along the streamlines.

In retrospect it is obvious without formal analysis that ideal flows in a rigid/plastic medium are compatible also with the specific elastoplastic response (2.3), at least locally. The reason is that the Jaumann flux of the stress deviator in any convected element is zero. This is because the same two fibres are perpetually aligned with constant principal values k, $-k$ as the element travels along a streamtube. Therefore, even if the response were elastoplastic, no elastic constituent would be induced. There is still the question though of compatibility with purely elastic streams at entry and exit (if that is the configuration); this is left for future investigation.

The flows (5.5) can be superimposed in arbitrary proportion along (x_1, x_2) respectively as joint components

$$w_1 = (p_0 + q_0) \cos\eta\exp(\sigma/2k), \quad w_2 = (q_0 - p_0)\sin\eta\exp(-\sigma/2k),$$

$$\tag{5.6}$$

where (p_0, q_0) are constants. Correspondingly, when (p,q) are restricted to be functions of σ alone, the general solution of (4.15) is

$$\left.\begin{array}{l} p = p_0 \cosh(\sigma/2k) + q_0 \sinh(\sigma/2k), \\[2mm] q = p_0 \sinh(\sigma/2k) + q_0 \cosh(\sigma/2k). \end{array}\right\} \tag{5.7}$$

Plainly, (p_0, q_0) are the uniform values of (p,q) along the locus $\sigma = 0$ in a plane of flow, if the field extends as far. Conversely, if these values are assigned on some segment of the locus, the present solution is uniquely determined within the domain of influence appropriate to the value of η (or k/μ). Plainly also, (p,q) take uniform values along any curve in the family parametrized by σ. As is well known, one set of diagonals of any *equiangular* net of sliplines belongs to this family. Since (w_1, w_2) also are uniform along such diagonals, so are the speed of flow and the streamline inclination to the trajectories of principal stress. There is no elastic contribution to the deformation: the relevant Jaumann flux

vanishes as before, by superposition. Correspondingly, a convected
element maintains its orientation relative to the local axes of
stress. This is formally apparent from (4.16), since the family
(5.7) has the distinctive property that

$$dp = qd\sigma/2k, \quad dq = pd\sigma/2k \tag{5.8}$$

for arbitrary (dx, dy), in particular in the directions (x_1, x_2).
Therefore any material fibre that coincides with a stress trajectory
at one instant is convected into another trajectory of the same
family. A like statement holds for an orthogonal grid of such fibres.

We turn lastly to flows compatible with any homogeneous field of
stress, where the sliplines and characteristics are rectilinear.
The solution of (4.11) is then a combination of non-uniform simple
shears, given by arbitrary functions $p(n_1)$ and $q(n_2)$, where n_1 and
n_2 are the perpendicular distances from some basal pair of characteristics. The material spin is $\frac{1}{2}\{q'(n_2) - p'(n_1)\}$ and generally
does not vanish; since convection now has no counterbalancing effect
on the objective stress, the deformation has an elastic constituent.
Take the case of a single non-uniform shear $p(n_1)$ and define $\gamma = \frac{1}{2}p'(n_1)$ locally. A straightforward calculation reveals that the
elastic contribution to the strain-rate tensor is a pure shear with
respect to the axes of stress (illustrating a general property
enunciated earlier) and that its magnitude is $\gamma k/\mu$. Further, this
elastic shear contributes a part $\gamma k^2/\mu^2$ of the overall component γ.
Finally, the active shear stress is $k(1 - k^2/\mu^2)^{\frac{1}{2}}$, while the associated direct stresses are $\sigma + k^2/\mu$ and $\sigma - k^2/\mu$ along and perpendicular
to the streamlines.

According to this analysis a concentrated band of simple shear is
conceivable and hence also, in the limit, a jump $[p]$ across a single
characteristic. Equally, such discontinuities are admissible across
a curvilinear characteristic in a general flow; then the normal
velocity $q\sin2\eta$ is continuous but the tangential velocity $p + q\cos2\eta$
is not. Since $[\partial p/\partial s_1] = 0$ by (4.11) or (4.15), $[p]$ is constant just
as in the classical case. However, in elastoplastic media, the admissible jumps are no longer across sliplines. Even for small k/μ
this difference may affect strongly both the local and global configurations in solutions to boundary-value problems.

6. CLOSING REMARKS

The present approach could be regarded in some sort as complementary
to the finite element method, which has been applied by Lee, Mallett,
and Yang (1977) to steady flows in elastoplastic media (albeit with
a degree of hardening). That qualification apart, the constitutive
bases and practical objectives are broadly similar, whereas the
mathematical inputs and levels of description are quite different.
The outcome of the present analysis suggests that steady flows in
non-hardening elastoplastic media are amenable to numerical methods
not far removed from those currently applied to slipline nets. The
separation of stress and velocity characteristics, though, does
require unfamiliar procedures in solving boundary-value problems.
There are also basic theoretical questions to be resolved in matching
a working zone to the adjoining elastic streams. Novel configurations

can be expected, even when k/μ is small. The limiting behaviours as $k/\mu \to 0$ should clarify the status of the rigid/plastic model and dispose of the effects of degeneracy.

REFERENCES

Alexander, J. M. (1955). A slipline field for the hot rolling process. *Proc. Inst. Mech. Engrs.*, <u>169</u>, 1021-1030.
Collins, I. F. (1968). The algebraic geometry of slipline fields with applications to boundary-value problems. *Proc. R. Soc.(London)*, <u>A303</u>, 317-338.
Collins, I. F. (1982). Boundary-value problems in plane-strain plasticity. In H. G. Hopkins and M. J. Sewell (Eds.), *Mechanics of Solids: The Rodney Hill 60th Anniversary Volume*. Pergamon Press, Oxford. pp. 135-184.
Dewhurst, P. (1978). On the non-uniqueness of the machining process. *Proc. R. Soc. (London)*, <u>A360</u>, 587-610.
Geiringer, H. (1937). Fondements mathématiques de la théorie des corps plastiques isotropes. *Mém. Sci. Math.*, <u>86</u>, 1-89.
Green, A. P. (1951). A theoretical investigation of the compression of a ductile material between smooth flat dies. *Phil Mag.*, <u>42</u>, 900-918.
Hill, R. (1947). Comments on a paper by E. Siebel. *J. Iron Steel Inst.*, <u>156</u>, 513-517.
Hill, R. (1950). *The Mathematical Theory of Plasticity*. Clarendon Press, Oxford. (Paperback 1983).
Hill, R. (1952). The theory of the extrusion of metals. In *Selected Government Research Reports*, Vol. 6, *Strength and Testing of Materials*, Part I. Her Majesty's Stationery Office, London. pp. 191-204.
Hill, R. (1954). On the limits set by plastic yielding to the intensity of singularities of stress. *J. Mech. Phys. Solids*, <u>2</u>, 278-285.
Hill, R. (1956). On the problem of uniqueness in the theory of a rigid/plastic solid. *J. Mech. Phys. Solids*, <u>4</u>, 247-255; <u>5</u>, 153-161.
Hill, R. (1966). A remark on diagonal streaming in plane plastic strain. *J. Mech. Phys. Solids*, <u>14</u>, 245-248.
Hill, R. (1967). Ideal forming operations for perfectly plastic solids. *J. Mech. Phys. Solids*, <u>15</u>, 223-227.
Hill, R. (1979). On the theory of plane strain in finitely deformed compressible materials. *Math. Proc. Cambridge Phil. Soc.*, <u>86</u>, 161-178.
Hill. R. and Hutchinson, J. W. (1975). Bifurcation phenomena in the plane tension test. *J. Mech. Phys. Solids*, <u>23</u>, 239-264.
Johnson, W. and Kudo, H. (1962). *The Mechanics of Metal Extrusion*. Manchester University Press.
Johnson, W. and Mamalis, A. G. (1977). *Engineering Plasticity: Theory of Metal Forming Processes*, Vol. 2. Springer-Verlag, New York.
Johnson, W. and Mellor, P. (1973). *Engineering Plasticity*. Van Nostrand-Reinhold, London.
Johnson, W., Sowerby, R. and Venter, R. D. (1982). *Plane-Strain Slipline Fields for Metal-Deformation Processes*. Pergamon Press, Oxford.
Lee, E. H., Mallett, R. L. and Yang, W. H. (1977). Stress and deformation analysis of the metal extrusion process. *Computer Methods in Applied Mechanics and Engineering*, <u>10</u>, 339-353.
Petryk, H. (1979). On the slipline field relations for steady-state and self-similar problems with stress-free boundaries. *Arch. Mech. Stos.*, <u>31</u>, 861-874.

Richmond, O. and Devenpeck, M. L. (1962). A die profile for maximum
    efficiency in strip drawing. *Proc. 4th U. S. National Congress
    of Applied Mechanics*, 2, 1053-1057.
Sokolovsky, V. V. (1962). Complete plane problems of plastic flow.
    J. Mech. Phys. Solids, 10, 353-364.

UPPER BOUND APPROACH TO METAL FORMING PROCESSES — TO DATE AND IN THE FUTURE

H. KUDO

Department of Mechanical Engineering, Yokohama National University, Japan

ABSTRACT

The development and use of the upper bound approach for analysing metal forming processes are surveyed. The uncertainties and unreliabilities inherent in the approach are highlighted with particular reference to interfacial friction, non-perfectly plastic material properties, uncontained large deformation, pressure distribution over tool surfaces and material failure. Nevertheless it is concluded that the approach, if modified with reference to experimental observations, should continue to be useful in teaching, workshop trials and in research and development of bulk forming processes because of its simplicity and the assistance it provides in visualizing the flow of the metal.

INTRODUCTION

Early in 1957 when the present author was conducting research work on cold bulk metal forming processes using an upper bound approach ("UBA") as a theoretical tool (Kudo, 1957, 1958, 1959), he communicated to W. Johnson, then at the University of Manchester, who had been publishing a series of theoretical (using the slip-line field method, "SLF") and experimental papers on cold extrusion. In reply to it, Johnson wrote that he had also begun to use UBA (Johnson, 1958, 1959a, 1959b, 1959c). This established the first contact between Johnson and the present author.

In September 1958, while staying in Zurich, the present author wrote to Johnson inquiring if there was any possibility of publishing his work in English in a scientific journal. After exchanging letters, Johnson informed the present author in November, 1958 that publication of the International Journal of Mechanical Sciences ("I.J.M.S.") had been formally decided and invited him therefore to contribute the relevant papers to this journal. The author not only accepted this invitation but also decided to work with Johnson in Manchester for the first half of 1959.

Thus the papers of the present author and some papers jointly written with Johnson which showed the method of UBA and its uses appeared in

the first volume of I.J.M.S. (1960). The name "the upper bound
approach" (UBA) was given by Johnson to the analytical method of
Kudo's papers published in the I.J.M.S. The name impressed the
author favourably since "approach" implied the minimization of the
upper bound solution rather than attaining the limit of the correct
solution from above.

The earlier theoretical methods for the analysis of bulk forming
processes at that time were SLF developed by Prandtl (1920, 1921),
Hencky (1923) and others, the elementary or slab method by Siebel
(1922, 1923, 1926), and von Kármán (1925), the ideal deformation
energy method by Siebel and Fangmeier (1931), and the semi-empirical
visio-plasticity method by Thomsen and Lapsley (1954) and Thomsen,
Yang and Bierbower (1954). These had been used extensively in metal
forming studies relating to bulk deformation.

Although the use of variational principles or the upper and lower
bound theorems in metal forming problems had been suggested as early
as 1951 by Hill (1951) and Green (1951), this seems not to have been
followed until late in the 1950's except in the U.S.S.R. (Tarnovskii,
Pozdeev and Garago, 1959). The requirement which is inherent in
this method is to determine both upper and lower bound solutions.
The latter are not as easy to determine as the former and the wide
gaps which usually lie between the upper and lower bound solutions
for problems associated with the Coulomb friction law (Drucker, 1954)
have made researchers reluctant to use the method.

Since the emergence of the initial series of UBA papers in I.J.M.S.
and other journals, accounts have appeared in the books describing
UBA written by Johnson and Kudo (1962) and Johnson and Mellor (1962).
These authors were engineers rather than mathematicians and did not
pay due regard to the lower bound solutions. Subsequently a number
of papers on UBA have been published by several authors world wide
and the development and use of UBA to-date has been recently reviewed
by Kudo (1983a, 1983b). The results of UBA provided by Johnson and
Kudo (irrespective of ignoring the lower bound solutions) by use of
simple kinematical models or velocity fields and of a friction factor
instead of a coefficient of friction, led to a fairly clear understanding of various forming processes and produced relatively exact
predictions of forming loads, macroscopic deformation and defect
formation; this, in turn, is believed to have encouraged others to
use UBA.

UBA is merging today with the finite element method (FEM)   as computers and their software develop and the use of sophisticated
velocity fields becomes feasible. On the other hand, relatively
plain flow models are still being applied to complex industrial
problems for designing operation sequences and toolings and developing novel forming processes, examples of which will be shown below.

The present chapter aims at suggesting ideas of how to overcome the
uncertainties inherent in UBA when treating problems involving
interfacial friction, work-hardening, strain-rate and temperature
dependence of work material, inertia effects, unconstrained large
deformation and material failure. In addition, ways of developing
the technique are described which should enable it to continue to
be useful in metal forming analysis alongside FEM.

TREATMENT OF INTERFACIAL FRICTION

As is well known, UBA requires calculation of the rate of plastic deformation work $\dot{W}_d^*$ within the deforming volume V, the rate of internal shear work $\dot{W}_s^*$ along the assumed surfaces of velocity discontinuity S_D^*, the rate of frictional work $\dot{W}_f^*$ over the tool-workpiece interface S_C and the rate of work $\dot{W}_t^*$ done by the specified external traction T_i by using assumed kinematically admissible velocity fields v_i^* that are compatible with the given boundary condition of velocity, i.e., the tool velocity V_i on S_C and the condition of volume constancy. When the frictional condition is specified in terms of a Coulomb coefficient of friction μ, the actual value and direction of the prevailing frictional stress T_t is unknown. However these are required in order to calculate $\dot{W}_f^*$.

As a consequence of this difficulty, the upper and lower bound theorems as proposed by Drucker (1954) were overcautious and gave too high an upper bound solution and too low a lower bound solution to provide an adequate assessment of the actual solution sought. Drucker's theorems were improved by Yamada (1958) so that the gap between the upper and lower bounds could be narrowed in some limited cases. A further improvement was attempted to lower the upper bound by Collins (1969) using an artificial velocity field. The present author is however uncertain whether the proposed upper bound is really the one appropriate to the prescribed boundary conditions. Moreover, as Collins himself remarked, the velocity field for the minimized upper bound can deviate significantly from the actual one.

The problems associated with friction were significantly reduced by the introduction of the friction law due first to Shield (1955):

$$|T_t| \leq f \frac{Y}{\sqrt{3}} , \tag{1}$$

where T_t stands for the tangential or shear stress exerted by the tool surface on the workpiece surface S_C. f denotes the friction factor which is assumed constant over S_C during forming and Y is the yield or flow stress of the workpiece material in tension or compression, the von Mises yield criterion being adopted.

Kudo and Takahashi (1964) proposed for Lévy-Mises' materials the following upper and lower bound theorems into which the friction law of eqn. (1) was introduced:

$$\int_{S_C} \sum_i T_i^* v_i \, dS \leq \int_{S_C} \sum_i T_i v_i \, dS$$

$$= \int_V Y \, \bar{\dot{\epsilon}} \, dV + \int_{S_D} \frac{Y}{\sqrt{3}} \Delta v_t \, dS + \int_{S_C} f \frac{Y}{\sqrt{3}} \Delta v_t \, dS - \int_{S_T} T_i v_i \, dS$$

$$\leq \int_V Y \, \bar{\dot{\epsilon}}^* dV + \int_{S_D^*} \frac{Y}{\sqrt{3}} \Delta v_t^* dS + \int_{S_C} f \frac{Y}{\sqrt{3}} \Delta v_t^* dS - \int_{S_T} T_i v_i^* dS. \tag{2}$$

Here T_i^* stands for the traction at S_C that is in equilibrium with
a statically admissible stress distribution σ_{ij}^* in V. This stress
field satisfies the equations of equilibrium, does not violate the
yield criterion and is consistent with eqn. (1) at S_C and the prescribed tractions T_i at S_T. $\bar{\varepsilon}^*$ is the equivalent strain-rate derived
from v_i^* through the compatibility relation. Δv_t^* is the absolute
difference in the tangential velocity components v_t^* across S_D which
represents the internal surfaces of velocity discontinuity or those
of the work material and tool at S_C. T_i, lying between the inequality
signs, represents the unknown true traction at S_C and the middle term
in the inequality represents the unknown true rate of work $\dot{W}_C$ being
done by the moving tool. On the right hand side of eqn. (2), which
is denoted by $\dot{W}_C^*$ below, the first term represents $\dot{W}_d^*$, the second $\dot{W}_s^*$,
the third $\dot{W}_f^*$ and the fourth $\dot{W}_t^*$.

The above bounding theorem can offer a sufficiently narrow gap between the upper and lower bounds provided that appropriate σ_{ij}^* and
v_i^* are chosen. This depends largely upon the experience and intuition of the research worker, although obtaining good lower bounds
is recognised to be difficult.

An attempt to correlate the friction factor f with the coefficient
of friction μ was made by Kudo (1957, 1958), who introduced the
concept of an equivalent coefficient of friction defined by

$$\mu_e = \frac{fY/\sqrt{3}}{p_{max}^*} , \qquad\qquad (3)$$

where p_{max}^* stands for the mean normal pressure which acts on the
flat area of S_C for which the upper bound pressure for a given f
shows the largest value. Although $\dot{W}_C^*$ was shown to give usually an
upper bound to the actual rate of work at S_C for $\mu = \mu_e$, a considerable overestimate of $\dot{W}_C^*$ and of working pressure was unavoidable in
certain cases, Fig. 1 (Kudo, 1959).

In some limited cases when the direction and magnitude of relative
sliding Δv_t^* is assumed not to change over the material-tool interface S_C which is flat and the contact pressure is obtainable from
an equilibrium condition using $\dot{W}_C^*$, it was shown by Shindo (1962)
and Oxley (1963) that introduction of μ instead of μ_e into eqn. (3)
yields an upper bound, $\dot{W}_C^*$, which may be reasonably close to $\dot{W}_C$.

As will be discussed later, UBA is essentially suitable for treating
forming problems in which a major portion of the workpiece surface

is constrained by tool surfaces, viz. the area S_C is much larger than S_T. It turns out that knowledge of the correct friction stress over the interface becomes indispensable for achieving a reasonable upper bound in such problems. However, the exact prediction of the μ or f value which generally varies with time and with position along the deformed workpiece during a forming operation, is practically impossible both theoretically and experimentally.

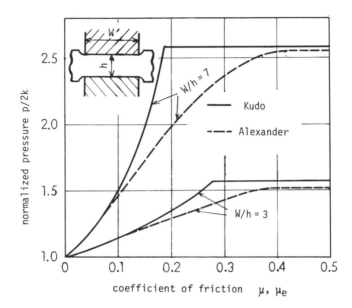

Fig. 1.   Comparison of normalized pressures required
           for plane-strain compression with Coulomb
           friction (μ) calculated by SLF method
           (Alexander, 1955) and UBA using an equivalent
           coefficient of friction ($μ_e$) (Kudo, 1959).

The present author is therefore inclined to use for metal forming studies the friction law of eqn. (1) rather than to adhere to the Coulomb law, both of which offer almost equal uncertainty. In many metal forming situations, the contact pressure decreases from the neutral area towards the tool opening or aperture where the inter-face expansion of the workpiece and the work-tool slip are more intense, thus resulting in increasing coefficients of friction μ and, accordingly, uniform f values over the entire interface (Rooyen and Backofen, 1960). Moreover, as will be suggested later, the most practical use of UBA is based on the determination of the hypothetical mean flow stress Y_m and friction factor f_m values for which are obtainable through comparison of predicted and observed working loads of deformations.

Another possible difficulty which may be created by the presence of interfacial friction is non-uniqueness of the solution. Kudo and Takahashi (1964) utilised the proof of uniqueness for solutions in plasticity problems as given by Hill (1950) and found that uniqueness does not hold in problems which include friction as a boundary condition even when it is prescribed in terms of the friction factor f. This again suggests the need for partial support for UBA from experiments.

INCORPORATION OF WORK-HARDENING

Inequality (2) indicates that a true upper bound to $\dot{W}_c^*$ can be obtained only when the exact current distribution of flow stress Y is known. This is only possible when the deformation history of the workpiece material is exactly determined since Y is a function of straining history. This indicates the reason why UBA has generally been applied to perfectly plastic materials.

In the UBA study of axisymmetric steady extrusion processes with work-hardening aluminium, Kudo (1958) adopted the mean flow stress Y_m which had been suggested by Hill (1950) and used by Johnson (1956) for modifying slip-line field solutions to assess the pressures needed for plane-strain extrusion of work-hardening materials. This idea is based on the assumption that the same mean equivalent strain $\bar{\varepsilon}_m$ is imparted to the extruded material whatever the work-hardening characteristics of a material. Thus, for a material with flow stress curve given by $Y = Y(\int \bar{\dot{\varepsilon}} dt) \equiv Y(\bar{\varepsilon})$, Y_m is expressed as

$$Y_m = \frac{\int_0^{p_e^*/Y_1} Y(\bar{\varepsilon}) \, d\bar{\varepsilon}}{p_e^*/Y_1} \,, \tag{4}$$

according to Hill (1950). Here p_e^* denotes an upper bound to the mean steady extrusion pressure per unit cross-sectional area of the container bore for perfectly plastic material having a constant flow stress Y_1. As Fig. 2 shows, the calculated steady extrusion pressure lie reasonably close to and above the experimental data.

The same assumption as above was also made by Kudo (1958) for non-steady forming processes. In this case, the sum $\dot{W}_d^* + \dot{W}_s^* + \dot{W}_f^*$ for the perfectly plastic material was uniformly distributed within the deforming volume V_d so that the mean equivalent strain $\bar{\varepsilon}_m$ and, accordingly, the mean flow stress Y_m were given by

$$Y_m = Y(\bar{\varepsilon}_m) = Y(\int \dot{\bar{\varepsilon}}_m \, dt)$$

$$= Y\left(\int_0^t \frac{\dot{W}_d^* + \dot{W}_s^* + \dot{W}_f^*}{Y_1 V_d} \, dt \right). \tag{5}$$

If the upper bound to the average pressure referred to the current cross-sectional area a of the deforming zone is denoted by p* and the (average) axial length of the deforming zone V_d by l, then eqn. (5) is rewritten in the absence of $\dot{W}_t^*$ as

$$Y_m = Y\left[\int_{l_0}^{l_e} \left(\frac{p^*}{Y} \right) \left(\frac{-dl}{l} \right) \right], \tag{6}$$

where l_0 and l_e stand for the initial and final lengths of the
deformation zone.

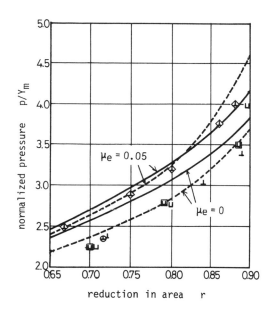

Fig. 2.   Steady backward rod and can extrusion
          pressures against reduction in area r
          obtained by UBA and experiment:
          ——, upper bound solution for rod
          extrusion; - - - - -, can extrusion; ⊥ and
          ⊔, experimental data for rod and can ex-
          trusion with hardened aluminium; ①, rod
          extrusion with annealed aluminium; ◈ , rod
          extrusion with lead; ⊡, can extrusion
          with hardened copper.

This criterion was found to result in good agreement between theory
and experiment when upset-forging processes were examined. However,
for post-steady extrusion processes in which the unextruded workpiece became very thin, the deviation between theory and experiment
increased as the process continued, see Fig. 3. This was the case
even in extrusion experiments carried out at a very slow speed in
which practically no temperature rise was expected. It was anticipated from hardness testing of extruded products that as the extreme
strains were accompanied by a considerable rotation of the principal
strain-rate axes, the actual work-hardening in the extrusion process
would not have been so high as that expected from the flow stress
curve obtained from compression test data and the isotropic work-
hardening criterion.

It is to be remarked here that in Kudo's study (Kudo, 1959), the
transition from steady to non-steady state extrusion was predicted
to occur earlier for work-hardening materials than for non-hardening

materials in accordance with experimental results. This was because the expansion of the volume of the deforming region at the transition gave rise to a reduction in the mean flow stress Y_m.

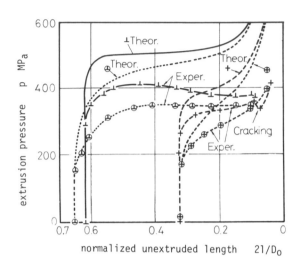

Fig. 3. Variations of extrusion pressure during
         punch travel obtained by UBA and experiment
         for an extrusion ratio of 3.52: ⊥ , rod
         extrusion with hardened aluminium; ⊕ , rod
         extrusion with annealed aluminium; ⊕ ,
         simultaneous rod-rod extrusion with annealed
         aluminium.

In a study of UBA for plane-strain extrusion of work-hardening materials through wedge-shaped dies, Halling and Mitchell (1964), using velocity fields consisting of rigid triangles, incorporated work-hardening along stream lines and determined optimum velocity fields (viz. minimum upper bound fields) by minimizing the right hand side of inequ.(2). This procedure may be regarded as the use of the inequality,

$$\int_{S_C} \sum_i T_i V_i \, dS = \int_V Y \, \dot{\bar{\varepsilon}} \, dV + \int_{S_D} \frac{Y}{\sqrt{3}} \Delta v_t \, dS + \int_{S_C} f \frac{Y}{\sqrt{3}} \Delta v_t \, dS -$$

$$\int_{S_T} \sum_i T_i V_i \, dS \leq \int_V Y^* \bar{\dot{\varepsilon}}^* dV + \int_{S_D} \frac{Y^*}{\sqrt{3}} \Delta v_t^* dS + \int_{S_C} f \frac{Y^*}{\sqrt{3}} \Delta v_t^* dS -$$

$$\int_{S_T} \sum_i T_i v_i^* dS, \tag{7}$$

where Y^* represents the variable flow stress determined from the assumed preceding velocity fields or strain history. In a discussion of the above paper, Haddow (1966) pointed out that inequ. (7) cannot be correct always since there is no guarantee that Y^* is nowhere less

than Y. Moreover the upper bounds to the extrusion pressure thus
calculated were practically the same as those obtained by using eqn.
(4).

Nevertheless, similar procedures as above have been followed by a
number of workers. It is likely that, when a kinematically admissible
velocity field v_i^* which somehow simulates the actual flow is chosen,

or when a velocity function which contains a sufficient number of
free parameters for optimization is adopted (as in FEM) for extrusion
and closed-die forging processes in which the major portion of the
surface of deforming zone of workpiece is constrained by tooling,
the calculated result is satisfactory when compared with experiment,
see e.g., Fig. 4 (Osakada and Niimi, 1975).

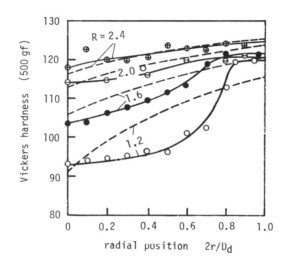

Fig. 4.   Comparison of experimental and calculated
          hardness distributions over cross section
          of forward extruded copper rod through 60°
          conical die with various extrusion ratios
          R: ———, experimental; ----, calculated
          using radial flow field.

INCORPORATION OF VISCOSITY, TEMPERATURE AND INERTIA EFFECTS

UBA for Visco-Plastic Workpieces

Generalizing the extremum theorems proposed by Prager (1954) and
other similar ones, Hill (1956) postulated a viscous material whose
stress components are derived from

$$\sigma_{ij} = \frac{\partial E}{\partial \dot{\varepsilon}_{ij}} ,$$ (8)

where E is, (in a similar manner to the concept of the plastic potential for non-viscous materials), called the "work function". The work

function is assumed convex so that the principle of maximum plastic work holds. If the flow stress is assumed to be given by

$$Y = Y_1 \, \bar{\varepsilon}^{\,m}, \tag{9}$$

where Y_1 and m are the material constants, the corresponding upper bound theorem is given by

$$\int_{S_C} \sum_i T_i v_i \, dS = Y_1 \int_V (\bar{\dot{\varepsilon}})^{1+m} dV - \int_{S_T} \sum_i T_i v_i \, dS$$

$$\leq \frac{Y_1}{1+m} \int_V (\bar{\dot{\varepsilon}}^*)^{1+m} dV + \frac{m \, Y_1}{1+m} \int_V (\bar{\dot{\varepsilon}})^{1+m} dV - \int_{S_T} \sum_i T_i v_i^* dS. \tag{10}$$

It is assumed that $T_t = 0$, so that the term which represents the rate of frictional work is excluded in inequ. (10), since no generally accepted friction law exists for the present visco-plastic materials. Note that the right hand side of inequ. (10) contains the actual unknown equivalent strain-rate $\bar{\dot{\varepsilon}}$.

Putting m = 0, i.e., $Y = Y_1$ = constant, inequ. (10) reduces to the upper bound theorem for perfectly-plastic materials given by inequ. (2) with f = 0. Otherwise, the right hand side of inequ. (10) yields an upper bound solution only if $\int_V \bar{\dot{\varepsilon}}^* dV$ is not less than $\int_V \bar{\dot{\varepsilon}} dV$. As in the case of work-hardening materials, this assumption may be reasonable when the selected velocity field is adequate, and the notion has been used repeatedly in the literature (Cristescu, 1975; Fenton, 1975; Lahoti and Altan, 1975).

The use of a mean flow stress Y_m based on a mean equivalent strain-rate $\bar{\dot{\varepsilon}}_m^*$ obtained from an admissible velocity field for a perfectly-plastic material, has also been adopted by Johnson and Kudo (1965).

Incorporation of Temperature Effects

Incorporation of the effect of temperature was first attempted by Nagpal, Lahoti and Altan (1978) in a study of the compression of a ring made of a work-hardening and strain-rate- and temperature-sensitive-material. After calculating the temperature distribution within the workpiece with a finite difference method using the preceding theoretical strain and temperature histories, they minimized the rate of energy dissipation $\dot{W}_C^*$ by varying parameters expressing the current velocity field v_i^*.

The general validity of UBA for temperature-affected problems based on an uncertain deformation history is again doubtful as in the cases of strain and strain-rate affected problems, in which the exact distribution of Y is not available.

Assessment of Inertia Effects with UBA

Avitzur, Bishop and Hahn (1972) were the first to incorporate inertia effects into UBA when solving an impact forming problem. Again, an implicit assumption was made, (whether the authors were aware of it or not, is not known to the present author) that the actual velocity v_i needed to calculate an upper bound was approximately the same as the assumed admissible velocity v_i^*.

Tirosh and Kobayashi (1976) formulated an upper bound in the absence of S_D as

$$\dot{W}_c \leq \int_V Y \, \overline{\dot{\epsilon}}^* dV + \int_{S_C} f \frac{Y}{\sqrt{3}} \, \Delta \, v_t^* dS - \int_{S_T} \sum_i T_i v_i^* dS + \int_V \sum_i p \dot{v}_i v_i^* dV,$$

(11)

and they introduced explicitly the following assumption needed to calculate the upper bound to $\dot{W}_c$

$$\int_V \sum_i p \dot{v}_i v_i^* dV \leq \int_V p \dot{v}_i^* v_i^* dV,$$

(12)

which may not necessarily be correct but which will not be totally unjustified. In this manner, however, they were able to present some interesting and useful conclusions regarding inertia effects.

UBA FOR UNCONTAINED LARGE DEFORMATION PROBLEMS

The present author should admit that the major weakness of UBA as a purely theoretical method is in its unreliability in treating un-contained large deformation problems, even if the work material is perfectly plastic and errors due to misestimate of the current flow stress distribution are excluded. This is because, when carrying out the calculation incrementally, the accumulation of possible errors in the predicted deformation increments may lead to consider-able deviations of the predicted current workpiece shape and tool-workpiece contact area from reality.

Nevertheless, a number of research workers including the present author have used UBA optimistically. This procedure is based on a conjectural upper bound theorem for non-steady processes which can be written as

$$\dot{W}_c = \int_V Y \, \overline{\dot{\epsilon}} dV + \int_{S_D} \frac{Y}{\sqrt{3}} \, \Delta \, v_t dS + \int_{S_C} f \frac{Y}{\sqrt{3}} \, \Delta \, v_t dS - \int_{S_T} \sum_i T_i v_i dS$$

$$\leq \int_{V^*} Y \, \overline{\dot{\epsilon}}^* dV + \int_{S_D^*} \frac{Y}{\sqrt{3}} \, \Delta \, v_t^* dS + \int_{S_C^*} f \frac{Y}{\sqrt{3}} \, \Delta \, v_t^* dS - \int_{S_T^*} \sum_i T_i v_i^* dS,$$

(13)

where V^* stands for the calculated current space occupied by the workpiece, S_C^* and S_T^* the calculated current work-tool interface and traction-prescribed surface and S_D^* the internal surface of tangential velocity discontinuity assumed within V^*, see Fig. 5.

In an UBA study of plane-strain deep indentation into a semi-infinite

perfectly-plastic workpiece, Johnson and Kudo (1964) used rigid-
triangle and centred fan velocity fields as shown in Fig. 6 and
optimized them so as to yield minimum upper bounds. In this solution,
the predicted bulge at the surface shown in Fig. 6, is quite unreal
compared with the broken line contour which represents a typical
actual profile. The actual indentation pressure would then be higher
than the present upper bound pressure since the velocity discontinuity
surface must reach the raised free surface resulting in a larger area
of the actual S_D than that of the assumed S_D^*.

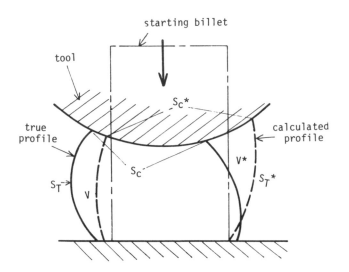

Fig. 5.   Possible difference in actual and calculated
           shapes of workpiece after uncontained large
           deformation.

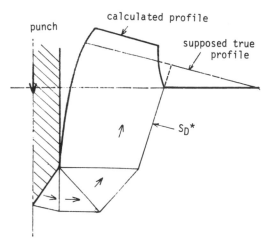

Fig. 6.   Rigid triangle velocity field used in UBA
           to plane-strain deep indentation of punch
           and resulting surface bulge.

In the final stages of die forging and coining, the major part of the workpiece surface becomes constrained by the tool surfaces. This implies that the errors in the calculated free surface profiles influence only slightly the upper bound values. It is expected therefore that reasonably good estimates of the final maximum loads can be obtained using UBA. These are most important in the design of tooling and in the selection of a forming machine, see Fig. 7 (Kudo and co-workers, 1980).

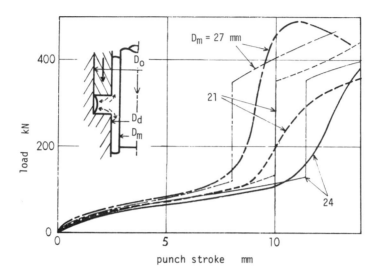

Fig. 7.   Comparison of experimental and calculated
          load-stroke diagrams for extrusion-forging
          of flanged hollow component from hollow
          billet of annealed aluminium:
          thick lines - experimental,
          thin lines - calculated.

When applying UBA to a pseudo-steady problem such as the indentation of a wedge-shaped or conical punch into a semi-infinite workpiece, the shape of the surface profile of the workpiece which is consistent with the adopted velocity field, should first be assumed.

In an attempt to correlate the work-hardening property of the workpiece with the pressure requirements and surface profile occurring in plane-strain wedge indentation processes, Matsubara and Kudo (1975) used the rigid-triangle velocity field and assumed first that the deformed workpiece surface near to the indenter is plane, see Fig. 8. They calculated changes in the flow stress along the trajectories of elements in the unit diagram (Hill, 1950) and minimized the upper bound loads by changing the configuration of the velocity field.

One of their noteworthy findings was that the minimized upper bound loads for a specified depth of indentation into a perfectly-plastic workpiece were lower than the slip-line field solutions by 0 to 5 percent depending on the wedge angle and friction factor. This indicates that minimization of the right hand side of inequ. (13)

does not always tend towards the true solution but could lead to a
deviation. The use of inequ. (13) should, therefore, be done carefully by checking the assumed surface profile with that experimentally
observed, especially when the area of the free or traction-prescribed
surface S_T is larger than the area of the workpiece-tool interface
S_C.

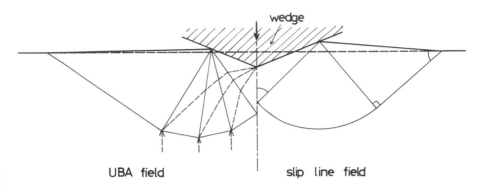

Fig. 8.   Comparison of optimum rigid triangle velo-
          city field with slip line field for plane-
          strain wedge indentation.

Such a check has specifically been carried out before choosing
families of the kinematically admissible velocity field v_i^* by Kudo

and Tamura (1969) and Usui and Masuko (1973) when studying V-groove
scratching and rolling processes in the surface of a semi-infinite
workpiece and in various three-dimensional machining processes respectively, see Figs. 9 and 10. The results show excellent agreement
between theory and experiment with regard both to the working load
and to the overall deformation.

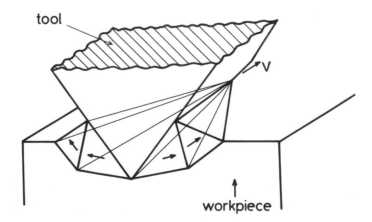

Fig. 9.   Block sliding velocity field assumed in UBA
          of surface scratching process.

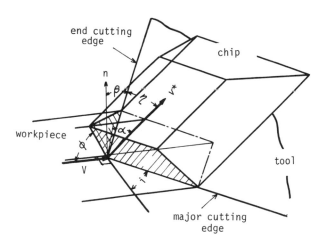

Fig. 10.   Block sliding velocity field assumed in
           UBA of three dimensional machining process.

The reader will readily appreciate the unreliability of inequ. (13)
if they think of a velocity field shown by the broken line in Fig.
11 in which S_c converges to a line and the rate of frictional work
$\dot{W}_f^*$ vanishes.   Although this field is certainly kinematically admissible, it yields a lower upper bound than the true working load.

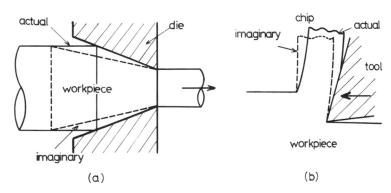

Fig. 11.   Unrealistic kinematically admissible velo-
           city fields for steady state processes:
           (a) rod drawing,   (b) machining.

A less empirical UBA of uncontained steady flow processes has been
proposed by Oh and Kobayashi (1975) taking the sideways spread in
single-pass plate rolling as an example.   Their line of thinking
will be more readily understood by following the description due to
Kato, Murota and Kumagai (1980).

Let us express a class of kinematically admissible steady velocity fields by $(v_i^*)_{steady} = \psi(x_i, c_s)$, where x_i represents the spatial co-ordinates and c_s a variable scalar parameter. Let us consider further another family of kinematically velocity fields given by $v_i^* = \psi(x_i, c)$ where $c_s \subset C$. This family is more comprehensive and does not necessarily represent steady state. For each chosen value of c_s, a $\dot{W}_c^*$-c relation given by inequ. (13) is obtained as shown by the thick-line curves, e.g., FDG in Fig. 12. The broken-line curve CAD shows the lowest line on the valley-shaped $\dot{W}_c^*$ versus c curves.

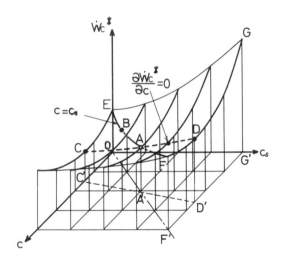

Fig. 12.   Description diagram for determining stable and optimum velocity field for steady state flow.

Whereas the optimum v_i^* should coincide with a steady state velocity field $(v_i^*)_{steady}$, viz. $(c)_{opt} = c_s$, the solution to the present problem is determined by the intersection of the $(c)_{opt} - c_s$ surface CADD'A'C' and the $c = c_s$ plane OEAFF'A'. Thus the stable steady state velocity field is given by the point A' of Fig. 12 and the associated total rate of work $\dot{W}_c^*$ at S_C by the point A. It should be remarked that the point A is not the lowest point among $\dot{W}_c^*$ given by the entire family of v_i^* which is kinematically admissible.

This procedure seems to be fairly reasonable. However, there is no sound proof that the true velocity field is approached by using an adequately big family of v_i^*. The choice of the family of velocity field should, therefore, be done again in the light of experimental observation.

ASSESSMENT OF PRESSURE DISTRIBUTION OVER TOOL SURFACES

The above upper bound theorems refer to the total rate of work being done by all moving tools. Either when the operation is performed by a single rigid tool or when multiple tools are arranged and operated symmetrically so that the rate of work for each moving tool can be identified, the upper bounds to the load and, consequently, the average pressure are obtainable. When the tools operate unsymmetrically, even the average pressures cannot be determined on the individual tool surfaces. It is also impossible for UBA to estimate the pressure distribution over a workpiece-tool interface. This information is often required in designing highly loaded tools, even when the total load or average pressure is known.

An attempt to obtain the average pressures on two unsymmetrically operating punch surfaces, see Fig. 13(a) was made by Kudo and Shinozaki (1974) in a study of a multi-ram extrusion process. They regarded one of the work-tool interface S_{C2} as S_T (i.e., the surface where the average traction T_{im} is prescribed), though actually the velocity V_2 is prescribed there instead of T_{im}. By assuming different values of T_{im} for the fourth term $(\dot{W}_t^*)$ on the right hand side of inequ. (2), optimum velocity fields $(v_i^*)_{opt}$ were determined which minimized $\dot{W}_C^*$. They assumed that the value of T_{im} which resulted in $(v_i^*)_{opt}$ which was consistent with V_2 was the average traction acting on S_{C2}.

Experimental results obtained in lateral extrusion tests on circular branches from a circular aluminium billet using two punches advancing at prescribed different speed ratios, are compared with the above upper bound solutions in Fig. 13(b). Although the present procedure has a theoretical basis it is found that T_{im} is not uniquely determined in certain cases, viz., different T_{im} values result in identical $(v_i^*)_{opt}$.

Johnson and Mamalis (1977) proposed a clever method of directly determining upper bound loads for plane-strain bulk forming processes of rigid-perfectly-plastic materials using force polygons associated with block sliding velocity fields. They also showed that the local average pressures over individual block-tool interfaces can be readily obtained with the force polygons. This method may provide us with an outline of the pressure distribution, however there is no guarantee that the actual local pressures are everywhere lower than the pressures thus determined.

Another proposal has been made by Kiuchi and Murata (1980) to obtain pressure distributions in axisymmetric problems in conjunction with UBA. They showed some examples in which the sum of the local pressures obtained equalled the load calculated from the total rate of work. It seems to the present author, however, that this method has not yet been generalized to make it applicable to general forming problems.

In order for UBA to be useful in predicting tool life and distortion, a general method of determining an upper bound to the local pressure distribution should be developed.

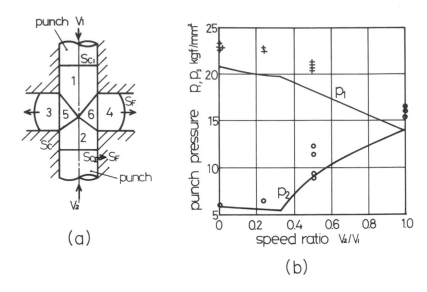

(a)

(b)

Fig. 13. Comparison of predicted and observed
          pressures on punch surfaces advancing at
          different speeds in multi-ram extrusion
          to form cross-shaped component from
          billet: (a) block sliding velocity
          field used for computation; (b) pre-
          dicted (line) and observed (plot) punch
          pressures for annealed aluminium.

PREDICTION OF MATERIAL FAILURE

Using the rigid triangle velocity field or block sliding model, Kudo
(1958) predicted the occurrence of the surface skin inclusion, (see
Fig. 14(a) and (d)) which leads to a partial decomposition of the
workpiece. He also suggested that this defect can be prevented or
minimized by the use of a rough or degreased tool surface which
anchors the surface skin. These observations have been confirmed
experimentally.

Sinking and cracking defects which produce a recess at a work-tool
interface have been suggested by Kudo (1958) and Johnson (1959), see
Fig. 14(b) and (c). They considered that when a velocity field which
results in local separation of the work material from tool surface
S_C consumes less work rate than does velocity fields within workpiece
which keep contact with the entire tool surface, the separation
should take place. The sinking shown in Fig. 14(b) has been attributed both by Kudo and Johnson to a simple moving away of a part of
workpiece surface from the tool surface. For a combined can-can
extrusion, Johnson proposed a sinking defect due to a necking mechanism shown in the top half of Fig. 14(c), while Kudo's mechanism
incorporated material cracking (lower half of Fig. 14(c)) in which
the energy consumption needed to create the new surface was disregarded. The choice between these mechanisms should be made

depending on the (residual) ductility of the work material.   The
predicted occurrence of those defects agreed qualitatively with
experimental observations made by Kudo and Johnson.

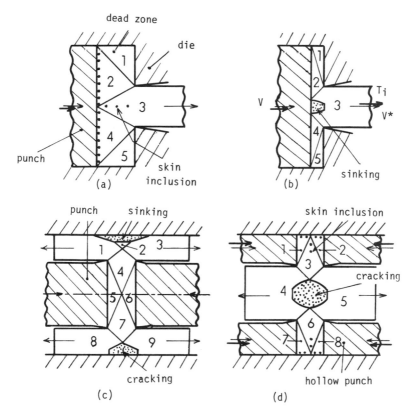

Fig. 14.   Occurrence of various material defects in
           extrusion predicted by triangle velocity
           field: (a) post steady state of rod ex-
           trusion, (b) final stage of rod extrusion,
           (c) final stage of simultaneous can-can
           extrusion, (d) final stage of simultaneous
           rod-rod extrusion.

The internal crack which often appears at the workpiece centre during
a simultaneous rod-rod extrusion, see Fig. 14(d), was attributed by
Kudo to the same mechanism as the rear centre sinking in the forward
rod extrusion shown in Fig. 14(b), the surface energy being again
disregarded.

Instead of introducing the energy required to create the new surface,
which would characterize the ductility of a material, Zimerman and
Avitzur (1970) incorporated work-hardening into an UBA of the central
bursting in rod drawing and extrusion.   They showed that materials
with a low hardening modulus are more susceptible to internal bursting.

This is in qualitative agreement with a number of experimental observations.

Both Kudo (1958) and Johnson (1959) mentioned that defects of the
type shown in Fig. 14(c) would be avoided or retarded if the friction over the container surface was increased. This suggestion was
correct since the third term of the right hand side of inequ. (2)
which expresses $\dot{W}_f^*$ increases with increasing Δv_t^* due to recess
formation. Thus the velocity fields in which no sinking or cracking
takes place result in lower upper bounds to $\dot{W}_c$.

The effect of the ambient hydrostatic pressure in increasing the
ductility of materials was discussed by Avitzur (1973) with reference
to composite materials. The effect is represented by the fourth
term of the right hand side of inequ. (2). It is obvious that the
formation of internal porosity or voids, as shown in Fig. 14(d),
gives rise to volume expansion and the rate of work done by the
ambient pressure $-T_i$, viz., $\int_{S_T} \sum_i T_i v_i^* dS$ becomes negative and increases the upper bound value $\dot{W}_c^*$. Thus velocity fields which result
in no volume expansion give lower upper bound, consequently ductility
is raised by the ambient hydrostatic pressure.

A criterion for successful extrusion to produce sound composite rods
was proposed by Osakada, Limb and Mellor (1973) based on UBA. It is
noteworthy because it has indicated an extended area of application
of UBA to instability problems.

The inverse process of void growth, i.e., void closure, has been
studied using UBA by Kiuchi and Hsiang (1981) who presented rolling
parameters which promote void closure.

It is to be remarked that although UBA does not give information on
stresses prevailing within the workpiece, it does predict occurrence
of various kinds of material defect at least qualitatively.

THE ROLE OF UBA: TO-DATE AND IN THE FUTURE

UBA is based upon visible deformation models. The deformation model
described by a kinematically admissible velocity field v_i^* which
satisfies the condition of volume conservation and the boundary condition V_i for velocity, allows simple superposition on other admissible fields as well as comparison with deformation patterns experimentally observed. It can be also described by block sliding models
which automatically satisfy the volume constancy condition and are
flexible enough to fit various two-dimensional as well as three-
dimensional problems (Gatto and Giarda, 1981).

Incidentally, the lower bound approach (LBA) uses a postulated
statically admissible stress field σ_{ij}^* which satisfies the equation
of stress equilibrium, a non-linear yield criterion and boundary
conditions given in terms of tractions T_i at S_T and friction factor
f at S_C. Thus σ_{ij}^* must satisfy more conditions than v_i^* and cannot
be simply superposed upon others. Moreover, σ_{ij}^* does not predict

deformation. Thus the easier use of UBA compared with LBA is obvious and, therefore, higher accuracies are generally anticipated with UBA solutions than with LBA solutions.

It is for the above reasons that UBA has been widely used to-date for the understanding of mechanisms that operate in various bulk forming processes and for the quantitative or at least qualitative decision making in workshop trials in spite of a number of weak points inherent in UBA, as pointed out in the preceding sections. The easy understanding of the mechanisms is accomplished when v_i^*

is relatively simple so that the upper bound solution is expressed in terms of the geometrical parameters, friction factor, flow stress, inertia effect etc.

The convenient flexibility of UBA is exemplified by simple conversion of the velocity field of a forward rod extrusion, Fig. 15(a), into that of a backward rod extrusion, Fig. 15(b), and that of a backward can extrusion, Fig. 15(c), by superposition of a uniform velocity V_p and by reversing the direction of v_2^* (Kudo, 1959).

This also enables conversion of the extrusion pressure from one process into other processes. It can thus be shown that the mean die and punch pressures for rod and can extrusion processes from the same billets are identical provided that the die aperture diameter D_d and the punch diameter D_p are the same. Fig. 16 confirms this conclusion experimentally.

UBA has also been utilized in designing forming operations and developing new forming processes. Kiuchi and Ishikawa (1981) formulated upper bound solutions for extrusion and drawing of eccentric pipes and on the basis of minimum work rate they proposed process parameters that can reduce the eccentricity.

Another noteworthy example is a novel precision cold forging process to produce thin toothed components shown in Fig. 17, that has been developed by Ohga, Kondo and Jitsunari (1984) with UBA as a guide. Starting from a circular blank, they managed to form almost complete teeth at a reasonably low working pressure with the aid of a central hole punched out from the blank to relieve excess pressure, see Fig. 17. This process is now being adopted in factory production of motorcar components.

It is true that in the future UBA itself must be refined so that, for example, reliable predictions of the uncontained large deformation of workpiece and of the pressure distribution on tool surfaces become possible. This will presumably require the use of more mathematically sophisticated velocity fields from which the results are obtained numerically at the expense of a large amount of computing time. The present value of UBA will then vanish and FEM may become preferable.

However, since knowledge of the flow stress Y, friction factor f and fracture criterion for materials deforming under actual forming conditions are rather poor, it seems to the present author that sophisticated computations are not worthwhile for technological and engineering purposes. He would suggest rather that theoretical methods are developed that can be used alongside and in conjunction with experimental observation. The use of an experimentally observed

current workpiece profile as the basis for calculation eliminates
the possible accumulation of errors in the profile and the need for
time consuming incremental or trial and error calculation.   This is
exemplified above concerning Figs. 9 and 10.

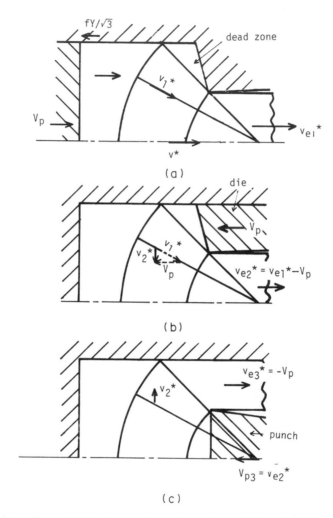

Fig. 15.   Conversion of radial flow field to various
           extrusion processes: (a) forward rod ex-
           trusion, (b) backward rod extrusion, (c)
           backward can extrusion.

The shortage of information about the flow stress and friction factor
can be supplied by a small number of experimental results.   Suppose
that an upper bound solution for the load, deformation mode or the
like is expressed as

$$\phi_i^* = \Phi(G_i, Y_m, f_m) \tag{14}$$

where G_i denotes the geometrical parameter of the i-th process configuration, Y_m and f_m the mean flow stress and mean friction factor
respectively. Y_m and f_m can be assessed from two experimental results on ϕ_i for two different values of G_i. The Y_m and f_m thus obtained may not be true. However if they are substituted into equ.
(14), fairly good prediction of other ϕ_i values are expected. In a
study of hot die forging, Kudo (1960) determined Y_m of steel in this
way and succeeded in predicting forging load and work done with a
satisfactory accuracy.

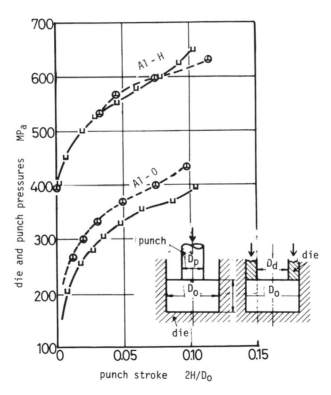

Fig. 16. Mean die and punch pressures during early
          stages of rod and can extrusion processes
          with the same die aperture and punch nose
          diameters:
          $D_p/D_0 = D_e/D_0 = 0.25$,
          ① rod extrusion,
          ⊔ can extrusion.

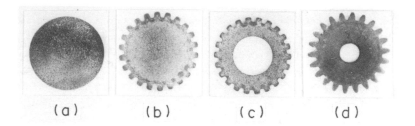

(a)   (b)   (c)   (d)

Fig. 17.   Cold precision forging of thin gears.
           (a) annealed 0.10% C steel blank of 19.5
               mm in diameter and 8 mm in thickness.
           (b) workpiece preformed in closed die
               under maximum pressure of 1560 MPa
           (c) pierced workpiece for pressure re-
               lieving.
           (d) finish-forged workpiece under maximum
               pressure of 1870 MPa.

Such usage of UBA as elucidated above will, the present author believes, be widely accepted among teachers, researchers and engineers
for the purpose of education, workshop trial and process design and
development in the future, perhaps in preference to FEM. Improvement
in UBA procedures which fit the above usage is highly desirable.

CONCLUSION

The upper bound approach has been surveyed above in relation to what
it could do and could not do and to what extent it has been shown to
be reliable and unreliable in its historical development. The present author is of the opinion that the so-called upper bound approach
should continue to be useful in teaching because of the way in
which it aids the understanding of the mechanism or essence of metal
forming processes. Moreover, if it is closely combined with experimental observations, it will continue to be a useful expedient for
decision making in workshop trials, the design of forming operations
and toolings and the development of novel forming processes.

One point of concern however is the name "upper bound approach".
As elucidated above, the so-called upper bound solution does not
necessarily offer an upper limit to the load or power needed for
a forming process. That is, the minimization of the upper bound
solution does not always mean the approach to the true solution.
Perhaps the name "lower energy method" (LEM) may be better given to
the procedure described above instead of UBA.

ACKNOWLEDGEMENT

The author wishes to express his thanks to Mrs. H. Kudo and Miss. Y.
Harada for their help in preparing the present manuscript.

REFERENCES

Alexander, J. M. (1955). Th effect of Coulomb friction in plane strain compression of a plastic rigid material. *J. Mech. Phys. Solids*, **3**, 233-245.

Avitzur, B. (1973). Tensile strength of composite materials - I. *Trans. A.S.M.E.*, **B95**, 827-834.

Avitzur, B., E. D. Bishop and W. C. Hahn (1972). Impact extrusion - Upper bound analysis of the early stage. *Trans. A.S.M.E.*, **B94**, 1079-1085.

Collins, I. F. (1969). The upper bound theorem for rigid/plastic solids generalized to include Coulomb friction. *J. Mech. Phys. Solids*, **17**, 323-338.

Cristescu, N. (1975). Plastic flow through conical converging dies using a viscoplastic constitutive equation. *Int. J. Mech. Sci.*, **17**, 425-433.

Drucker, D. C. (1954). Coulomb friction, plasticity and limit loads, *Trans. A.S.M.E.*, *J. Appl. Mech.*, **21**, 71-74.

Fenton, R. G. (1975). Effects of ram speed and size on the required extrusion pressure. *Proc. 3rd N.A.M.R.C.*, 41-51.

Gatto, F. and A. Giarda (1981). The characteristics of the three-dimensional analysis of plastic deformation according to spatial elementary rigid region method. *Int. J. Mech. Sci.*, **23**, 129-148.

Green, A. P. (1951). A theoretical investigation of the compression of a ductile material between smooth flat dies. *Phil. Mag.*, **42**, 900-918.

Haddow, J. B. (1966). Comments on "An upper-bound solution for axi-symmetric extrusion". *Int. J. Mech. Sci.*, **8**, 145.

Halling, J. and L. A. Mitchell (1964). Use of upper bound solutions for predicting the pressure for the plane strain extrusion of materials, *J. Mech. Eng. Sci.*, **6**, 240-249.

Hencky, H. (1923). Über einige statisch bestimmte Fälle des Gleichgewichts in plastischen Körpern. *Z. angew. Math. Mech.*, **3**, 241-251.

Hill, R. (1950). *The Mathematical Theory of Plasticity*. Clarendon Press, Oxford.

Hill, R. (1951). On the state of stress in a plastic-rigid body at the yield point. *Phil. Mag.*, **42**, 868-875.

Hill, R. (1956). New horizons in the mechanics of solids. *J. Mech. Phys. Solids*, **5**, 66-76.

Johnson, W. (1956). Experiments in plane-strain extrusion. *J. Mech. Phys. Solids*, **4**, 269-282.

Johnson, W. (1958). Over-estimates of load for some two-dimensional forging operations. *Proc. 3rd. U.S. Congr. Appl. Mech.*, 571-579.

Johnson, W. (1959a). Estimation of upper bound load for extrusion and coining operations. *Proc. Inst. Mech. Engrs.*, **173**, 61-72.

Johnson, W. (1959b). An elementary consideration of some extrusion defects. *Appl. Scient. Res.*, **A8**, 52-61.

Johnson, W. (1959c). Cavity formation and enfolding defects in plane-strain extrusions using a shaped punch. *Appl. Scient. Res.*, **A8**, 228-236.

Johnson, W. and H. Kudo (1962). *The Mechanics of Metal Extrusion*, Manchester University Press.

Johnson, W. and H. Kudo (1964). Plane-strain deep indentation. *Proc. 5th. Int. M.T.D.R. Conf.*, 441-447.

Johnson, W. and A. G. Mamalis (1977). Force polygons to determine upper bounds and force distribution in plane strain metal forming processes. *Proc. 18th. Int. M.T.D.R. Conf.*, 11-25.

Johnson, W. and P. B. Mellor (1962). *Plasticity for Mechanical Engineers*, van Nostrand Reinhold, London.

Kármán, T. von (1925). Beitrag zur Theorie des Walzvorganges. *Z. angew. Math. Mech.*, **5**, 139-141.

Kato, K., T. Murota and T. Kumagai (1980). An analysis of flat
    rolling of bar by the energy method - I, *J. Jap. Soc. Technol.
    Plasticity*, <u>21</u>, 359-367.
Kiuchi, M. and S. -H. Hsiang, (1981). Study on application of limit
    analysis to rolling process - I. *J. Jap. Soc. Technol. Plasticity*,
    <u>22</u>, 927-934.
Kiuchi, M. and M. Ishikawa (1981). Study on non-symmetric extrusion
    and drawing of pipe, *Seisan-Kenkyu*, Res. Inst. Indust. Sci., Tokyo
    University, <u>33</u>, 473-476.
Kiuchi, M. and Y. Murata (1980). Study on application of UBET.
    Proc. 4th. Int. Conf. Prod. Eng., Tokyo, 66-71.
Kudo, H. (1957). A computation of required pressure for extrusion-
    forging circular shells. *Proc. 7th. Jap. Nat. Cong. Appl. Mech.*
    57-62.
Kudo, H. (1958). Study on forging and extrusion processes, I -
    Analysis on plane-strain problems. *Kokenshuho, Tokyo University*,
    <u>1</u>, 38-96; II - Experiment on plane-strain problems. *ibid.*, 131-
    150.
Kudo, H. (1959). Study on forging and extrusion processes, III -
    Analysis on axisymmetrical problems. *Kokenshuho, Tokyo University*,
    <u>1</u>, 212-246; IV - Experiments on axisymmetrical problems, *ibid.*,
    247-299.
Kudo, H. (1960). An upper bound approach to simple axisymmetric
    closed-die forging. *Proc. 10th. Jap. Nat. Cong. Appl. Mech.*,
    145-150.
Kudo, H. (1983a). Teoriya Plasticheskikh Deformacii Metallov.
    Mashinostroenie, Moscow. pp. 280-337.
Kudo, H. (1983b). A review of development and use of upper bound
    approach to metal forming processes. *Grundlagen der Umformtechnik.
    Ber. Inst. Umformtech., Univ. Stuttgart*, <u>74</u>, 57-81.
Kudo, H., B. Avitzur, T. Yoshikai, J. Luksza, M. Moriyasu and S. Ito
    (1980). Cold forging of hollow cylindrical components having an
    intermediate flange - UBET analysis and experiment. *Ann. C.I.R.P.*,
    <u>29</u>, 129-133.
Kudo, H. and K. Shinozaki (1974). Investigation into multiaxial
    extrusion process to form branched parts. *Proc. 1 st. Int. Conf.
    Prod. Engg.*, Tokyo, 314-319.
Kudo, H. and H. Takahashi (1964). On some complete solutions for
    steady state extrusion in plane strain, *J. Jap. Soc. Technol.
    Plasticity*, <u>5</u>, 237-243; 464; see also Discussion to B. Avitzur
    (1964). Analysis of wire drawing and extrusion through conical
    dies of large cone angle, *Trans. A.S.M.E.*, <u>B86</u>, 305-316.
Kudo, H. and K. Tamura (1969). Analysis and experiment in V-groove
    forming, *Ann. C.I.R.P.*, <u>17</u>, 297-305.
Lahoti, G. D. and T. Altan (1975). Prediction of temperature dis-
    tributions in axisymmetric compression and torsion, *Trans. A.S.M.E.*,
    <u>H97</u>, 113-120.
Matsubara, S. and H. Kudo (1975). Wedge indentation into work-
    hardening materials. *Proc. Spring Conf. Plastic Working*, Tokyo.
    265-268.
Nagpal, V., G. D. Lahoti and T. Altan (1978). A numerical method
    for simultaneous prediction of metal flow and temperature in up-
    set forging of rings. *Trans. A.S.M.E.*, <u>B100</u>, 413-420.
Oh, S. I. and S. Kobayashi (1975). An approximate method for a
    three-dimensional analysis of rolling. *Int. J. Mech. Sci.*, <u>17</u>,
    293-305.
Ohga, K., K. Kondo and T. Jitsunari (1984). Research of precision
    die forging utilizing divided flow - V. *Trans. J.S.M.E.*, <u>C50</u>,
    (to be published).
Osakada, K., B. Limb and P. B. Mellor (1973). Hydrostatic extrusion
    of composite rods with hard cores. *Int. J. Mech. Sci.*, <u>15</u>, 291-
    301.

Osakada, K. and Y. Niimi (1975). A study on radial flow field for extrusion through conical dies. *Int. J. Mech. Sci.*, 17, 241-254.

Oxley, P. (1963). Note: Allowing for friction in estimating upper bound loads. *Int. J. Mech. Sci.*, 5, 183-184.

Prager, W. (1954). On slow visco-plastic flow. *R. von Mises Presentation Volume*. Academic Press, N.Y., pp. 208- .

Prandtl, L. (1920, 1921). Über die Härte plastischër Korper. *Nachr. Ges. Wiss. Göttingen, Math.-Phys. Kl.*, 74; *Z. angew. Math. Mech.*, 1, 15-20.

Rooyen, G. T. van and W. A. Backofen (1960). A study of interface friction in plastic compression. *Int. J. Mech. Sci.*, 1, 1-27.

Shield, R. T. (1955). Plastic flow in a converging conical channel. *J. Mech. Phys. Solids.*, 3, 246-257.

Shindo, A. (1962). General consideration on the compression of a wedge by a rigid flat die - I. *Bull. J.S.M.E.*, 5, 21-29.

Siebel, E. (1922, 1923). Grundlagen zur Berechnung des Kraft-und Arbeitsbedarfs beim Schmieden und Walzen. *Ber. Walzwerkausschuss. V.D.E.*, 28; *Stahl u. Eisen*, 43, 1295-1298.

Siebel, E. and E. Fangmeier (1931). Untersuchungen über den Kraftbedarf beim Pressen und Lochen, *Mitt. Kais.-Wilh.- Inst. Eisenforsch.*, *Düsseld.*, 13, 29-41.

Tarnovskii, I. Ya., A. A. Pozdeev and O. A. Ganago (1959). *Deformatsii i Usiliya pri Obrabotkye Metallov Davleniem*. Mashinostroitel'noi Literatur', Moscow.

Thomsen, E. G. and J. T. Lapsley (1954). Experimental stress determination within a metal during plastic flow. *Proc. Soc. Exptl. Stress Anal.*, 11, 59-68.

Thomsen, E. G., C. T. Yang and J. B. Bierbower (1954). An experimental investigation of the mechanics of plastic deformation of metals. *Univ. California*, *Pub. Engg.*, 5, 89-144.

Tirosh, J. and S. Kobayashi (1976). Kinetic and dynamic effects on the upper bound loads in metal-forming processes. *Trans. A.S.M.E.*, E43, 314-318.

Usui, E and M. Masuko (1972). Fundamental study on three-dimensional machining. *Trans. J.S.M.E.*, 38, 3255-3271.

Yamada, Y. (1958). Yield point load of rigid-plastic body - I, *Kikai-no-Kenkyu, Tokyo*, 10, 621-628.

Zimerman, Z. and B. Avitzur (1970). Analysis of the effect of strain hardening on central bursting defects in drawing and extrusion. *Trans. A.S.M.E.*, B92, 135-145.

ROTARY FORMING OF RINGS UNDER KINEMATIC CONSTRAINTS

Z. MARCINIAK

Technical University of Warsaw, Poland

ABSTRACT

Rotary upsetting of a circular flange surrounding a central rigid hub is analysed theoretically. Also the process of rotary forging of a ring placed tightly in the hole of a die is investigated. In both cases the mean pressure exerted by the punch is evaluated by applying upper and lower bound methods. It is concluded that different flow patterns may take place during forming depending on the outer to inner diameter ratio. If the inner diameter of a ring is small enough, and outer diameter is kept constant a plastic wave is developed in front of the contact area. The in-plane twisting of the ring during rotary forging is also analysed.

INTRODUCTION

The ring rolling operation, in which the roll force is exerted radially by two rolls was analysed by Johnson and his co-workers (1968, 1973, 1976). It was observed that outside of the forming region a plastic hinge is developed, which makes the spread of the ring possible. A similar flow pattern is also observed in rotary upsetting of rings as this process can also be regarded as a kind of axial rolling. Hawkyard et al. (1979) proposed a model which employs a plastic hinge to analyse this process. Recently the rotary upsetting of rings was experimentally investigated by Wang and co-workers (1982).

In the analysis presented below two particular cases of rotary forming of rings are considered. It is assumed that due to the appropriate kinematic boundary conditions either the inner or outer diameter remains constant. The first case, in which the inner diameter of the ring remains constant is often encountered in practice. It takes place when a thin circular flange is rotary upset while the central thick hub is free of pressure and behaves as a rigid body (Fig. 1). This process has been investigated by Nakane et al (1982) and Hawkyard and Moussa (1982). In order to calculate the axial force and plastic work Hawkyard (1982) assumed a simple plane strain indentation model, for which the axial pressure equals to $2k(1+\frac{\pi}{2})$.

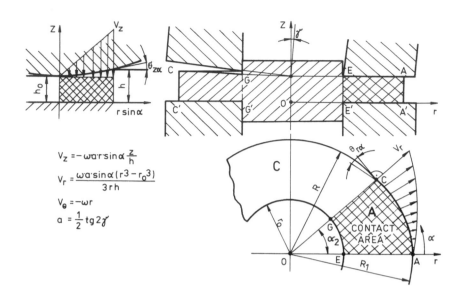

$$V_z = -\omega a r \sin\alpha \frac{z}{h}$$

$$V_r = \frac{\omega a \sin\alpha (r^3 - r_0^3)}{3 r h}$$

$$V_\theta = -\omega r$$

$$a = \frac{1}{2} tg 2\gamma$$

Fig. 1.   Rotary upsetting of a flange surrounding
a central rigid hub.

The second case, for which the outer diameter of the ring remains
constant due to the kinematic constraints, refers to the rotary
upsetting of rings places tightly in the die cavity, as shown in
Fig. 4.

In both cases under consideration the expansion of the ring is constrained and plastic hinges cannot develop. The purpose of this
paper is to analyse the flow patterns, which are possible under the
assumed kinematic conditions and to investigate the influence of the
forming conditions and geometric parameters of the ring on the mode
of deformation.

ROTARY UPSETTING OF A CIRCULAR FLANGE

The approximate analysis of the process shown in Fig. 1 is based on
the following simplifying assumptions:

1. The workpiece-die contact area is restricted to the section AEGC
    in Fig. 1. It is equivalent to the assumption, that the die
    closing rate is zero and the initial cross-section of the ring
    CC'GG' is deformed into the final rectangle AA'EE'.

2. The central hub, $2r_0$ in diameter, is free of pressure and does
    not undergo plastic deformation.

3. The friction forces are neglected.

4. Material obeys the Huber-Mises yield criterion and the associated
    flow rules.

Since the analysis is to be a qualitative one the load bounding approach will be used for assessment of the most probable flow pattern.

Upper Bound Approximation

Using cylindrical coordinates r, α, z; the z axis coincides with the axis of the press and the axis of the upper die, inclined by the angle γ, lies in the z-r plane. The distance h between the conical surface of the upper die and the flat surface of the lower one is the following function of coordinates r and α:

$$h = h_0 + ar\,(1-\cos\alpha) \qquad \text{where,} \qquad a = \frac{1}{2}\tan 2\gamma$$

The system of coordinates rotates about the z-axis with the angular velocity ω, and the upper die rotates about its own axis with the velocity $-\omega$. Thus the axial velocity at the point lying on the surface of the upper die is

$$v_z = \omega\ a\ r\ \sin\alpha \tag{1}$$

Let us assume, that material flows radially in the contact area A, (see Fig. 1) while it is rigid outside this region. The velocity field in region A is then determined by the velocity components

$$v_z = -\omega\ a\ \sin\alpha\ \frac{r\ z}{h}$$

$$v_\theta = -\omega\ r \tag{2}$$

$$v_r = \omega\ a\ \sin\alpha\ \frac{r^3 - r_0{}^3}{3rh}$$

In consequence, the strain rate components are:

$$\dot\varepsilon_z = -\omega\ a\ \sin\alpha\ \frac{r}{h} \qquad\qquad \dot\gamma_{r\alpha} = \omega\ a\ \cos\alpha\ \frac{r^3 - r_0{}^3}{3r^2 h}$$

$$\dot\varepsilon_r = \omega\ a\ \sin\alpha\ \frac{2r^3 + r_0{}^3}{3r^2 h} \qquad \dot\gamma_{\alpha z} = -\omega\ a\ \cos\alpha\ \frac{z}{h} \tag{3}$$

$$\dot\varepsilon_\alpha = \omega\ a\ \sin\alpha\ \frac{r^3 - r_0{}^3}{3r^2 h} \qquad \dot\gamma_{zr} = -\omega\ a\ \cos\alpha\ \frac{z}{h}$$

where partial derivatives $\frac{\partial h}{\partial r}$ and $\frac{\partial h}{\partial \alpha}$ have been neglected as being very small. The rate of equivalent strain $\dot\varepsilon$ is thus

$$\dot\varepsilon = \sqrt{\frac{2}{3}\ \dot\varepsilon_z^2 + \dot\varepsilon_r^2 + \dot\varepsilon_\alpha^2 + \frac{1}{2}\left(\dot\gamma_{r\alpha}^2 + \dot\gamma_{\alpha z}^2 + \dot\gamma_{zr}^2\right)} \tag{4}$$

If V denotes the volume of the deformation region, the rate of energy $\dot W_V$ dissipated in this volume because of the plastic deformation is thus

$$\dot W_V = \int_V \sigma_p\ \dot\varepsilon\ dV \tag{5}$$

It is to be noted that along the plastic-rigid boundary (plane CG) a discontinuity in the velocity field occurs. Streamlines bend through an angle θ on crossing a plane of discontinuity for which $\alpha = \alpha_A$. The components of shear deformation, occurring in the r-α and z-α planes are:

$$\tan\theta_{r\alpha} = \frac{v_r}{\omega r} \qquad\qquad \tan\theta_{r\alpha} = \frac{r^3 - r_0^3}{3r^2 h}\, a\sin\alpha_A$$

$$\tan\theta_{z\alpha} = \frac{v_z}{\omega r} \qquad\qquad \tan\theta_{z\alpha} = \frac{z}{h}\, a\sin\alpha_A \tag{6}$$

The rate of energy $\dot{W}_{CG}$ dissipated along the discontinuity plane CG can be determined from the integral

$$\dot{W}_{CG} = \frac{\sigma}{\sqrt{3}} \int_{S_{CG}} \sqrt{(\tan\theta_{r\alpha})^2 + (\tan\theta_{z\alpha})^2}\ r\omega\ dS_{CG}. \tag{7}$$

Another discontinuity in velocity field occurs along the cylindrical section GE. The rate of plastic work $\dot{W}_{GE}$ done by the shear stresses $\tau = \sigma_p/\sqrt{3}$ along this surface can be determined from the formula

$$\dot{W}_{GE} = \frac{\sigma}{\sqrt{3}} \int_{S_{GE}} v_z\, dS_{GE}. \tag{8}$$

The mean pressure p, exerted by the upper die on the contact area can be determined by comparing the rate of work done by the pressure with the sum of the energy rate dissipated in the material

$$p\int_{S_{AGCE}} v_z\ dS = \dot{W}_v + \dot{W}_{CG} + \dot{W}_{GE}. \tag{9}$$

The results of a numerical computation are displayed in Fig. 2 by the line b. This calculation was made for $h = 0.2r_0$ and $\alpha_2 = 40°$.

In the previous consideration, a uniform distribution of radial velocity v_r along the thickness of the flange was assumed. However, if the outer to inner diameter ratio R/r_0 is small enough, the non-uniform distribution of velocity v_r shown on the left hand side of Fig. 2 gives a lower pressure p (represented by the line C in Fig. 2). This velocity field results in the cross-section of the flange having a *fishtail* shape - a case often met in practice.

Lower Bound Approximation

In order to estimate a lower bound on the axial pressure let us assume that in the forming region ACGE the stress field is described by the principal stresses:

$$\text{radial:} \qquad \sigma_{2A} = \frac{\sigma_p}{\sqrt{3}} \ln \frac{(\sqrt{1 + 3\rho^2} + 1)\rho_1}{(\sqrt{1 + 3\rho_1^2} + 1)\rho} \tag{10}$$

circumferential: $\quad \sigma_{1A} = -\dfrac{\sigma_p}{\sqrt{3}} \dfrac{1 + 3\rho}{\sqrt{1 + 3\rho^2}} + \sigma_{2A}$

axial: $\quad\quad\quad\quad\quad \sigma_{3A} = -\dfrac{\sigma_p}{\sqrt{3}} \dfrac{1 + 3\rho}{\sqrt{1 + 3\rho^2}} + \sigma_{2A}$

where, $\quad \rho = \left(\dfrac{r}{r_0}\right)^2 \quad$ and $\quad\quad \rho_1 = \left(\dfrac{R}{r_0}\right)^2 .$

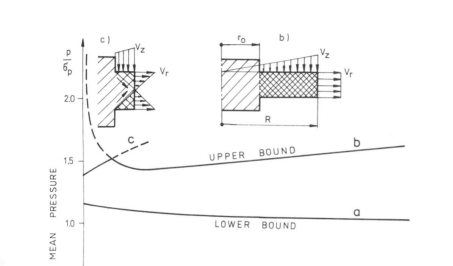

Fig. 2.   Mean pressure p versus R/r_0 ratio.   Inner
diameter of the ring is constant.

This stress distribution represents the known solution of the axially
symmetric problem of a ring subjected to an axial frictionless compression between two parallel plates under the following boundary
conditions:

1. The outer cylindrical surface of the ring is stress free ($\sigma_2 = 0$
    for $r = R$).

2. The inner diameter of the ring $2r_0$ is kept constant during defor-
    mation ($\varepsilon_1 = 0$ for $r = r_0$).

It is also assumed that in the remaining part of the ring, i.e. outside the contact area, the axial stress $\sigma_{3C} = 0$, while the radial
and circumferential stresses are the same as in the forming region
$\sigma_{2A} = \sigma_{2C}$, $\sigma_{1A} = \sigma_{1C}$ (see Fig. 3).

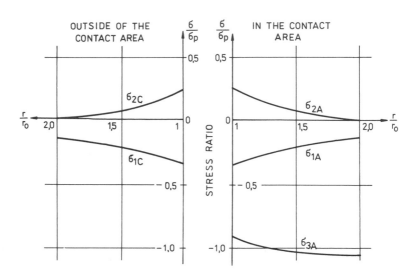

Fig. 3.   Stress distribution across the flange.

The stress field under consideration is one of the admissible ones
as it satisfies the condition of equilibrium, the yield criterion
and the static boundary conditions. Also the yield criterion is
not violated in the rigid part C.

By integrating σ_{3A} over the whole contact area ACGE the total axial
force and the mean axial pressure p can be determined. The result
of the calculation is represented by the line a in Fig. 2. The
large distance between the lower bound (a) and upper bound lines
(b, c) indicates that both the velocity and stress fields assumed
above are far from real fields.

ROTARY UPSETTING OF RINGS INSIDE A DIE

Let us consider the forming process shown in Fig. 4. The ring subjected to deformation is tightly placed inside the cylindrical
cavity of a die.   Under the axial force exerted by the conical
nose of the punch, the initial, trapezoidal cross-section CGC'G'
is deformed into the final rectangle AEA'E'. It is assumed that
the outer diameter of the ring $2r_0$ remains constant during deformation due to the kinematic constraints imposed by the die.

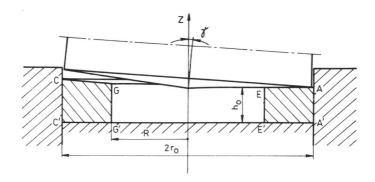

Fig. 4.   Rotary upsetting of a ring placed inside
          of the die.

Upper Bound Approximation

In order to estimate the possible flow pattern of the material for
different values of the outer to inner diameter ratio of the ring,
the upper bound method is applied.   The three velocity fields, shown
in Fig. 5, are taken into consideration.

The first velocity field (Fig. 5a) is based on the assumption that
tha material in the forming region flows radially inwards ($v_\alpha = 0$).
The forming region coincides with the workpiece-die contact area
ABEF.   The remaining part C of the ring is assumed to be rigid.   As
a result of the radial flow, the final inner radius R_1 of the ring
is smaller than the initial one R.   The thickness of the ring is
also reduced ($h_0 < h$).   The velocity and strain rate components are
described by eqs. (2) and (3) in which $r < r_0$ is taken.   The results
of numerical calculations are shown by the line DC in Fig. 6.   In
the calculation, the rate of work done by friction forces acting at
the cylindrical inside surface of the die has been neglected.   It
is seen that the mean pressure p increases rapidly as the inner to
outer diameter ratio decreases.

The second velocity field considered is shown in Fig. 5b.   It is
assumed that in the contact region A material flows both in the
radial and circumferential directions in such a way, that along the
outer arc AB ($r = r_0$) the radial and circumferential strain rates
are equal to each other $\dot\varepsilon_r = \dot\varepsilon_\alpha = -\frac{1}{2}\dot\varepsilon_z$.   The second assumption is
that no twisting of radial lines occurs, so that the circumferential
velocity v_r is proportional to the radius r.   The two assumptions
are satisfied by the velocity field described by the components:

$$v_{zA} = -\omega\ a\ \sin\alpha\ \frac{r \cdot z}{h}$$

$$v_{\alpha A} = \omega\ a(1 - \cos\alpha)\ \frac{r_0 \cdot r}{2h} - \omega\ r \qquad (11)$$

$$v_{rA} = \omega\ a\ \sin\alpha\ \frac{4r^3 - 3r_0 r^2 - r_0^3}{12r \cdot h}$$

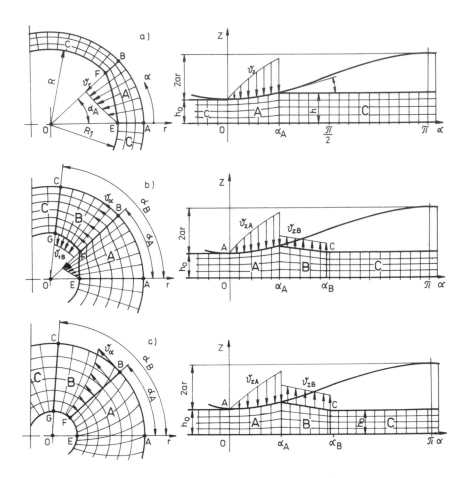

Fig. 5.   Modes of deformation of a ring whose outer
          diameter is constant.

Thus, the corresponding strain rate components are

$$\dot{\varepsilon}_z = -\omega \; a \; \sin\alpha \; \frac{r}{h} \qquad\qquad \dot{\gamma}_{r\alpha} = \omega \; a \; \cos\alpha \; \frac{4r^3 - 3r_0 r^2 - r_0^3}{12 r^2 h}$$

$$\dot{\varepsilon}_\alpha = \omega \; a \; \sin\alpha \; \frac{4r^3 + 3r_0 r^2 - r_0^3}{12 r^2 h} \qquad\qquad \dot{\gamma}_{\alpha z} = -\omega \; a \; \cos\alpha \; \frac{z}{h} \qquad (12)$$

$$\dot{\varepsilon}_r = \omega \; a \; \sin\alpha \; \frac{8r^3 - 3r_0 r^2 + r_0^3}{12 r^2 h} \qquad\qquad \dot{\gamma}_{zr} = -\omega \; a \; \sin\alpha \; \frac{z}{h}$$

Let us notice that at the border BF (Fig. 5b) of the contact region
A ($\alpha = \alpha_A$) the material flows in the circumferential directions with
an angular velocity $\omega_A = \frac{v_\alpha}{r}$ which is given by

$$\omega_A = \omega \; a(1 - \cos\alpha_A) \; \frac{r_0}{2h} \; - \; \omega$$

or $$\omega_A = -(\omega - \Delta\omega)$$

where $$\Delta\omega = \omega \; a(1 - \cos\alpha_A) \; \frac{r_0}{2h}. \tag{13}$$

The decrement $\Delta\omega$ in the angular velocity results in the appropriate increment in the cross-section area in the plane BB'FF' compared with the initial area of the cross-section CC'GG' which rotates with higher angular velocity ω. This follows directly from the incompressibility condition. It is assumed that the passage from the initial velocity field of the rigid part C ($v_{\alpha C} = -\omega \; r$, $v_{rc} = v_{zc} = 0$) to the velocity field existing at the border BF takes place within an additional plastic region B which is formed in front of the contact region A as is shown in Fig. 5b. This additional plastic region B may be regarded as a plastic wave which precedes the forming region A. The velocity field in the region B may be described by the components

$$v_{\alpha B} = -\omega \; r + \Delta\omega r \; \frac{\alpha_B - \alpha}{\Delta\alpha_B}, \qquad\qquad v_{zB} = \frac{\Delta\omega z}{2\Delta\alpha_B}$$

$$v_{rB} = \frac{\Delta\omega}{4\Delta\alpha_B} \; (r_0^2 - r^2) \tag{14}$$

where $\Delta\alpha_B = \alpha_B - \alpha_A$ is the angle corresponding to the region B. The components of the strain rate field in the region B are thus

$$\dot{\varepsilon}_{zB} = \frac{\Delta\omega}{2\Delta\alpha_B}$$

$$\dot{\varepsilon}_{rB} = \frac{\Delta\omega}{2\Delta\alpha_B} \; [\frac{1}{2}(\frac{r_0}{r})^2 + \frac{1}{2}]$$

$$\dot{\varepsilon}_{\alpha B} = -\frac{\Delta\omega}{2\Delta\alpha_B} \; [\frac{1}{2}(\frac{r_0}{r})^2 + \frac{3}{2}] \tag{15}$$

$$\dot{\gamma}_{r\alpha} = \dot{\gamma}_{\alpha z} = \dot{\gamma}_{rz} = 0.$$

The shear strains which take place at the border between regions A and B (plane BB'FF'), where a discontinuity in velocity field occurs, can be determined from the formulae

$$\tan\theta_{r\alpha} = \frac{v_{rA} - v_{rB}}{v_\alpha}$$
$$\qquad\qquad\qquad \text{for } \alpha = \alpha_A \tag{16}$$
$$\tan\theta_{z\alpha} = \frac{v_{zA} - v_{zB}}{v_\alpha}$$

By analogy, the shear strains produced at the border between rigid region C and plastic region B are

$$\tan\theta_{r\alpha} = \frac{v_{rB}}{\omega \; r} \quad \text{and} \quad \tan\theta_{z\alpha} = \frac{v_{zB}}{\omega \; r} \quad \text{for } \alpha = \alpha_B \tag{17}$$

Applying the previously described method, the rate of energy dissipated in the material due to its plastic deformation and mean punch
pressure p can be determined as a function of radius ratio R/r_0.
The results of numerical calculations performed for $h/r_0 = 0.2$ and
$\Delta\alpha_B = 60°$ are presented by the line CB in Fig. 6.

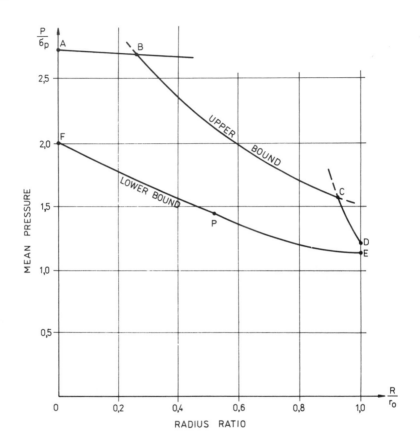

Fig. 6.   Mean pressure p versus R/r_0 ratio.   Outer
          diameter of the ring is constant.

The third velocity field shown in Fig. 5c is based on the assumption
that no radial flow occurs.   In the contact region A the velocity
field is thus described by components

$$v_{zA} = -\omega\ a\ \sin\alpha\ \frac{r \cdot z}{h}$$

$$v_{rA} = 0 \qquad\qquad\qquad\qquad\qquad\qquad (18)$$

$$v_{\alpha A} = \omega\ a(1 - \cos\alpha)\ \frac{r^2}{h} - \omega\ r$$

The corresponding strain rate components are

$$\dot{\varepsilon}_{zA} = -\omega \, a \, \sin\alpha \, \frac{r}{h} \qquad\qquad \dot{\gamma}_{r\alpha} = \omega \, a \, (1 - \cos\alpha) \frac{r}{h}$$

$$\dot{\varepsilon}_{rA} = 0 \qquad\qquad \dot{\gamma}_{zA} = -\omega \, a \, \cos\alpha \, \frac{r}{h} \qquad\qquad (19)$$

$$\dot{\varepsilon}_{\alpha A} = \omega \, a \, \sin \, \frac{r}{h} \qquad\qquad \dot{\gamma}_{zr} = -\omega \, a \, \sin\alpha \, \frac{r}{h}$$

At the border BF of the region A (for $\alpha = \alpha_A$) the circumferential velocity v_α equals to

$$v_{\alpha A} = \omega \, a \, (1 - \cos\alpha_A) \, \frac{r^2}{h} - \omega \, r$$

Note that this hoop velocity is not proportional to the radius r, which results in the in-plane twisting of the ring. So the initially straight radial lines are deformed into curves, as it is seen in the grid in Fig. 5c. In addition to twisting, a decrease of circumferential velocity of the material also takes place at the border BF. In consequence, the thickness of the ring h in this place is larger than the initial thickness h_0. So a plastic wave B arises naturally in front of the contact region A.

Assuming that the central angle of the arc B equals $\Delta\alpha_B = \alpha_B - \alpha_A$, the velocity field in the additional plastic region B, which satisfies the kinematic boundary conditions along the borders CG and BF, can be described by its components

$$v_{\alpha B} = \omega \, a \, (1 - \cos\alpha_A) \, \frac{r^2}{h} \, \frac{\alpha_B - \alpha}{\Delta\alpha_B} - \omega r$$

$$v_{\iota B} = 0 \qquad\qquad (20)$$

$$v_{zB} = \omega \, a \, (1 - \cos\alpha_A) \, \frac{r \cdot z}{\Delta\alpha_B h}$$

The corresponding strain rate components are then

$$\dot{\varepsilon}_{\alpha B} = -\omega \, a \, (1 - \cos\alpha_A) \, \frac{r}{\Delta\alpha_B \cdot h} \qquad\qquad \dot{\gamma}_{r\alpha} = \omega \, a \, (1 - \cos\alpha_A) \, \frac{\alpha_B - \alpha}{\Delta\alpha_B}$$

$$\dot{\varepsilon}_{rB} = 0 \qquad\qquad \dot{\gamma}_{\alpha z} = 0 \qquad\qquad (21)$$

$$\dot{\varepsilon}_{zB} = \omega \, a \, (1 - \cos\alpha_A) \, \frac{r}{\Delta\alpha_B h} \qquad\qquad \dot{\gamma}_{rz} = \omega \, a \, (1 - \cos\alpha_A) \, \frac{z}{h \cdot \Delta\alpha_B}$$

Taking into account the energy dissipation in the volume V_B of the region B and in the volume V_A of the region A, as well as the energy dissipated on the discontinuity planes CG and BF, the total rate of plastic work and the mean punch pressure p can be calculated. The angle $\Delta\alpha_B$ of the plastic wave (region B), can be chosen so as to minimize the total rate of plastic work. The results of calculations, performed for $h_0 = 0.2 \, r_0$; $\alpha_A = 40°$; $\Delta\alpha_B = 60°$, are presented by the almost horizontal line AB in Fig. 6.

Lower Bound Approximation

The statically admissible stress field is chosen in an analogous
manner to the previously discussed case of upsetting the external
flanges. However, the two cases should be distinguished when this
method is applied to the ring with the inner surface free of pressure. If the ratio of inner to outer radius of the ring is sufficiently large, the assumed stress field is that represented in
Fig. 7b by the principal stresses σ_{1A}, σ_{2A}, σ_{3A}. The yield condition is satisfied in the contact region A only, whereas in the
remaining region C the stresses σ_{1C} and σ_{2C} satisfying the condition
of equilibrium are too small to satisfy the yield criterion. There-
fore, the ring remains rigid. The mean pressure p is then represented by the line AP in Fig. 6. However, both parts of the ring
are plastic when the inner to outer radius ratio is smaller than
the abscissa of the point P. The stresses σ_{1A}, σ_{2A}, σ_{3A} shown in
Fig. 7a, produced in the contact region A, as well as the stresses
$\sigma_{1B} = \sigma_{1A}$, $\sigma_{2B} = \sigma_{2A}$, $\sigma_{3B} = 0$ in the remaining part, satisfy both
the yield and equilibrium conditions. The mean pressure p versus
R/r_0 ratio is then presented by the line PF in Fig. 6.

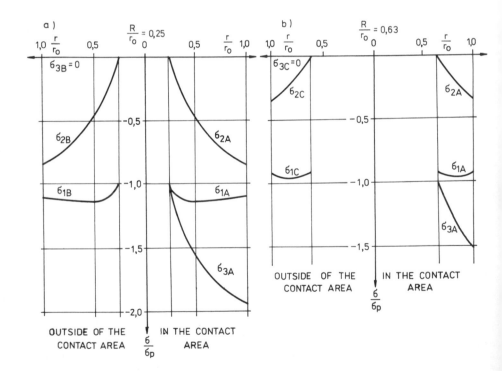

Fig. 7.   Stress distribution in the ring. Inner
surface is free of radial stresses.

CONCLUSIONS

By applying the upper bound method the assumed flow patterns can be assessed by minimizing the rate of plastic work. When an external flange surrounding a rigid central hub is subjected to rotary up-setting, the predominant flow pattern is radially outwards flow of the material particles. When the width to thickness of the flange is small enough, a non-uniform radial velocity distribution is pre-ferred which results in the fishtail shape of the cross-section. If the angle α_A of the forming region is sufficiently large, no plastic wave arises in front of the contact region.

When the outer diameter of the ring remains constant due to kinematic constraints, the flow pattern depends significantly on the inner to outer radius ratio. If this ratio is large enough, the predominant flow pattern is radially inwards flow within the workpiece-die con-tact region. The remaining part of the ring in then rigid. If the inner diameter is smaller, the material outside of the forming region starts to be also plastically deformed. A plastic wave arises in front of the contact area in which the thickness of the ring increases. If the R/r_0 ratio is very small, or when a central mandrel is placed in the hole of the ring to prevent material from inwards radial flow, the plastic wave forms. Moreover, the material in the plastic regions A and B is subjected to in-plane twisting. As a result straight radial lines marked on the material surface are deformed into the curves shown in Fig. 5c. It is worth noting that this twisting vanishes when the thickness of the ring, h, is proportional to radius r.

REFERENCES

Hawkyard, J. B., W, Johnson, J. Kirkland and E. Appleton (1973). Analyses for roll force and torque in ring rolling with some supporting experiments. *Int. J. Mech. Sci.*, 15, 873-893.
Hawkyard, J. B. and J. B. Ingham (1979). In Proc. of First Int. Conf. on Rotary Metalworking Processes. IFS Publications (Ed). An investigation into profile ring rolling.
Hawkyard, J. B. and G. Moussa (1982). In Proc. of 2nd Int. Conf. on Rotary Metalworking Processes. IFS Publications (Ed). Rotary forging of a component with a non circular flange-control of forging force and energy dissipation. pp. 73-80.
Johnson, W., J. MacLeod and G. Needham (1968). An experimental investigation into the process of ring or metal tyre rolling. *Int. J. Mech. Sci.*, 10, 455-468.
Johnson, W. (1982). In the Rodney Hill 60th Anniversary Volume. The mechanics of some industrial pressing, rolling and forging processes. H. G. Hopkins and M. J. Sewell (Ed) . Pergamon Press Oxford. pp. 303-356.
Mamalis, A. G., J. B. Hawkyard and W. Johnson (1976). Spread and flow patterns in ring rolling. *Int. J. Mech. Sci.*, 18, 11-16.
Nakane, T., M. Kobayashi and K. Nakamura (1982). In Proc. of 2nd Int. Conf. on Rotary Metalworking Processes. IFS Publications (Ed). Deformation behaviour in simultaneous backward extrusion - upsetting by rotary forging. pp. 59-71.
Pei Xinghua, Zhou Decheng and Wang Zhonren.(1982). In Proc. of 2nd Int. Conf. on Rotary Metalworking Processes. IFS Publications (Ed). A study of the rotary forging. pp. 91-100.

A STUDY OF THE CONTACT
ZONE AND FRICTION
COEFFICIENT IN HOT-ROLLING

P. S. THEOCARIS

Section of Mechanics, Department of Engineering Science,
The National Technical University of Athens,
5 Heroes of Polytechnion Avenue,
Zographou, GR-157 73 Athens, Greece

ABSTRACT

It is a common practice to accept in the theory of rolling that the
coefficient of friction along the contact arc remains constant and
the length of the roll-throat may be evaluated from the initial
radii of the rolls. Both these assumptions are only approximate
and only in special cases approach reality. In order to obtain
more realistic results an experimental technique was developed,
based on the optical method of reflected caustics, which was used
to investigate the variation of the coefficient of friction in the
roll gap during strip rolling and to define the extent of the contact arc. Tests were carried out on an experimental rolling mill
using lead at room temperature to obtain quantitative information
about the pressure distribution and friction relevant to hot rolling,
i.e. when frictional effects are pronounced. The rolls were made of
perspex. The normal and tangential pressure distribution and thence
the coefficient of friction along the arc of contact at the external
lateral faces of the strip were calculated from the caustic curves
formed because of the elastic deformation of the rolls, although
the rolled strip was deformed plastically.

For *strip rolling*, where plane-strain conditions prevail, it
was sufficient to evaluate from the elastic deformation of the external faces of the rolls, the distribution of normal, N, and tangential, T, stresses along the contact arc. For *ingot rolling*, where
a three-dimensional state of stress is dominant in the rolls, it was
necessary to derive information on the N and T distributions at
various sections inside the transparent rolls, which, in this case,
were made by sandwich plates held together by friction.

For the evaluation of the contact length the size and the relative
position of the caustics formed at the contact zone were used. The
coordinates of the extremities of their almost circular branches were
used to define the exact value of the contact arc.

It was found that there is an almost linear decrease of the friction
coefficient from a maximum value at the entry plane to the roll gap
to zero at the neutral plane. It then increases in a parabolic,
tending to linear, manner towards the exit plane. This is signifi-

cantly different from the usual assumption that the coefficient of
friction remains constant during hot rolling.

The various applications of the method to different aspects of strip
rolling indicate that the method constitutes a potentially valuable
method for evaluating accurately the characteristics of many types
of rolling process.

INTRODUCTION

Knowledge of the distribution of forces over the zone of contact
between rolls and stock is of great importance in metal forming.
Several assumptions have been made when constructing theories of
rolling, since the problem is so complicated and it is unlikely
that any theory can lead to a complete solution of the problem.

Cold rolling is much simpler than hot strip-rolling since the process is essentially independent of the rate of deformation and the
frictional forces between roll and stock are rather low. Nevertheless the process is analysed by making a series of basic assumptions,
which are partly justified in practice.

First of all, certain limits have been imposed from practical considerations on the ratio of the strip thickness to the roll diameter.
Furthermore, a basic assumption is that the ratio of the strip width
to thickness ratio should be large enough to ensure that the non-
plastically deformed material of the stock, outside the roll-gap,
prevents any lateral spread of the material of the stock passing
through the roll gap. The flow is considered to be everywhere
parallel to a given plane. Therefore, conditions of overall plane-
strain may be assumed for the study of cold-rolling phenomena. In
practice the lateral spreading of the stock material seldom exceeds
1 to 2 percent.

The hot-rolling case is much more complicated because in this case
the rolled material must be taken as strain-rate dependent. The
frictional forces at the interfaces between rolls and stock are in
this case very high, whereas for cold rolling they are rather limited.
Owing to the difficulty of the problem, no complete theory has been
devised so far for the case of considerable lateral spread, which
happens in hot-rolling. In this case, the plane-strain assumption
is no longer valid and the case should be treated as a three-dimensional problem of plasticity for the stock and of elasticity for the
rolls.

The early assumptions made in the theories of rolling were summarized
by Orowan (1943). These assumptions for cold rolling are: i) The
rolled material obeys the Mises yield criterion, ii) During rolling,
plane sections perpendicular to the rolling direction remain plane,
which implies the existence of slipping friction along the contact
arc, iii) The contact arc is assumed circular, iv) The coefficient
of friction over the contact arc is assumed to be constant, v) There
is no lateral spreading of the rolled material and vi) The elastic
compression of the rolled material is negligible. The second assumption is important and constitutes the basis of definition of the
limits and the mode of deformation in the plastic zone between rolls.

Concerning friction between the roll and stock, it is worthwhile
mentioning that the kind of friction which is generated between two
bodies, one of which is submitted to large plastic deformations,
cannot be assumed to be the same as the frictional conditions between

two elastic bodies. This fact, which was mentioned by Ford (1957),
is generally overlooked. For hot rolling, where the frictional
forces are high, this fact plays a considerable role in the evaluation of the friction coefficient between rolls and stock.

Smith, Scott and Silwestrowicz (1952) studied the distribution of
pressure over the contact arc by the pressure-pin method, but they
considered that the coefficient of friction could not be evaluated
by the rather inaccurate measurements made using this technique.
Smith et al. studied the problem of correcting for the finite size
of the pin in the pressure-pin technique introduced by Siebel and
Lueg (1933). According to this technique a square pressure-trans-
mitting pin is placed in a radial cavity of the rolls. When it
comes in contact with the rolled material it presses against a piezo-
electric-force transducer during rolling, thus allowing the pressure
at the roll faces to be measured.

Another early method was described by Whitton and Ford (1955), where
an equivalent friction coefficient is determined by rolling at steady
speed a strip at any suitable pass-reduction and measuring continuously the roll force and torque. An additional applied back-tension
to the strip displaces the neutral point to the exit, so ensuring
that the friction along the whole contact zone is in the same direction. The pin-load cell method was further studied by Mamalis,
Johnson and Hawkyard (1976a) and details of the transducer were
given. The method was applied to measure the pressure distribution
along the rolling surface for ring rolling (1976a) or, additionally,
the roll force and torque in cold ring-rolling (1976b).

On the other hand, resistance strain gauges were used by McGregor
and Palme (1948), whereas Van Rooyen and Backofen (1957) used this
method for evaluating the friction coefficient along the contact
arc in cold rolling by measuring normal and tangential force distributions. Similar applications of strain gauges in strip, ingot and
section rolling are given in the papers by Mamalis, Johnson and
Hawkyard (1976a, b).

A variation of the pin load-cell method is the membrane method,
where the pin of the transducer comes in contact with a thin cover
on the external surface of the roll left during drilling of the
cavity for the transducer, instead of coming directly into contact
with the rolled stock.

All these mechanical gauges, although they have been continuously
improved over the years, present inherent difficulties and drawbacks
which are accentuated because of their having to function in hostile
environments where large forces are applied, considerable deformations
take place, and, furthermore, high temperature atmospheres eventually
exist, which make difficult the accurate measurements of the force
distribution along the rolling zone. This has as a result to lead to
the simplifying idea that the coefficient of friction during the
rolling process may be considered to be constant.

In this work a new method having considerable potential is described and reviewed. It enables the normal and tangential forces
developed in the roll gap during any type of rolling process to be
accurately evaluated from the sides of the elastically deformed rolls.
The method is a combination of the optical methods of *caustics* - for
the points and areas where stress-singularities are developed, because of loading-jumps or variations of slope and curvature of the
force distribution in the contact zone - and *pseudocaustics* - for
the remaining non-singular zones of the roll gap. It is shown,

further, that the method of pseudocaustics can be applied, not only
to the contact surfaces between rolls and stock but also to any concentric internal surface of the roll where improved measurements may
be executed. Furthermore, the method can be used to define the length
of contact of the flattened rolls in strip rolling, a quantity which
to date has proved difficult to measure or calculate in a reliable
way by any other experimental or theoretical method.

THE EXPERIMENTAL METHOD OF CAUSTICS AND PSEUDOCAUSTICS

The method of reflected caustics has been developed in the last
fifteen years (Theocaris, 1970; 1979; 1981). It is an optical
method, using the reflections of light rays impinging along a singular
zone of the stress field, or, for the method of pseudocaustics (Theocaris, 1973a, b; Theocaris and Razem, 1977; 1978; 1979) along a
boundary, or any line marked on the deformed body. The reflected
rays yield accurately the real length of contact of the two bodies
and the normal and tangential force distribution along the contact
zone.

The method has been used to determine the stresses resulting from a
concentrated load (Theocaris, 1973a) or from a uniformly distributed
load (Theocaris, 1973b) in a half-plane in plane stress. More important is the inverse problem, where there is an arbitrary distribution of load along a straight boundary. By using data from the
caustics created by the singularities (Theocaris and Razem, 1978)
and the pseudocaustics generated by the deformed boundary of the
half-plane (Theocaris and Razem, 1977), the end-values of the load
and its distribution along the boundary can be found. This may be
achieved by using a hybrid method of solving an ordinary integral
equation, in which data from the pseudocaustics along the contact
zone are utilized. The method of caustics was also used for studying
the intensity, slope, and curvature-discontinuities in loading distributions at the contact of two plane bodies (Theocaris and Razem, 1979),
while the method of pseudocaustics has been successfully used for
the experimental solution of the general elastic plane-stress contact
problem of two bodies of different materials and arbitrary shapes,
one of which is finite while the other is infinite (Theocaris, 1977).

The method of caustics has also been used for the experimental determination of the contact length between two elastic discs (Theocaris
and Stassinakis, 1978) and the coefficient of friction in the contact
zone between an elastic disk and a body which has deformed plastically
(Theocaris and Stassinakis, 1982). Moreover, using data from pseudocaustics, a practical application of the method was made to define
successfully the roll pressure distribution and the coefficient of
friction in hot-rolling, where the assumption of plane-strain conditions in the rolls and the stock is valid (Theocaris, Stassinakis
and Mamalis, 1983). Finally, the three-dimensional deformation of
rolls and stock for the ingot hot-rolling was examined in a recent
paper (Theocaris, Mamalis and Stassinakis, 1984). It can be seen
from the foregoing discussion that the various quantities involved
in contact problems can be found from measurements of caustics or
pseudocaustics. The principal pseudocaustics (i.e. those formed
from real boundaries of the deformed body) are used to define the
contact length and the distribution of forces, if the nature of the
deformation of the contacting boundaries allows the distinct creation
of sharp illuminated curves. In cases of high loading, where plastic
deformation of the one or both bodies distorts considerably the
contact boundaries, *secondary pseudocaustics*, that is pseudocaustics
formed from lines traced inside the stress field, combined together

with a cubic-spline polynomial representation of the stresses inside the body, may yield the coefficient of friction along the contact zone (Theocaris and Stassinakis, 1984).

The use of cubic spline-polynomials is not applicable for discontin- uous stress distributions (Theocaris and Razem, 1979). In these cases the areas of intermediate discontinuities of loading are readily detected, because they must create caustics lying between the two extreme ones. In these cases the contact problem may be solved piecewise between successive pairs of caustics (Theocaris and Razem, 1979).

Caustics and pseudocaustics are governed by the same physical prin- ciples and the method for each is developed by the same theory as for caustics. The method will first be applied, after a brief introduction of the method, to the classical problem of contact between a disc and a plane or between two cylinders, where first of all the contact length will be evaluated.

It has been shown that the equation of caustics is expressed by (Theocaris 1970; 1979; 1981)

$$W = \lambda_m [z + C \overline{\Phi'(z)}] \tag{1}$$

where z is the complex variable in the plane $z = x + iy$ representing the middle plane of the specimen, W is the complex expression for the caustic ($W = X + iY$) on the plane of the reference screen, λ_m is the magnification ratio of the optical setup, C is a constant, which depends on the mechanical and optical properties of the material and the characteristics of the optical set-up, and $\overline{\Phi(Z)}$ is the complex-conjugate of Muskhelishvili's stress-potential function.

In contact problems two caustics are created by the two singular ends of the contact area and a pseudocaustic is formed by the boundary of the contact zone (Theocaris and Razem, 1977; 1978). Generally, it is impossible to establish a correspondence between points z_i of the initial curve and W_i of the caustic, with the exception of a few characteristic ones, such as the extreme points. On the other hand, the pseudocaustic extends between the two caustics and it is a well-defined curve representing the deformed contact boundary. It is suitable for establishing a relationship between the points z_i of the specimen and the points W_i of the plane of the caustic.

The pseudocaustic depends on the intensity of the applied load, as well as on the contact-length. If the contact-length is reduced, or the intensity of load is increased, the pseudocaustic disappears, being absorbed by the two expanding caustics, which tend to coalesce (Theocaris and Razem 1977; 1978). Any arbitrary point z of the specimen and its corresponding point W on the caustic satisfies Eq. (1). Hence, if we choose points near the contact-boundary, but away from the caustics, to avoid their interference, we can have in all cases pairs of points, which satisfy the above equation. For this purpose a fine grid consisting of two orthogonal families of curves is scribed at the area of contact, thereby defining with high accuracy the coordinates of the points z_i, while the distorted grid on the caustic-plane gives the coordinates of the corresponding points W_i.

P. S. Theocaris

If we consider a half-plane, or an elastic disc, on which a load distribution is applied, the complex potential functions of Muskhelishvili are given by (Fig. 1)

$$\Phi(z) = \frac{1}{2\pi i} \int_a^b \frac{N(x)+iT(x)}{x-z}\, dx \tag{2}$$

$$\Phi(z) = \frac{1}{2\pi i} \int_{t_1}^{t_n} \frac{N(t)+iT(t)}{t-z}\, dt + B_0 \tag{3}$$

respectively.

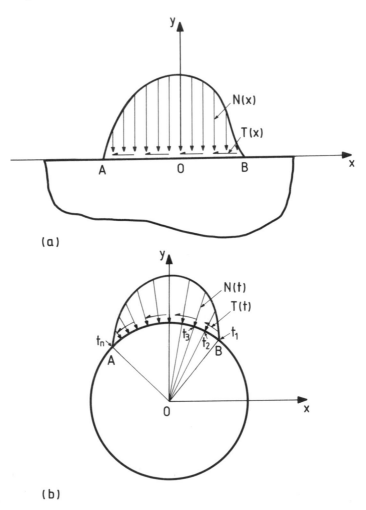

(a)

(b)

Fig. 1.   Arbitrary normal and tangential pressure distributions along the boundaries of (a) a half-plane and (b) a circular disc.

The limits a, b define the contact length between the points A, B
(Fig. 1(a)), whereas complex variables t_1, t_n in Eq. (3) define the
contact arc between the points A, B (Fig. 1(b)). The constant B_0 is
the value of $\Phi(z)$ at infinity and it is independent of the variable
z, whereas the functions N(t) and T(t) represent the normal and tangential forces along the contact length. Since in the equation for
the caustics we use the first derivatives of equations (2) and (3)
with respect of z, we can neglect the constant B_0 and we can use a
unique expression for equations (2) and (3) in either case.

Differentiating relation (3) with respect to z, yields

$$\Phi'(z) = \frac{1}{2\pi i} \int_{t_1}^{t_n} \frac{N(t)+iT(t)}{(t-z)^2} dt \qquad (4)$$

and from Eq. (1) it follows that

$$\overline{\Phi'(z)} = \frac{W-\lambda_m z}{\lambda_m C}, \qquad (5)$$

Combining equations (4) and (5) we obtain

$$\frac{1}{2\pi i} \int_{t_1}^{t_n} \frac{N(t)+iT(t)}{(t-z)^2} dt = \frac{\overline{W-\lambda_m z}}{\lambda_m C}. \qquad (6)$$

By solving the integral equation (6) we can determine the normal and
tangential stresses applied along the contact surface between the
bodies.

Application of Caustics to Contact Forces

We consider the elastic disc shown in Fig. 1(b), which is loaded by
a complex load distribution N(t)+iT(t) along an arc defined by the
points t_1, t_n. The function F(t)=N(t)+iT(t) can be approximated by
the polynomial

$$F(t) = a+b(t-t_1)+c(t-t_1)^2 + \sum_{i=1}^{j} d_i(t-t_i)^3 \text{ for } t_j \leq t \leq t_{j+1}$$
$$(7)$$

which is continuous, as are its first and second derivatives, between
the points t_1, t_n. In this relation the coefficients $a,b,c,d,...,d_{n-1}$,
and the variables $t_1,t_2,...,t_{n-1}$, which correspond to discrete points
on the arc of the disc between the points t_1 and t_n are complex.

From equations (3) and (7) we have

$$2\pi i\Phi(z) = a\int_{t_1}^{t_n} \frac{dt}{t-z} + b\int_{t_1}^{t_n} \frac{t-t_1}{t-z} dt + c\int_{t_1}^{t_n} \frac{(t-t_1)^2}{t-z} dt +$$

$$+ \sum_{j=1}^{n-1} \int_{t_j}^{t_n} \frac{(t-t_j)^3}{t-z} dt \qquad (8)$$

Putting

$$H_a(r_1,r_2,z) = \int_{r_1}^{r_2} \frac{dt}{t-z}, \quad H_b(r_1,r_2,z) = \int_{r_1}^{r_2} \frac{t-r_1}{t-z} dt$$

$$H_c(r_1,r_2,z) = \int_{r_1}^{r_2} \frac{(t-r_1)^2}{t-z} dt, \quad H_d(r_1,r_2,z) = \int_{r_1}^{r_2} \frac{(t-r_1)^3}{t-z} dt$$

(9)

Eq. (8) becomes

$$2\pi i \Phi(z) = aH_a(t_1,t_n,z) + bH_b(t_1,t_n,z) + cH_c(t_1,t_n,z) +$$

$$+ \sum_{j=1}^{n-1} d_j H_d(t_j,t_n,z).$$

(10)

Combining the first derivative of Eq. (10) with Eq. (6) we derive the relation

$$aH_a'(t_1,t_n,z) + bH_b'(t_1,t_n,z) + cH_c'(t_1,t_n,z) + \sum_{j=1}^{n-1} d_j H_d'(t_j,t_n,z)$$

$$= \frac{\overline{W - \lambda_m z}}{\lambda_m z} \frac{1}{2\pi i}.$$

(11)

Putting

$$P_1 + iP_2 = \frac{\overline{W - \lambda_m z}}{\lambda_m C}, \quad \alpha = \alpha_1 + i\alpha_2, \quad \alpha = a,b,c, \quad d_i = d_{i,1} + id_{i,2},$$ (12)

$$i = 1,2,\ldots,n-1$$

Eq. (12) yields

$$a_1 \text{Re}[H_a'(t_1,t_n,z)] - a_2 \text{Im}[H_a'(t_1,t_n,z)] + b_1 \text{Re}[H_b'(t_1,t_n,z)] -$$

$$b_2 \text{Im}[H_b'(t_1,t_n,z)] + \ldots\ldots + d_{n-1,1} \text{Re}[H_d'(t_{n-1},t_n,z)] -$$

$$d_{n-1,2} \text{Im}[H_d'(t_{n-1},t_n,z)] = -\frac{P_2}{2\pi}$$

(13a)

and

$$a_1 \text{Im}[H_a'(t_1, t_n,z)] + a_2 \text{Re}[H_a'(t_1,t_n,z)] + b_1 \text{Im}[H_b'(t_1,t_n,z)] +$$

$$b_2 \text{Re}[H_b'(t_1,t_n,z)] + \ldots\ldots + d_{n-1,1} \text{Im}[H_d'(t_{n-1},t_n,z)] +$$

$$d_{n-1,2} \text{Re}[H_d'(t_{n-1},t_n,z)] = -\frac{P_1}{2\pi}$$

(13b)

where the symbols Re and Im represent the real and imaginary parts of the complex quantity.

Equations (13a) and (13b) have as unknowns the real coefficients $a_1, a_2, b_1, b_2, \ldots, d_{n-1,1}$ and $d_{n-1,2}$, which number $(2n+4)$. In order to

define these coefficients we need to define n+2 points on the pseudocaustics, which form a linear system of (2n+4) equations with (2n+4)
unknowns. Solving this system we calculate the coefficients $a_1, a_2,$
$b_1, b_2, \ldots, d_{n-1,1}, d_{n-1,2}$ and derive the unknown stress-function F(t).

In order to check the method, it is applied to the problem of contact
of two discs. Two elastic and isotropic discs with radii R and r,
respectively, are in elastic contact and they are subjected to point-
loads at diametrically opposite points. Although the initial contact
area is a point where the applied load is insignificant, when the load
is increased the contact zone is extended and a distribution of
stresses is developed. If we consider one of them, we can state that
in this disc a distribution of stresses equal to the developed contact
stresses is applied along an arc of its boundary, which is equal to
the contact arc. For this problem the function $\Phi(z)$ is given by Eq.
(3).

From the form and the size of the caustics appearing at the ends of
the loading zone we can extract all the necessary information about
the slopes and the values of stresses at the two ends of the loaded
zone (Theocaris and Razem, 1977).

In order to calculate the normal loads developed along the contact
zone we used six points from the second horizontal line of the scribed
grid. The corresponding coordinates W_i were measured on the screen
and these data were introduced into (12) and (13).

In Fig. 2a, where the caustics are shown in both discs, it is clear
that there are no separated caustics corresponding to loading jumps
at the ends of the contact zone. Therefore, the normal load distribution in the contact zone is smooth, starting from zero loads at the
extremities of the contact.

By solving the respective system of equations we obtained the values
of the normal loads represented in Fig. 2b by dots. The differences
between the applied loads and calculated ones were small, with a
maximum deviation of 9 percent at the end-points of the loading zone.

In order to calculate the extent of the real contact zone between
the discs the deformed patterns of the pseudocaustics were recorded
for each step of the loading. The contact length may be calculated
by using the relation (Theocaris and Stassinakis, 1978)

$$\bar{\ell}^2 = \frac{2Rr}{\pi(R+r)} KP \tag{14}$$

where $\bar{\ell}$ is the contact length, as it is calculated from the distance
of the extremities of the caustic formed by the contact of the two
bodies and their elevations from the image of the reference boundary
of the discs (their common tangent). P is the applied load between
the contacted bodies, R and r are the radii of the discs and K denotes
a real constant given by

$$K = \frac{k_1+1}{4\mu_1} + \frac{k_2+1}{4\mu_2} \tag{15}$$

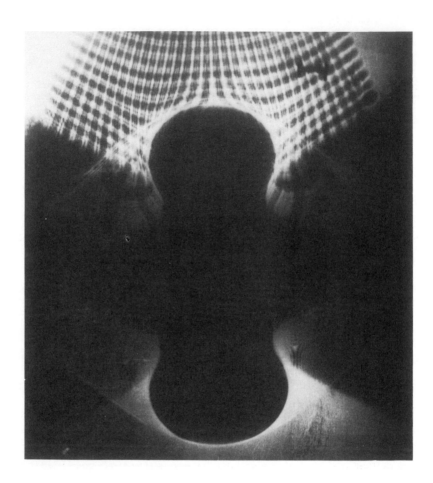

Fig. 2.   (a) The caustics formed along the contact
arc of two elastic discs.

where $\mu_{1,2}$ are the shear moduli of the materials of the discs and
$k_{1,2}$ are expressed by

$$k_{1,2} = (3-4\nu_{1,2}) \qquad \text{and} \qquad k_{1,2} = (3-\nu_{1,2})/(1+\nu_{1,2}) \qquad (16)$$

for plane-strain   for plane-stress.

Figure 2(a) presents the caustics and pseudocaustics formed from the
reflected rays from the discs. The applied load was 3500N, the value
of constant K was $1.33 \times 10^{-3} N^{-1} m^2$, and the contact length ℓ_c was found
to be 0.0045m. The length calculated by relation (14) was $\bar{\ell}_c = 0.0042m$,
so we have a discrepancy of only 7 percent.

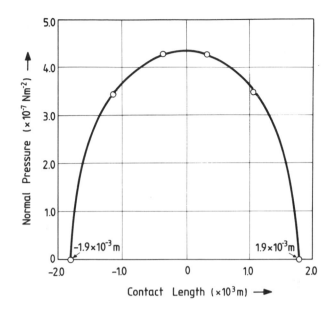

Fig. 2.   (b) The normal pressure distribution along
          their contact arc.

Using data from the fifth concentric circle of the scribed fine grid
on the lower disc and following the aforementioned analysis we calculated the stresses at six points in the contact zone.   The calculated
values are represented in Fig. 2(b) by dots, whereas the continuous
line represents the *Hertzian elliptic stress distribution*.   The
differences between the calculated values and the corresponding value
of the elliptic distribution are small;   the experimental distribution
resembles a Hertzian one.

The method may be also applied with equally satisfying results to
solve contact problems in which one of the bodies is deformed plastically, by using data from the body which is deformed only elastically.   This idea may be applied to evaluate the variation of the
coefficient of friction along the contact arc in rolling of thin
hard strips as, for example, in the rolling of stainless and razor-
blade steels.

For thin-strip rolling, there is a flattening of the rolls, with a
resulting increase in the contact-arc, which ceases to remain cylindrical.   Along the contact zone there is a flat contact, whose length
is unknown.   The method described above allows the accurate evaluation
of the real contact length ℓ, which may then be introduced into previously established relations for the definition of the variation of
the coefficient of friction along the contact arc.   A related problem
to *thin-strip rolling* was solved by the method of caustics by Theocaris and Stassinakis (1982), where the variation of the coefficient
of friction between a cylindrical indenter and a plastically deformed
plate, representing rolled stock, was studied.   In this work is was

assumed that the cylindrical indenter which deforms elastically,
presses the rolled stock into the form of a plate which is deformed
plastically. The normal and tangential loads were assumed to follow
a Coulomb-type relationship.

As a first step, the average value of the coefficient of friction
may be found from the shape and size of the caustic formed at the
contact zone and from the value of the average length of contact.
This average value of the coefficient of friction, together with the
actual contact arc allow a sound simplified consideration of the problem of cold and hot rolling.

However, for an exact and detailed solution of the problem of stress
and strain distributions at the contact arc it is necessary to undergo
the whole analysis from data taken from the caustics and pseudocaustics.

Finally, it is worthwhile pointing out that the solution of the plate
plastically deformed by a cylindrical indenter does not correspond
exactly to the case of thin-strip rolling, since in this problem the
strip should be very thin so that the influence of the pressure exerte
by either roll on the other cannot be neglected. However, although
the case of thin-strip rolling presents some complications in the
analysis, it can be treated by the method of caustics and may yield
more accurate results than those which exist at present which are
based on the Karman assumptions and the solution forwarded by Stone
(1953). In this the contact arc is evaluated by Hitchcock's formula
(Hitchcock, 1935).

THE VARIATION OF COEFFICIENT OF FRICTION IN HOT-ROLLING

Crude assumptions are made at the present time concerning the change
in the coefficient of friction over the arc of contact between stock
and rolls in flat rolling. Rolling theory assumes that the coefficien
of friction is constant (Orowan, 1943), but this assumption has been
questioned on theoretical grounds by many researchers (see for instanc
Ford, 1957).

The measurement of the coefficient of friction has always been one
of the most difficult problems in the development of flat-rolling
theory. This difficulty stems from the fact that, in rolling, the
friction force varies from point to point and also changes direction
at the neutral point and, therefore, it cannot be determined by
measuring overall quantities along the contact region. However, the
accurate determination of an average value of this coefficient by the
optical method of caustics, as it was developed in the previous section, constitutes a first important step towards an accurate solution
of the rolling problem. Therefore, it is desirable to measure the
normal and tangential stress distribution over the region of contact
between the rolls and the workpiece and subsequently to determine the
nature of the friction condition (Wusatowsky, 1969). In this section
the optical method of reflected caustics will be used to measure the
variation of normal and tangential stress distribution along the arc
of contact between the roll stock and the rolls. The latter are made
of Perspex, whilst in order to simulate hot-rolling conditions, lead
is used as a test material, and we concentrate on the case of plane-
strain strip rolling. Assuming that Coulomb-friction prevails during
rolling the variation of the coefficient of friction along the arc of
contact may be determined.

In the method of reflected caustics a light beam impinges on the

specimen which in this case is the end face of one of the rolls, and
the reflected light rays from this face are deviated and concentrated
along a highly illuminated surface in space, which forms the caustic
surface. When the caustic surface is cut by a reference screen, placed
at some distance z_0 from the specimen, a singular curve, called the

caustic, is formed on it. Any caustic created by such a procedure
of reflection or refraction can be classified into one of two categories, depending on the properties of its generatrix curve on the
specimen: i) caustics, closely related to the characteristic parameters of singular stress fields, or ii) pseudocaustics, depicting
any displacement distribution along the boundaries or any lines in
the interior of non singular stress fields. Figure 3 presents a
schematic diagram of the experimental arrangement for detecting the
reflected caustics from the upper roll in the close vicinity of the
respective contact arc. The Oxyz-Cartesian coordinate system refers
to the roll specimen at a generic instant of the rolling procedure,
whereas the O'x'y'z'-frame corresponds to the reference frame (a
ground-glass plate), where the caustics are formed.

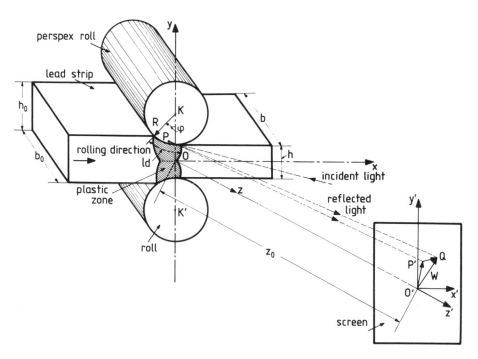

Fig. 3.   Schematic diagram of the experimental set-
           up in rolling tests.

Relations (3) and (4) yield the $\Phi(z)$-stress function and its complex
derivative for the normal, $N(t)$, and tangential, $T(t)$, distribution
of the contact stresses. If φ is the angle of contact and R the radius
of the undeformed roll, putting $t=Re^{i\varphi}$ into Eq. (4) and substituting
$\Phi'(z)$ in Eq. (1) we obtain for the total angle of contact φ_d the
relation

$$\int \frac{N(\varphi) + iT(\varphi)}{(Re^{i\varphi} - z)^2} \, e^{i\varphi} d\varphi = \frac{2\pi}{R} \frac{\overline{\lambda_m^{z-W}}}{\lambda_m C} \tag{17}$$

Equation (17) is solved numerically for the normal and tangential stress distribution. The values of the displacements W were experimentally determined along the principal pseudocaustic or a secondary one, in order to check the experimental data taken from the pseudocaustics along the contact zone. Introducing these experimental values into Eq. (17) the $N(\varphi)$ and $T(\varphi)$-values were determined.

The experimental rolling mill used for the present investigation was a small two-high, single-stand mill, convenient to dispose rolls of 75-125mm dia. and of a barrel length 120mm. The rolls were driven by Cardan shafts from an enclosed gear box, the roll-necks being fitted with needle bearings. The necessary drive power was provided by a variable three-phase motor and was transmitted through a fly-wheel.

Rolling was carried out at a constant speed of 4m/min. between two rolls made of perspex. The rolls consisted of six perspex discs of 89mm diameter and 10mm thickness, the overall barrel-length being 60mm approximately. The elastic discs were assembled on the roll-shaft and held together using steel spacers. Their working surface was finished by grinding. The mechanical properties of perspex were elastic modulus $E = 1217 N/mm^2$, Poisson's ratio $\nu = 0.33$, and its global constant $C = 1.05 \times 10^{-10} m^4 N^{-1}$.

A coarse grid of concentric circles and radial lines was accurately scribed on to the surface of the outer disc of the upper roll. This surface and the corresponding one of the lower roll in the vicinity of the roll gap were illuminated by a coherent light beam, emitted from a He-Ne laser, (Fig. 3). The elastic deformation of the rolls, due to plastic deformation of the rolled strip, resulted in the formation of front caustics and pseudocaustics extending both sides from the extremities of the caustics as shown in Fig. 4, which presents in an axonometric picture the zone around the contact arc at the rolling mill and the respective caustics and pseudocaustics formed on the ground glass of the reference screen placed at a distance z_0 from the front face of the rolled stock and rolls.

The distortion of the grid lines shown in the upper part of Fig. 4b allowed the determination of the stress distribution along each such line. The magnification ratio was always $\lambda_m = 4.0$.

The present optical set-up enables the examination of the roll deformation only at the external face of the roll, i.e. at the edge of the strip. Therefore plane-strain conditions must prevail, so that the stress distribution obtained at the edge should approximate the stress distribution across the strip-width. For a strip width/thickness (h_0/b_0)-ratio larger than 6 spread is usually negligible and rolling problems are solved as ones of plane strain.

The initial length of the lead-strips was kept constant and equal to 0.158m. The remaining initial and final dimensions of the rectangula lead strips are given in Table 1 together with other characteristic values derived from the experiments.

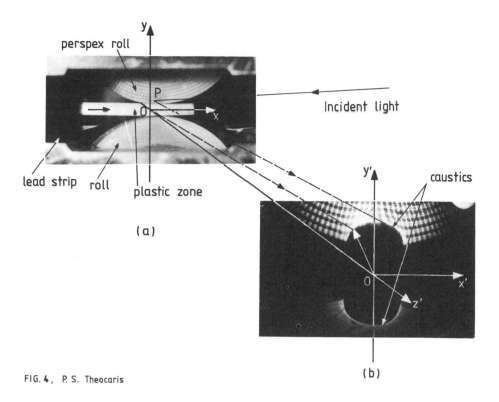

(a)

(b)

FIG. 4, P. S. Theocaris

Fig. 4. (a) An axonometric arrangement of the experimental rolling-mill and rolled stock and (b) the caustics formed at the reference plane at a distance z_0 from the mill.

TABLE 1

Specimen	Initial strip dimensions (mm)		Final strip dimensions (mm)		$\frac{b_0}{h_0}$	$\bar{\varepsilon}$ (s^{-1})	Total reduction $\Delta h/h_0$ (%)
	h_0	b_0	h	b			
1	5.9	38.2	5.5	38.4	6.5	1.11	7
2	6.0	38.2	5.2	38.3	6.4	1.53	13
3	6.0	38.0	4.5	38.5	6.3	2.21	25
4	4.0	26.0	2.2	26.7	6.5	4.04	45

Specimen	Total angle of contact $\varphi_d(0)$	Total length of contact $l_d = R\varphi_d$ (nm)	Maximum normal pressure N_{max} (N/mm²)	Angles of friction	
				$\alpha(0)$	$\beta(0)$
1	8	6.21	35.5	53	65
2	11	8.54	44.5	45	53
3	14	10.87	51.0	39	53
4	16	12.43	51.5	39	43

P. S. Theocaris

Rolling was conducted always in one pass and total reductions $\Delta h/h_0$ of the strip varied between 7 percent and 45 percent in four steps (7, 13, 25 and 45 percent). Lead in the *as cast* condition at room temperature was used as the test material to simulate steel at elevated temperatures. The stress-strain curve for lead, as obtained from a quasi-static compression test, resembled a typical stress-strain curve of an elastic-almost perfectly plastic material, with insignificant hardening at large deformations.

Hot-worked metals are strain rate-dependent. An estimate of the mean strain rate in the pass during hot-rolling, assuming sticking friction, is given by (Ford, 1957).

$$\bar{\dot{\varepsilon}} = \frac{2\pi N_r}{60} r (1-3r/4) \left(\frac{2rR}{2h(1-r)} \right)^{\frac{1}{2}} \tag{18}$$

where $\bar{\varepsilon}$ represents the mean strain rate in the pass, N_r is number of revolutions per minute of the rolls, and r, is the pass reduction, expressed as the fraction of $(h_0-h)/h_0$, and h is the thickness of the outgoing strip. Figure 5 presents a schematic of the rolling procedure of the strip with the geometry of the entering and outgoing strip and the roll gap. Calculated values of the mean strain rate $\bar{\varepsilon}$ in our tests are given in Table 1.

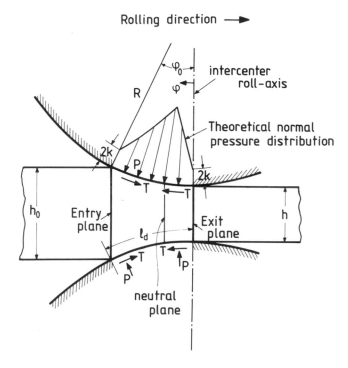

Fig. 5. Schematic diagram of the idealized geometry along the roll-gap in strip rolling.

The mechanical properties of the annealed lead used in the tests
were as follows: yield stress and strain in compression $\sigma_0=29.40\text{Nmm}^{-2}$
and $\varepsilon_0=0.2$ and maximum stress and strain $\sigma_{max}=32.60\text{Nmm}^{-2}$ and $\varepsilon_{max}=0.6$
measured at a strain rate of $\dot{\varepsilon}=5\times10^{-3}\text{s}^{-1}$.

Typical caustics due to the elastic deformation of the perspex rolls
are shown in Fig. 6 for pass-reductions of 7, 13, 25 and 45 percent
respectively. From the caustics and pseudocaustics of these figures
the normal, N, and tangential, T, distributions along the respective
contact arcs, versus the angle φ of contact were calculated and
plotted in Figs. 7 and 8 respectively. In the same figures and for
the sake of comparison the yield stress 2k-curves are also plotted,
using the Tresca yield criterion, where k is, as usual, the yield
stress of the material in pure shear.

If we assume now that there are no sticking phenomena during rolling
(there being no evidence to the contrary) and therefore *Coulomb
friction* prevails between rolls and stock, the coefficient of friction in the contact arc and its variation may be readily determined
from the values of N- and T-distributions. The coefficient of friction, μ, and its variation along the contact arc versus the angle of
contact was plotted in Fig. 9. It can be deduced from Figs. 7 and
8 that a smooth variation of the normal and tangential contact
stresses exists between the rolls and the stock.

Moreover, the coefficient of friction, μ, is not constant over all
of the contact arc, but it varies almost linearly from a maximum
value at the entry plane to a zero value at the neutral point, and
then it starts to increase again linearly up to a maximum appearing
at the exit plane.

Since it has been shown (Theocaris and Razem, 1978; 1979) that the
method of caustics constitutes a very sensitive gauge, indicating
in a clear way the existence of jumps in loading, as well as abrupt
variations in slope and curvature of the loading function in a two-
dimensional elastic problem, where the contact of two bodies is
studied, and since in all our tests no such caustics appeared along
the contact arc for all pass-reductions, it would appear that such
phenomena do not exist in hot-rolling and the distributions given
in Figs. 7 to 9 are the real ones.

Before commenting on the experimental results it may be pointed out
that the most commonly used theory of rolling, originally due to
Orowan (1943) and adapted by Bland and Ford (1948), predicts the
friction-hill type of normal pressure distribution, as shown in Fig.
5, assuming that the coefficient of friction, μ, remains constant
throughout the roll gap. At the neutral plane, a reversal of the
direction of the tangential frictional forces T occurs.

From our tests it is clear that the normal pressure increases slowly
at the entry plane to the roll gap to reach a peak in the region of
the neutral plane and afterwards falls more abruptly towards the
exit plane. The neutral plane appears to be close to the intercenter
roll-axis (ira), which always coincides with the zero abscissa in
Figs 7-9.

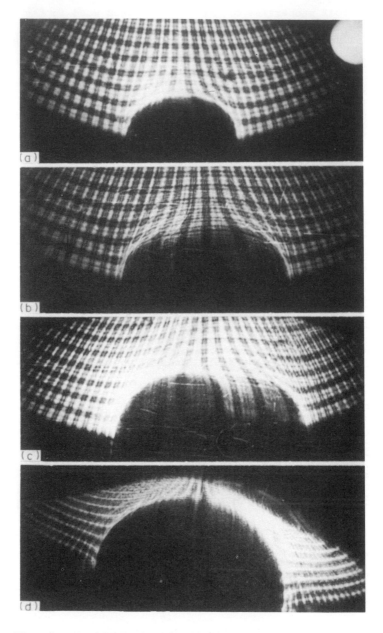

Fig. 6.   Caustics formed on the perspex rolls during
          assimilated strip hot-rolling for total
          reductions (a) 7 percent (R/h_0=7.6), (b)
          13 percent (R/h_0=7.4), (c) 25 percent
          (R/h_0=7.4), and (d) 45 percent (R/h_0=11.1).

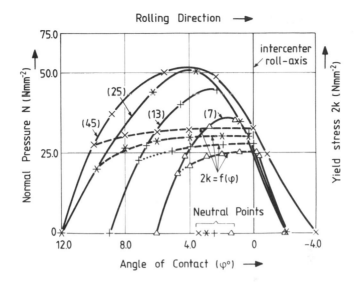

Fig. 7.   Normal pressure distribution in strip hot-
          rolling for total reductions (a) 7 percent
          (Δ), (b) 13 percent (+), (c) 25 percent
          (*), and (d) 45 percent (x).

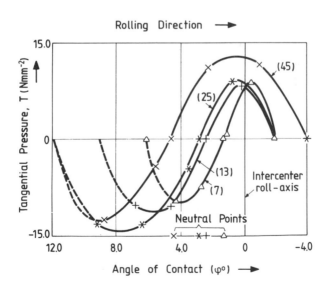

Fig. 8.   Tangential pressure distribution in strip
          hot-rolling for total reductions (a) 7
          percent (Δ), (b) 13 percent (+), (c) 25
          percent (*), and (d) 45 percent (x).

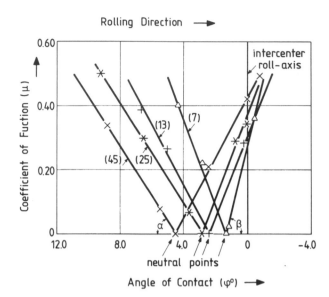

Fig. 9.   The friction-coefficient along the contact
          arc in strip hot-rolling for total reductions
          (a) 7 percent (Δ), (b) 13 percent (+), (c)
          25 percent (*), and (d) 45 percent (x).

As reduction increases, the position of the pressure peak, i.e. the
neutral plane, moves towards the entry plane. Note, however, that,
in all the cases examined, the experimental curve is rounded at the
region of the neutral plane and does not come to a point-type peak,
because of the non-deforming zone.

The pressure near the intercenter roll-axis is equal to 2k. At this
point a change in curvature of the pressure curve occurs, indicating
the end of the plastic-deformation régime. Beyond the intercenter
roll-axis and up to the exit plane, deformation is due to the elastic
recovery of the strip. However, this deformation region is small,
in general, probably because of the softness of the material, which
produces only small elastic deformations. Note also that the total
length of contact, including both the elastic and plastic deformation
zones, increases with increasing reduction. Figure 10 shows the
variation of the peak of the normal pressure, versus the total reduction in strip thickness. The peak pressure increases continuously
until a certain reduction. From this curve it can be concluded that
the envelope of all these maxima for the normal pressure distribution
curves in relation to total reduction is a curve similar to the flow
stress-strain curve of the material.

On the other hand, according to rolling theory, tangential stresses
appear along the contact zone. They reverse their direction at the
neutral plane. The stress direction between the entry and neutral
planes is in accordance with the roll surface moving faster than the
strip, whilst beyond the neutral plane the strip is supposedly moving

faster than the rolls. The experimental results described above
show that the tangential pressure distribution curve has a sinusoidal-
like shape. The tangential pressure increases rapidly from the entry
plane to the roll gap, to reach a peak and remains almost constant
for a certain contact, halfway to the neutral plane. Afterwards it
decreases slowly, towards the neutral plane, where it falls to zero.
At the neutral plane, the tangential pressure reverses its direction
and increases slowly with almost the same slope as its previous decrease before the neutral plane, until it reaches another peak very
close to the intercenter roll-axis. Thereafter it falls more abruptly
towards the exit plane. The position of the early peak, close to the
entry plane, and that of the peak near the intercenter roll-axis
correspond to the positions where plastic deformation starts and ends,
i.e. they define the *plastic deformation zone*. As mentioned above,
at the same positions, a change in curvature of the normal distribution curve occurs.

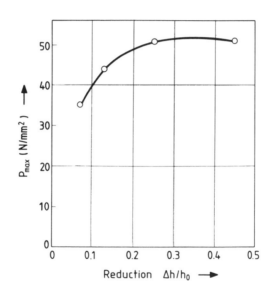

Fig. 10.   Variation of the normal pressure peak with
           total reduction of lead strip during rolling.

The following remarks may be made regarding the coefficient of friction. If Coulomb friction prevails, $T=\mu N$, then it holds that the
distribution of the coefficient of friction along the arc of contact
is determined from the experimental normal and tangential stress
distribution curves and plotted in Fig. 9. The existing methods
for determining the coefficient of friction are based, one on the
maximum angle of bite, and the other of the maximum value of forward
slip. However, in both methods it is assumed that the coefficient
of friction is constant along the arc of contact and therefore only
a mean-value of it can be calculated (Wusatowsky, 1969). The same
assumption is also made in the rolling theory of Mamalis, Johnson
and Hawkyard (1976).

Contrary to this assumption, from the experimental results obtained,

it has been shown consistently that an almost linear variation of the
coefficient of friction μ along the arc of contact is derived. The
coefficient μ falls according to a parabolic law, which may be approximated by a linear variation, from a maximum value, at the entry plane,
to zero, at the neutral plane. Thereafter, it increases again approximately linearly as the exit plane is approached. This variation of
μ=f(φ) is depicted in Fig. 9. This mode of variation of the coefficient of friction versus angle of contact suggests that μ is *speed-
dependent*.

In Table 1 the values of the angles α and β, were included, indicating
the slope of two almost linear branches of the distribution of μ.
The slope of its linear average decrease from entry to the neutral
plane is always smaller than the slope of its respective average
linear increase from the neutral to the exit plane for all cases
examined. With increasing total reduction of strip-thickness both
angles α and β decrease.

In absolute value, μ is in the region $0.5 \leftrightarrow 0 \leftrightarrow 0.5$, whilst an average
value of it is about 0.3. These values are in the same region as
the values of μ used in hot-rolling practice (Wusatowsky, 1969).
However, when strip is hot rolled the coefficient of friction between
the rolls and the stock may be high enough to cause shearing of the
metal in contact with the rolls. For the case of hot rolling with
sticking friction over the whole arc of contact, slip-line field and
upper-bound solutions are available (Johnson and Mamalis, 1977).

From the experimental results reported here it is evident that in
every case $\mu N/k<1$, implying that there is no-sticking along the interface. However, the highest values of $\mu N/k$, approaching unity were
found close to the entry plane indicating that probably some sticking
may occur.

Summarizing the main features of the results when rolling lead-strip,
some..of which are novel, the following conclusions may be made:
(i) A friction hill-type of normal pressure distribution was found
with the peak in the region of the neutral plane and close to the
intercenter roll-axis. The envelope of all these maxima for the
normal pressure distribution curves, in relation to total reduction,
was found to be a curve similar to the flow stress-strain curve of
the material.
(ii) The tangential stress distribution curve has an almost sinusoidal
shape, reversing its direction at the neutral plane, where it falls
to zero.
(iii) Two peaks were found in the tangential-pressure distribution
curve, one close to the entry plane, and another near the intercenter
roll-axis, indicating the start and the end of the plastic deformation
zone. In these positions the normal pressure equals 2k.
(iv) There is an average linear variation of the coefficient of friction along the arc of contact. In reality, this coefficient drops
according to a parabolic law tending to a straight line from a maximum
value at entry to zero at the neutral plane and then increases again
almost linearly, towards the exit plane. This is contrary to the
assumption of a constant coefficient of friction made in rolling
theory since in this theory only average values of friction coefficient
could be measured. However, the method of reflected caustics, which
allows a more detailed examination of the distribution of normal and
shearing forces along the contact zone, yields a more meticulous and
precise distribution of this coefficient.
(v) In general, Coulomb friction, with no-sticking along the interface, prevails for the cases examined.

THE COEFFICIENT OF FRICTION IN INGOT HOT-ROLLING

The experimental technique used in the previous section was limited in this paper to a study of the stress distribution at the outer surface of the strip and therefore only plane-strain conditions could be encountered by this arrangement. Indeed, in our tests the strip-width to thickness ratio of the stock material was taken greater than six, assuming approximate plane-strain conditions with insignificant lateral barrelling. However, the method can be readily applied to any depth along the thickness of the rolls, by considering sandwich rolls with an appropriate number of slices, where the caustics and pseudocaustics may be studied. In order to find the distinguishing features of the ingot rolling process, fairly thick and narrow speci-mens with an initially square section having thickness to width ratio equal to unity, for which plane-strain conditions cannot be assumed to prevail, were rolled. For various rolling conditions, the normal and tangential stress distribution and, thence, the coefficient of friction along the arc of contact and across the width of the ingot were calculated from the caustics, formed due to elastic deformation of the rolls, whilst the rolled stock was deformed plastically.

The difficulty in evaluating the coefficient of friction in plane-strain strip-rolling stemmed from the fact that this quantity was variable along the contact arc and, therefore, its calculation could not be based on measuring the overall normal and tangential forces acting upon the interface in the contact arc. In ingot rolling, the N and T stresses are neither constant along the contact-arc, nor are they constant along the width of the rollers. The problem of evalu-ating μ in this general case of rolling becomes extremely difficult.

In order to study the variation of the N and T stresses across the rolls we need to obtain caustics in planes parallel to the end faces of the rolls. For this reason, as has been already described pre-viously, the rolls consisted of two perspex discs of a diameter D=0.089m and a thickness t=0.018m so that the overall width of the barrel was 0.036m. The cylindrical slices of each roll were scribed with a fine grid, consisting of isometric concentric circles and radii of equal angle, held together by friction through a central steel pin and bolts to ensure complete conformity of the elements under the effect of friction. The laser-light beam, impinging on the vicinity of the contact arc was focussed on each reflective intermediate sur-face of the layers of the roll so that a series of superimposed caustics and pseudocaustics could be obtained, corresponding to different depths inside the roll. In this way the variation of N and T stresses inside the contact cylindrical surfaces could be evalu-ated.

However, for better results and for higher accuracy we limited our study to the external lateral face of the rolls and only the first internal face displaced by 0.018m inside the roller and constituting the middle planes of the rolls.

The number of lateral faces of each roll, consisting of the two ex-ternal and two internal lateral faces was sufficient for checking the symmetry of the loading and studying the variation of forces across the width of the roll. Otherwise, the method of caustics, as it has been described for plane-strain strip-rolling, was entirely sufficient.

In order to take into consideration the features of ingot rolling, fairly thick and narrow bars made of annealed lead were used. The initial lengths of the bars were always equal to 0.150m, whereas

their other initial and final dimensions were tabulated in Table 2.
Again, for ingot rolling the procedure was conducted in a single
pass for four different reductions of the thickness, $\Delta h/h_0 = 7, 13, 19$
and 29 percent. Annealed lead in the *as cast* condition at room
temperature was again used as the test material to simulate steel
rolling at elevated temperatures.

TABLE 2

No of specimen	Initial bar dimensions (mm) h_0	b_0	Final bar dimensions (mm) h	b	Total reduction $\Delta h/h_0$ (%)	Total angle of contact $\varphi_d(0)$	Total length of contact ℓ_d (mm)
1	16.0	15.9	14.9	16.2	7	8	6.21
2	15.9	16.1	13.9	16.7	13	12	9.32
3	15.8	15.8	12.8	16.9	19	18	13.98
4	15.8	15.9	11.2	17.8	29	22	17.09

$\ell*$ (mm)	$\dfrac{h_0+h}{2\ell}$	$\dfrac{b_0+b}{2\ell}$	Maximum normal pressure P_{max} (N/mm²) At Mid-plane	At External face	Coefficient of friction (0) α_f	α_m	β_f	β_m
4.66	3.32	3.42	39.1	34.3	78	81	55	63
7.77	1.92	2.11	55.5	44.1	78	73	55	56
10.87	1.32	1.50	67.4	54.0	72	70	60	55
12.43	1.09	1.36	84.0	63.8	75	61	65	55

* ℓ is the roll-throat deformation length

Typical caustics formed from reflections at different depths are presented in Fig. 11. From these caustics and by using the analytic
procedure developed previously the normal, N, and the tangential, T,
stress distributions were evaluated and plotted in Figs. 12 and 13
respectively, against φ, the angle of contact. In the same figures
and for the sake of comparison the yield stress, 2k, versus φ was
also plotted using the Tresca yield condition.

Assuming again that Coulomb friction prevails during the rolling process in the roll gap, which implies that no sticking phenomena occur
in this zone, the variation of the friction coefficient along the
contact zone can be evaluated and it is given in Fig. 14. In Figs.
12 to 14 two groups of curves were plotted. The one group correspond
to data extracted from the lateral faces of the rolls and corresponding to the lateral faces of the rolled stock whereas the other group
corresponded to the same quantities at the middle plane of the rolls
and stock. It is clear from these two different groups that significant difference exists outside and inside the rolled stock during
ingot rolling.

Since in hot-rolling the flattening and the elastic distortion of
rolls is generally insignificant (Wusatowsky, 1969) it was assumed
in our calculations that the radius of the roll was equal to its
initial value R. Then, the total extent of the deformed zone or the

total contact length ℓ_d was given by

$$\ell_d = R\varphi_d$$

where φ_d is the total angle of contact, measured experimentally from the caustics. The values of φ and ℓ_d, as they have been experimentally defined are included in Table 2.

Fig. 11.   Typical caustics formed from reflections
at the contact zone on the lateral faces
and the middle planes of the upper roll
in ingot-rolling (the rollers are made
by contact of two identical discs pressed
together).

From this experimental study of ingot-rolling the following results may be derived. The normal pressure inside the roll gap increases rapidly from the entry plane gap to reach a peak in the vicinity of the neutral plane and afterwards falls slowly towards the exit plane. The formation of the early peak near the entry-plane is different from that observed in flat-strip rolling, where the peak appears some times close to the exit plane. With increasing reduction, the position of the neutral plane on the contact arc moves towards the entry plane.

All of the experimental curves are rounded in the vicinity of the neutral plane and do not present a point-type peak as would be expected because of the non-deforming zone (Orowan, 1943). The pressure near

the intercenter roll-axis for the pressure-distribution at the middle
plane of the rolls and the rolled bar is approximately equal to 2k,
during rolling, whereas it is reduced for a large reduction well
below this value in the early stages of rolling. At the external
lateral faces of the rolled bar the normal pressure near the intercenter roll-axis is also well below 2k. At this point, a change in
the curvature of the pressure curve occurs, indicating the position
where plastic deformation occurs. Beyond the intercenter roll-axis
and up to the exit-plane, the deformation is due to elastic recovery
of the bar.

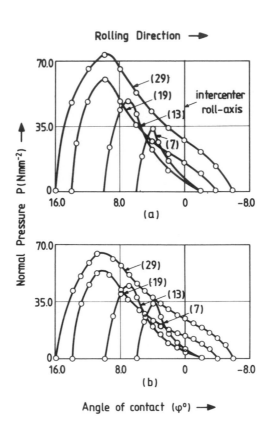

Fig. 12.   Normal pressure distribution in ingot
           rolling for total reduction (i) 7 percent
           (ii) 13 percent, (iii) 19 percent and (iv)
           29 percent (a) at the middle plane and (b)
           at the lateral faces of the rolled bar.

A similar feature, that is the appearance of a peak near the entry
point, has been observed by McGregor and Palme (1959) and Vater,
Nebe and Petersen (1966) during rolling of large aluminium ingots.
The remarks of Tselikov (1967) regarding ingot rolling are also
similar and compare well with similar conclusions drawn from hot-
rolling of tellurium lead rings (Mamalis, Johnson and Hawkyard, 1976).

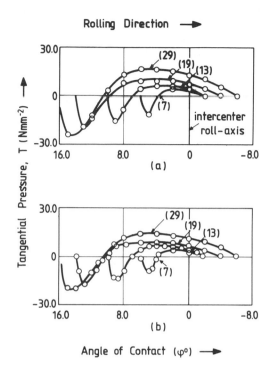

Rolling Direction ➔

(graph a axis labels) Tangential Pressure, T (Nmm⁻²) ; values 30.0, 0, -30.0 ; horizontal axis 16.0, 8.0, 0, -8.0 ; labels (29), (19), (13), (7), intercenter roll-axis ; (a)

(graph b) 30.0, 0, -30.0 ; 16.0, 8.0, 0, -8.0 ; (29), (19), (13), (7) ; (b)

Angle of Contact (φ°) ➔

Fig. 13.   Tangential pressure distribution in ingot
            rolling for total reduction (i) 7 percent,
            (ii) 13 percent, (iii) 19 percent and (iv)
            29 percent (a) at the middle plane and (b)
            the lateral faces of the rolled bar.

The effect of the friction-hill and the expected lateral spread for
a given amount of compression in the case of conventional rolling
has been examined experimentally by several authors. The friction
hill and the associated metal flow-patterns have been suggested in
Tselikov (1967) and Johnson (1971). In the case of ingot rolling,
since the length of contact is generally small in relation to the
mean-ingot thickness, friction would be expected to have a small
influence. By referring to the slip-line field solution for the
indentation problem as given by Hill (1950), frictional forces might
have a greater effect near the "conventional" rolling régime i.e.
for values of h_m/ℓ, where h_m is the mean-ingot thickness and ℓ is
the roll-throat deformation extent between the entry point and intercenter roll-axis, which are of the order of unity (Mamalis, Johnson
and Hawkyard, 1976).

Examination of our experimental results concerning the tangential
force distribution shows that the tangential stress increases abruptly
from the entry plane to the roll gap, to reach a peak, very close to
the entry plane, and then decreases with almost the same slope, towards the neutral plane, where it falls to zero. At this plane the

tangential stress reverses its slope, increasing slowly, to reach
an almost constant value at the intercenter roll-axis. Thereafter,
it falls slowly towards the exit plane. The positions of the early
peaks close to the entry plane and those near the intercenter roll-
axis correspond to the positions where plastic deformation starts
and ends, i.e. to the plastic deformation zone, which may be defined
in this way. At the same positions changes of curvature of the
normal pressure distribution occur.

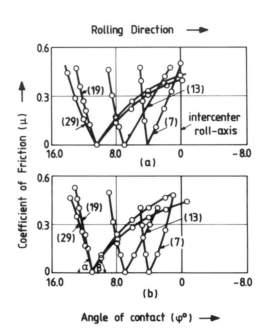

Fig. 14.   The friction coefficient along the contact-
            arc in ingot rolling for total reduction
            (i) 7 percent, (ii) 13 percent, (iii) 19
            percent and (iv) 29 percent (a) at the
            middle plane and (b) the lateral faces of
            the rolled bar.

Contrary to the assumption made in rolling theory that the friction
coefficient is constant along the arc of contact, from the experimental results obtained for ingot rolling it is again evident that
the friction coefficient μ varies with the angle φ of contact in a
manner which suggests that μ is speed-dependent. The variation of
μ with the contact angle φ is shown in Fig. 14. The coefficient μ
falls according to an almost linear law from a maximum value of the
entry-plane to the role gap, to zero, at the neutral-plane. Afterwards, it increases again according to a parabolic law, which may be
approximated to a linear variation, as the exit plane is approached.
The angles α and β shown in Fig. 14 whose values are given in Table
2 indicate the average slope of the two almost linear branches of
the distribution of the friction coefficient along the contact arc.
With increasing total reduction of thickness of the rolled stock the
angle α remains almost constant, while the angle β increases slightly

The coefficient μ is in the region between 0.5↔0↔0.5. From the experimental results reported it is evident that generally μN/k<1, implying that there is no-sticking along the interface. The highest values of μN/k, which approach unity, were found close to the entry-plane indicating that some sticking may occur there.

CONCLUSION

In conclusion it has been shown that the optical method of caustics and pseudocaustics can be applied in an elegant and accurate way to problems of flat strip cold and hot rolling, where plane strain conditions prevail and also to ingot rolling, where three-dimensional problems are encountered. Since one is using a whole-field analysis, it is expected that the results of the method are closer to reality and important novel conclusions have been derived concerning the extent of the contact and the variation of the coefficient of friction along the plastic zone of deformation of the stock. It is reasonable to assume that the method can be applied successfully to other problems of rolling and, in general, to the contact of two bodies and will yield reliable results.

REFERENCES

Bland, D. R. and H. Ford (1948). The calculation of roll-force and torque in cold strip rolling with tensions. *Proc. Inst. Mech. Engrs.*, *159*, 144-153.
Ford, H. (1957). The theory of rolling. *Metall. Reviews*, *2*, 1-28.
Hitchcock, J. (1935). Elastic deformation of rolls during cold rolling. *Roll-neck Bearings ASME*, N. York Appendix I.
Hill, R. (1950). The Mathematical Theory of Plasticity, Clarendon Press, Oxford.
Johnson, W. (1971). The mechanics of metal working plasticity. *Appl. Mech. Rev.*, *24*, 977-981.
Johnson, W. and A. G. Mamalis (1977). Engineering Plasticity: Theory of Metal Forming Processes, Springer Verlag (CISM)
Mamalis, A. G., W. Johnson and J. B. Hawkyard (1976a). On the pressure distribution between stock and rolls in ring rolling. *J. Mech. Engrg. Sci.*, *18*, 184-195.
Mamalis, A. G., W. Johnson and J. B. Hawkyard (1976b). Pressure distribution, roll force and torque in cold ring rolling. *J. Mech. Engrg. Sci.*, *18*, 196-209.
McGregor, C. W. and R. B. Palme (1948). Contact stresses in the rolling of metals I, *J. Appl. Mech.*, *Trans ASME*, *70*, 297-302.
McGregor, C. W. and R. B. Palme (1959). The distribution of contact pressures in the rolling of metals. *J. Basic Engrg.*, *Trans ASME*, *79*, 669-680.
Orowan, E. (1943). The calculation of roll pressure in hot and cold flat rolling. *Proc. Inst. Mech. Engrs.*, *150*, 140-167.
Siebel, E. and W. Lueg (1933). Untersuchungen uber die Spannungsverteilung in Walzplat. *Mitt. Kön. Wilhelm Inst. Eisenforsch.*, *15*, 1-13.
Smith, C. L., F. H. Scott and W. Sylwestrowicz (1952). Pressure distribution between stock and rolls in hot and cold flat rolling. *J. Iron Steel Inst.*, *170*, 347-359.
Stone, M. D. (1953). *Iron Steel Eng.*, *30* (2), 61-72.
Theocaris, P. S. (1970). Local yielding around a crack tip in plexiglas. *Jnl. Appl. Mech.*, *Trans. ASME*, *37*, 409-415.
Theocaris, P. S. (1973a). Stress singularities at concentrated loads. *Exp. Mech.*, *13*, 511-518.

Theocaris, P. S. (1973b). Stress singularities due to uniformly
   distributed loads along straight boundaries. *Intern. J. Solids
   and Struct.*, 9, 655-670.
Theocaris, P. S. (1977). The contact problem by the method of
   caustics. *Proc. 3rd Bulgarian Nat. Cong. Theor. Appl. Mech.*, *Varna*,
   1, 263-268.
Theocaris, P. S. (1979). The method of caustics applied to elasticity problems. Developments in Stress Analysis - 1. G. Holister
   Editor, Appl. Sci. Publ., England, chapt. 2, 27-63.
Theocaris, P. S. (1981). Elastic stress intensity factors evaluated
   by caustics, *Experimental Determination of Crack Tip Stress Intensity
   Factors, Mechanics of Fracture*, vol. VII, G. C. Sih Editor, M.
   Nijhoff Publ., Hague, chapt. 3, 189-252.
Theocaris, P. S., A. Mamalis and C. Stassinakis (1984). On the
   pressure distribution and the friction coefficient between stock
   and rolls in ingot hot-rolling. *Proc. First Intern. Conf. on
   Technology of Plasticity, Tokyo Japan 1984*, 1, 30-36.
Theocaris, P. S. and C. Razem (1977). Deformed boundaries determined
   by the method of caustics. *J. Strain Anal.*, 12, 223-232.
Theocaris, P. S. and C. Razem (1978). The end-values of distributed
   loads in half-planes by caustics. *J. Appl. Mech.*, *Trans ASME*, 45,
   313-319.
Theocaris, P. S. and C. Razem (1979). Intensity, slope and curvature
   discontinuities in loading distributions at the contact of two
   plane bodies. *Intern. J. Mech. Sci.*, 21, 339-353.
Theocaris, P. S. and C. Stassinakis (1978). The elastic contact of
   two discs by the method of caustics. *Exp. Mech.*, 18, 409-416.
Theocaris, P. S. and C. Stassinakis (1982). The determination of the
   coefficient of friction between a cylindrical indenter and a plastically deformed body by caustics. *Intern. J. Mech. Sci.*, 24,
   717-727.
Theocaris, P. S.,C. Stassinakis and A. Mamalis (1983). Roll-pressure
   distribution and coefficient of friction in hot rolling by caustics
   Intern. J. Mech. Sci., 25, 833-844.
Theocaris, P. S. and C. Stassinakis (1984). The contact problem
   studied by pseudocaustics formed at the vicinity of the contact
   zone. *J. Engrg. Mat. and Technology*, 106, 235-241.
Tselikov, A. I. (1967). Stress and Strain in Metal Rolling, MIR Publ
   Moscow.
Van Rooyen, G. T. and W. A. Backofen (1957). Friction in cold rollin
   J. Iron Steel Institute, 186, 235-244.
Vater, M., G. Nebe and J. Petersen (1966). Die Druckverteilung im
   Waltzspalt beim Kaltwalzen von Bändern mit und ohne Langszugspannun
   Stahl and Eisen, 86, 710-720.
Whittow, P. W. and H. Ford (1955). Surface friction and lubrication
   in cold-strip rolling. *Proc. Inst. Mech. Eng.*, 169, 123-133.
Wusatowsky, Z. (1969). *Fundamentals of Rolling*, Pergamon Press,
   Oxford.

MICRO COMPUTER PROGRAMS FOR ROLLING AND EXTROLLING

J. M. ALEXANDER

Department of Mechanical Engineering, University of Surrey,
Guildford, Surrey GU2 5XH, UK

ABSTRACT

The paper gives a discussion of the application of desk-top micro-computers to determine roll pressure distribution and other main parameters in the processes of hot and cold strip rolling and in a novel process of combined rolling and extrusion, called "extrolling", invented by Avitzur (1975).

The soft-ware is based on a previous FORTRAN program for a main-frame computer (Alexander (1972)). This developed a comprehensive solution to von Karman's basic linear first-order differential equation for rolling, using a fourth order Runge-Kutta procedure. The mixed boundary condition of $\tau=\mu s$ or k, whichever is the smaller, was used together with the effect of back and front tensions (or compressions), both entry and exit elastic arcs and variation of the flow stress through the arc of contact.

NOTATION (see Fig. 1)

a'	constant in Johnson's equation for plane strain extrusion
A_3	extruded area of product in extrolling
b'	constant in Johnson's equation for plane strain extrusion
B	constant in constitutive equation
C	constant in constitutive equation
E	Young's modulus
G	roll torque due to plastic arc of contact, per unit width
G_{e1}	roll torque due to entry elastic arc, per unit width
G_{e2}	roll torque due to exit elastic arc, per unit width
G_T	total roll torque, per unit width
h	local thickness
h_0	annealed thickness
h_1	entry thickness
h_2	exit thickness (=thickness entering extrusion die in extrolling)
k	yield shear stress at any section
m	constant in constitutive equation

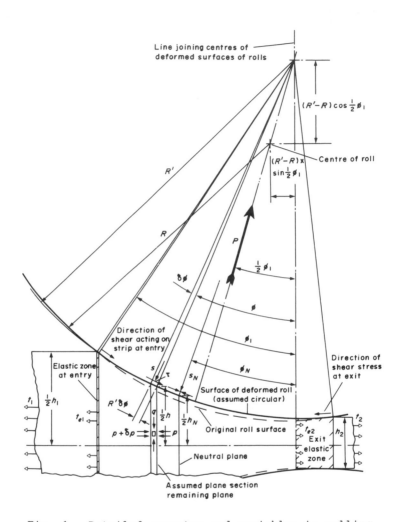

Fig. 1.   Detailed geometry and variables in rolling.
         (after Alexander (1972)).

n	constant in constitutive equation
p	horizontal compressive stress at any section
P	roll force due to plastic arc of contact, per unit width
P_{e1}	roll force due to entry elastic arc, per unit width
P_{e2}	roll force due to exit elastic arc, per unit width
P_T	total roll force, per unit width
q	vertical compressive stress at any section
Q	activation energy
R	original radius of roll, (also extrusion ratio), (also universal gas constant)

R'   radius of the deformed arc of contact (assumed circular)
r   reduction in plane strain extrusion
s   local normal pressure on the deformed roll surface
t_1 back tension stress
t_2 front tension stress
t_{e1} back tension stress actually transmitted to the plastic zone
t_{e2} front tension stress actually transmitted to the plastic zone
T   absolute temperature
w_1 entry width of strip (for extrolling)
Y   uniaxial yield stress ($=\sqrt{3}k$) at uniaxial effective strain $\bar{\varepsilon}$.
Y_0 uniaxial yield stress at zero strain (constant in constitutive
            equation

        GREEK

δ 'draft' = $h_1 - h_2$
$\bar{\varepsilon}$ uniaxial effective strain
$\dot{\bar{\varepsilon}}$ uniaxial effective strain rate
μ Coulomb coefficient of friction
ν Poisson's ratio
τ surface shear stress at any section
ϕ angle of considered section (see Fig. 1) (plane sections
            assumed)
ϕ_N angle of neutral plane (see Fig. 1)
ϕ_1 angle of entry plane (see Fig. 1).

        INTRODUCTION

The micro-computer programs developed in this chapter are based on
a FORTRAN program for strip rolling written for a main-frame computer
by Alexander (1972). The original program was intended to remove
many of the approximations which had been made by previous investigators who had developed numerical solutions for rolling, e.g. Orowan
(1943), Bland and Ford (1948)(1952), Bland and Sims (1953), Cook and
McCrum (1958), Ford, Ellis and Bland (1951), Lianis and Ford (1956)
and Sims (1954) as discussed in the review by Ford (1957).

The most comprehensive of these earlier theories was that of Orowan
(1943), who developed a 'homogeneous graphical method' of solution,
including an attempt to allow for the inhomogeneity of deformation
throughout the volume of the plastically deforming material in the
arc of contact. All other theories, including the original basic
theory of von Karman (1925) assumed that 'plane sections remain plane'
during passage of the strip through the arc of contact. Also, apart
from Orowan, all other researchers had used approximations such as
$\sin\phi \simeq \tan\phi \simeq \phi$ and $(1-\cos\phi)=0$ or $\phi^2/2$ and that a mean flow stress could
be used through the contact arc. This was understandable, in view
of the complexity of Orowan's homogeneous graphical method.

With the advent of the modern electronic digital computer, it seemed
sensible to set up a numerical solution on the lines of Orowan's
approach and obtain a more accurate solution than the approximate
methods just described. That was the reason for developing the main-
frame computer program previously mentioned (Alexander, 1972) and
comparisons were made in that paper between the computer solution

and the approximate methods, for a few typical examples, revealing significant differences, especially in the estimation of roll torque, which is one of the more sensitive parameters. Unlike Orowan's method, however, no allowance was made in the computer program for inhomogeneous deformation through the arc of contact, although the present writer has previously attempted to obtain an estimate of such effects by using slip line field theory (Alexander, 1955). More recently, Venter and Abd-Rabbo (1980) have shown that the inclusion of Orowan's 'w' function or inhomogeneity factor in the author's original computer solution does make a slight difference to the predicted values of parameters such as roll force and roll torque, but at the expense of considerable added complexity to the computer program and in its execution time. Since the inhomogeneity factor introduced by Orowan can at best be only an approximation to the real situation of unknown accuracy it does not appear worthwhile to try to include it in the computer solution. The only way in which a true prediction of the inhomogeneity of deformation can be achieved is either by development of slip line field solutions with their attendant approximations or by using finite element methods, also prone to considerable approximation unless very fine meshes requiring large computer capacity can be employed.

For these reasons it seems desirable to develop the original computer program in order to give reasonably accurate prediction of rolling parameters for use in planning manufacturing sequences and to adapt it to the flexible system characterised by desk-top micro-computers. That is the main objective of the present study. A further objective is to assist in the development of on-line computer control systems for rolling mills but that would probably require the use of larger computing facilities. Yet another is to obtain an accurate prediction of the main parameters in extrolling, a novel process combining extrusion and rolling, invented by Avitzur (1975).

THEORY

In essence, the basic equation for rolling developed by von Karman (1925) can be written thus:-

$$\frac{d}{d\phi}\ [h(s\ -\ 2k\ \pm\ \tau\tan\phi)]\ =\ 2R'(s\sin\phi\ \pm\ \tau\cos\phi) \tag{1}$$

where the upper sign refers to the exit side and the lower sign refers to the entry side of the neutral plane.

Also, from the geometry of the deformed circular arc of contact illustrated in Fig. 1,

$$h\ =\ h_2\ +\ 2R'(1-\cos\phi) \tag{2}$$

A mixed boundary condition on the shear stress is used in the computer solution, namely $\tau=\mu s$ or k, whichever is the smaller.

The von Mises yield criterion is q-p=2k.

To allow for work-hardening Swift's equation may be used, viz:-

$$Y\ =\ Y_0\,(1+B\bar{\varepsilon})^n \tag{3}$$

In plane strain compression the yield stress is $2k\ =\ 2Y/\sqrt{3}$, and the total effective strain is $\bar{\varepsilon}\ =\ (2/\sqrt{3})\ln(h_0/h)$, (where h_0 is the thickness of the strip at its last annealing).

Thus, the compressive flow stress in plane strain at any thickness h is given by the expression

$$2k = \frac{2}{\sqrt{3}} Y_0 \; (1 + \frac{2}{\sqrt{3}} B \; \ln \frac{h_0}{h})^n \tag{4}$$

The two frictional conditions which can exist at the roll-workpiece interface are either $\tau = \mu s$ or $\tau = k$, as already mentioned. Summarizing the theoretical development of von Karman's basic equation, then:-
(i) If $\tau = \mu s$, from equation (1),

$$\frac{ds}{d\phi} = g_{1(\phi)} s + g_{2(\phi)} \tag{5}$$

where $g_{1(\phi)} = \pm \; \mu \sec\phi \; (\frac{2R'}{h} + \sec\phi) / (1 \mp \mu\tan\phi)$

$$g_{2(\phi)} = \{ \frac{2R'}{h} \; 2k\sin\phi + \frac{d(2k)}{d\phi} \} / (1 \mp \mu\tan\phi)$$

(ii) If $\tau = k$,

$$\frac{ds}{d\phi} = g_{3(\phi)} \tag{6}$$

where $g_{3(\phi)} = 2k\{ \frac{2R'}{h} \sin\phi(1 \pm \tan\phi) \pm (\frac{R'}{h}\cos\phi + \frac{1}{2}\sec^2\phi) \} + (1 \pm \frac{1}{2}\tan\phi)$
$\frac{d}{d\phi}(2k)$.

As before, in all these equations the upper sign refers to the exit side, the lower sign to the entry side of the neutral plane. For a complete solution, the mixed boundary condition $\tau = \mu s$ or k, whichever is the smaller, must be used.

The horizontal pressure at any section is $p = s - 2k \mp \tau\tan\phi$. At entry to the plastic arc where $\phi = \phi_1$, $2k = 2k_1$, and $p = -t_{e1}$. The roll pressure at entry is, therefore,

$$s_1 = 2k_1 - t_{e1} - \tau\tan\phi_1 \tag{7}$$

(or $s_1 = (2k_1 - t_{e1}) / (1 + \mu\tan\phi_1)$ if $\tau_1 = \mu s_1$)

and at exit, where $\phi = 0$,

$$s_2 = 2k_2 - t_{e2} \tag{8}$$

In all these equations the derivative $\frac{d}{d\phi}(2k)$ is given by

$$\frac{d}{d\phi}(2k) = -\frac{8}{3} Y_0 n B R' \sin\phi (1 + \frac{2}{\sqrt{3}} B \; \ln\frac{h_0}{h})^{n-1} /h \tag{9}$$

Neglecting the elastic arcs, the roll force per unit width is given by

$$P = R' \int_0^{\phi_1} s\cos(\phi - \frac{1}{2}\phi_1) d\phi + R' [\int_{\phi_N}^{\phi_1} \tau\sin(\phi - \frac{1}{2}\phi_1) d\phi - \int_0^{\phi_N} \tau\sin(\phi - \frac{1}{2}\phi_1) d\phi]$$

$$\tag{10}$$

Neglecting elastic arcs, the roll torque per unit width is given by

$$G = R'(R'-R)\int_0^{\phi_1} s\sin(\phi-\tfrac{1}{2}\phi_1)d\phi + R'\int_{\phi_N}^{\phi_1}[R'\tau-(R'-R)\tau\cos(\phi-\tfrac{1}{2}\phi_1)d\phi$$

$$- R'\int_0^{\phi_N}[R'\tau-(R'-R)\tau\cos(\phi-\tfrac{1}{2}\phi_1)]d\phi \tag{11}$$

It is possible to obtain an analytical solution when $\tau=k=constant$ throughout the arc of contact and this is given in Alexander (1972) but the expressions do have to be evaluated numerically, even in that case.

Deformed Arc of Contact

Neglecting the elastic arcs of contact, Hitchcock's formula is often used in the form

$$R'/R = 1 + 0.000334 \ P/\delta \tag{12}$$

where δ is the 'draft' and $\delta=h_1-h_2$ and 0.000334 is the appropriate factor for steel rolls (in dimensional units of in^2/Tf).

Ford, Ellis and Bland (1951) showed that the entry and exit elastic arcs could be allowed for by modifying this equation to

$$\frac{R'}{R} = 1 + \frac{0.000334P}{\{\sqrt{(\delta+\delta_{e2}+\delta_t)}+\sqrt{\delta_{e2}}\}^2} \tag{13}$$

where $\delta = h_1-h_2,$

$$\delta_{e2} = \frac{(1-\nu^2)}{E}(2k_2-t_{e2})h_2,$$

$$t_{e2} = t_2 - 2\mu P_{e2}/h_2,$$

$$\delta_t = \frac{\nu(1+\nu)}{E}(h_2t_2-h_1t_1).$$

Elastic Arcs of Contact

Ford, Ellis and Bland (1951) developed the following equations for estimating the contributions of the elastic arcs of contact to both roll force and roll torque.

$$P_{e1} = \frac{(1-\nu^2)h_1}{4}\sqrt{(R'/\delta)}\ \frac{(2k_1-t_{e1})^2}{E} \quad \text{(Entry elastic arc)} \tag{14}$$

where $t_{e1}=t_1-2\mu P_{e1}/h_1$

$$P_{e2} = \frac{2}{3}(2k_2-t_{e2})^{\frac{3}{2}}\sqrt{\{\sqrt{R'}h_2(1-\nu^2)/E\}} \quad \text{(Exit elastic arc)} \tag{15}$$

$$G_{e1} = \mu R P_{e1} \quad \text{(Entry elastic arc)} \tag{16}$$

$$G_{e2} = -\mu RP_{e2} \quad \text{(Exit elastic arc, negative since strip is moving faster than the roll surface at exit)} \quad (17)$$

Thus the total roll force and roll torque per unit width are

$$P_T = P + P_{e1} + P_{e2} \tag{18}$$

$$G_T = G + G_{e1} + G_{e2} \tag{19}$$

Slab Entry Condition

For a given slab entry thickness h_1, h_2 cannot be less than

$$h_{2min} = h_1 - 2R\{1 - \cos(\arctan\mu)\}. \tag{20}$$

Otherwise, the frictional drag of the roll on the slab will be insufficient to draw it into the roll gap.

The computer program is arranged to replace h_2 with h_{2min} if $h_2 < h_{2min}$.

Tension Conditions

If the back tension stress $t_1 > 2k_1$, either the roll pressure at entry would become negative or the strip would become unstable due to yielding in longitudinal tension before entry into the roll gap; similarly at exit if $t_2 > 2k_2$. The computer program is arranged to make $t_1 = 2k_1$ and $t_2 = 2k_2$ if bigger tensions are applied.

For the extrolling process both tensions may well be negative (i.e. compressive). To cater for this, the computer program is arranged to replace t_1 with $2k_1 t_1 / |t_1|$, if $|t_1| > 2k_1$, thereby catering for both positive and negative tensions at entry to the roll gap. The exit tension t_2 will always be negative (compressive) in extrolling and *can* be greater than $2k_2$ in absolute magnitude.

Approximate Theories

In practice, the most well-known and well-used theories have been those of Bland and Ford (1948) for cold rolling and Sims (1954) for hot rolling. Comparisons between the predictions of the computer program under discussion and those theories were made by Alexander (1972).

COMPUTER PROGRAM

Computer solutions for equation (1) and also for the various approximate theories were given by Alexander (1972). In this paper attention will be concentrated on the complete solution of equation (1). In general, capital letters in the computer program alpha-numeric listing have been chosen to comply with the notation used in the theory already

presented. Since the symbols I,J,K,L,M,N can only be used for integer
numbers the notation XMU has been chosen to represent μ and X2K2 to
represent $2k_2$, for example, since such quantities are floating point
numbers, in general. A complete copy of the program for rolling forms
Appendix 1, entitled ROLLNG.

The main problem is to determine the pressure distribution s. Thus
S(I) is the computer program notation for the M ordinates of pressure
distribution over the arc of contact, M being an odd number; e.g. if
the arc of contact is divided into 1000 equal segments, M=1001. The
reason for having an odd number of ordinates is to cater for the easy
application of Simpson's rule for integration.

The values of $E,\nu,Y_0,B,n,M,\mu,R,h_0,h_1,h_2,t_1$ and t_2 are entered, in the
case of rolling, and their values written out. Tests are then made,
to increase h_2 if the reduction is too great for slab entry and to
decrease the tensions (whether positive or negative) if they exceed
$|2k_1|$ or $|2k_2|$ (lines 24 and 25 of Appendix 1). The final values of
h_2, t_1 and t_2 which are permissible are then written out. In the case
of extrolling, t_2 is not known beforehand and the additional parameters
w_1 and A_3 are entered and written out. In PROGRAM ROLLNG, w_1 is
added to lines 10 and 11, with FORMAT 7F10.5 in lines 10 and 12,
whilst A_3 replaces T_2 (lines 14 and 15). The value of t_2 for extrolling is negative (compressive) since it is the pressure required
to extrude the rolled strip through the extrusion die, as described
by Alexander and Tilakasiri (1979) and Alexander and Hatakeyama (1983).
The equation used for extrusion is derived from the well-known expression put forward by Johnson (1956) for plane strain conditions,
viz:-

$$p = 2k\{a'+b'\ln 1/(1-r)\}, \text{ or, for this specific application,}$$

$$t_2 = -2k_2[0.47+1.2\ln(h_2w_1)/A_3] \qquad (21)$$

which is expressed in FORTRAN thus:-

        T2 = - X2K2*(0.47+1.2*ALOG(H2*W1/A3))

and is inserted in place of line 25 which is not applicable to extrolling, for the following reason:-

In equation (21) the parameter h_2w_1/A_3 is the extrusion ratio R and
it can be seen from that equation that $t_2=-2k_2$ when $(0.47+1.2\ln R)=1$,
i.e. when R=1.555. This is a relatively low extrusion ratio so that
the restriction $|t_2|<|2k_2|$ is not applicable to this case (for ext-
rolling). Of course, it is possible to select values other than 0.47
and 1.2 for the empirical constants given in equation (21), from
Johnson's comprehensive results.

An approximate value of the roll force is calculated to provide an
initial realistic value of R'(RD in the program) and hence ϕ_1(PHI1).
The values of S(I) and S(M) at the entry and exit planes may then be
estimated from equations (7) and (8) and hence the values of k over
the arc of contact (XK(I) in the program).

It is then possible to integrate equation (1) over the whole arc of contact, working inwards from both ends, using a fourth order Runge-Kutta process as defined in the sub-routine RUNGE. The stress ordinates from S(1) up to S(M) working from entry up to exit are denoted by SEN(I), whilst those from S(M) at exit back to S(1) are denoted by SEX(I). The actual value of S(I) is then the lesser of the two as designated by the IF-statements (lines 67 and 68)

```
67:      IF(SEX(I).LE.SEN(I))  S(I)  = SEX(I)
68:      IF(SEN(I).LE.SEX(I))  S(I)  = SEN(I)
```

The mixed boundary condition that $\tau=\mu s$ or k whichever is the smaller is ensured by the IF-statements (lines 91 and 92)

```
91:      IF(XMU*S(I).LT.XK(I))  TAU(I)  = XMU*S(I)
92:      IF(XMU*S(I).GE.XK(I))  TAU(I)  = XK(I).
```

which have also been incorporated in the sub-routine RUNGE in lines 258 et seq. Having found SEN(I) and SEX(I) across the arc, the ordinates between which SEN(I) and SEX(I) intersect can be determined by the logical IF-statement (line 71)

```
71:      IF(SEX(I).GE.SEN(I).AND.SEN(I+1).GE.SEX(I+1))  GO TO 16
```

Statement 16 is line 74, viz

```
74:      16   X=DPHI/(1.0+(SEX(I)-SEN(I))/(SEN(I+1)-SEX(I+1)))
              et seq:,
```

which gives the fraction of the interval dϕ (DPHI) in which ϕ_N(PHIN)

occurs; in the event that that logical IF-statement is not satisfied then the program is arranged to state that "no neutral point can be found, therefore a feasible solution does not exist for these conditions" and to print out the distributions of SEN(I) and SEX(I).

Having calculated ϕ_N it is then possible to calculate the roll pressure at the neutral plane (SN) and hence integrate the pressure by the trapezoidal rule (sub-routine TRAPEZ) and Simpson's rule (sub-routine SIMPSN) to determine the roll force P (equation (10)) due to the plastic arc of contact. The entry and exit elastic zones can then be estimated, to give PE1 and PE2, their contributions to the total roll force. Statement 40 (line 122) then gives the *current* estimate of the total roll force per unit width, viz

```
122:     40   P=P+PE1+PE2.
```

This whole procedure is then repeated until compatible values of the roll force and the radius of the deformed (assumed circular) arc of contact are obtained (i.e. J iterations to give $|P-P1|/P1<10^{-8}$, where P1 is the latest value of P) or until J>10, according to line 130, viz:-

```
130:     IF(ABS((P-P1)/P1).LT.I.OE.-8.OR.J.GT.10) GO TO 17.
```

Statement 17 et seq then arranges for calculation of the elastic arcs contributions to the torque and hence GAC12, the most accurate value of torque GAC from the plastic arc (according to equation (11)), with GE1 and GE2 added as set out in line 153, viz:-

```
153:     GAC12=GAC+GE1+GE2
```

Finally, the program is arranged to write out all the parameters of
interest, together with the values of s,sϕ,2k,τ and k spaced out at
equal intervals across the contact arc.

TYPICAL RESULTS

A typical computer print-out is shown in Appendix 2, for the program
ROLLNG and in Appendix 3 for the program EXTROL. Considering the
extrolling problem first, the material chosen was hot aluminium with
a Young's Modulus E of 4600 T/in², a Poisson's Ratio ν of 0.34 and a
constant flow stress Y_0 of 1.65 T/in² (i.e. B=n=0). In order to try
to predict the *experimental* results derived previously (Alexander and
Hatakeyama (1983)), the roll radius chosen for this example was R =
3.375 in., h_0=h_1=0.625 in, h_2=0.5 in, t_1=-0.5 T/in², A_3=0.375 in, and
w_1=1 in. Thus the extrusion ratio (R) was $h_2 w_1/A_3$ = 0.5/0.375 = 1.333
and the extrusion pressure therefore $-t_2 = \frac{2}{\sqrt{3}}Y_0(0.47+1.2\ell nR)$ = 1.65x
(0.47+1.2ℓn1.333)x2/$\sqrt{3}$ = 1.5532 T/in².

The result of inputting the values assumed into the program is shown
in Appendix 3 - the input data is printed out and the tests on h_2,t_1
and t_2 are made and the resulting values are printed out also. It
will be observed that the exit tension stress is t_2 = -1.5532 T/in²,
as expected.

All parameters are printed out - the initial approximate value of roll
force was PAP = 2.44367 T/in width. The final value achieved after
J = 4 iterations was P = 2.11621 T/in. On the main-frame computer
with 1501 ordinates (M=1501) this program would have taken about 50
seconds actual computer processing time to give a result of greater
precision, having 1501 ordinates (M) of pressure distribution. By reducing
the number of ordinates (M) to 31 only, a less precise result is
achieved on the desk-top computer using the program described here,
in about 3 minutes computer processing time. Previous experience in
running these programs on a desk-top computer has shown that, to
achieve the same accuracy as the main-frame computer (with M=1501)
requires about 5 hours. The computer actually used was an ACT(I)
Sirius 16-bit microprocessor (equivalent to a Victor 9000 micro-
computer in the U.S.A.), with a RAM of 256k bytes and MSDOS operating
system for the FORTRAN 77 language.

The same data was fed into the program ROLLNG and exactly the same
results were obtained, as expected and as indicated in Appendix 2.
The back tension was -0.5 T/in² for both ROLLNG and EXTROL, as was
the front tension of -1.5532 T/in². The neutral point occurred at
an angle of ϕ_n = 0.0252582 radians, the total angle of contact arc
being ϕ_1 = 0.192258. Thus the maximum pressure was quite near to the
exit point, achieving a value for SN of approximately 3.948 T/in².
By reducing the area of the extruded product A_3, possibly a slightly
bigger extrusion ratio could be achieved and this could quickly be
investigated with this computer program. It is easy to see that not
much more extrusion reduction could be achieved before the neutral
plane would coincide with the exit plane, unless a higher compressive
stress could be imposed on the strip entering the roll gap.

Investigations of this type can be made quickly with this computer program, enabling the optimum extrolling conditions to be established for any given material and rolling/extrusion geometries, as discussed and demonstrated previously by Alexander and Hatakeyama (1983). Although the program described here contains only the relatively simple constitutive relation of Swift (as exemplified by equation (3) and lines 22 and 23 of the computer program itself), it would be a simple matter to include well-known factors such as $(C\bar{\epsilon}^m)$ and $\exp\left(-\frac{Q}{RT}\right)$ to allow approximately for the effects of strain rate and temperature to be included. Such effects are doubtless of much greater importance in affecting the rolling mill and extrusion parameters than is either the effect of inhomogeneity of deformation through the roll gap or the precise shape of the deformed arc of contact.

REFERENCES

Alexander, J. M. (1955). A slip line field for the hot rolling process. *Proc. I. Mech. E.*, (169), 1021-1030.

Alexander, J. M. (1972). On the theory of rolling. *Proc. Roy. Soc. Lond. A.*, 326, 535-563 and 329, 493-496.

Alexander, J. M. and K. Hatakeyama (1983). On the hot extrolling of aluminium. *Proc. 24th Int. M.T.D.R. Conference*. Macmillan.

Alexander, J. M. and A. N. Tilakasiri (1979). A study of the process of extrolling. *Proc. 20th Int. M.T.D.R. Conference*. 93-102. Macmillan.

Avitzur, B. (1975). Extrolling: combining extrusion and rolling. *Wire Journal*, July, 73-80.

Bland, D. R. and H. Ford (1948). The calculation of roll force and torque in cold strip rolling with tensions. *Proc. I. Mech. E.*, 159, 144.

Bland, D. R. and H. Ford (1952). Cold rolling and strip tension - Part III: An approximate treatment of elastic compression of the strip in cold rolling. *J. Iron and Steel Inst.*, 171, 245.

Bland, D. R. and R. B. Sims (1953). A note on the theory of rolling with tensions. *Proc. I. Mech. E.*, 167, 371.

Cook, P. M. and A. W. McCrum. (1958). *The calculation of load and torque in hot flat rolling*, B.I.S.R.A., London.

Ford, H., F. Ellis and D. R. Bland (1951). Cold rolling with strip tension - Part I: A new approximate method of calculation and a comparison with other methods. *J. Iron and Steel Inst.*, 168, 57.

Ford, H. (1957). The theory of rolling. *Metallurgical Reviews*, 2, 1-28.

Johnson, W. (1956). Experiments in plane strain extrusion, *J. Mech. Phys. Solids*, 4, 269.

Karman, T. von (1925). Beitrag zur Theorie des Walzvorganges. *Z. Angew. Math. Mech.*, 5, 139.

Lianis, G. and H. Ford (1955-56). A graphical solution of the cold rolling problem, when tensions are applied to the strip. *J. Inst. of Metals*, 84, 299.

Orowan, E. (1943). The calculation of roll pressure in hot and cold flat rolling, *Proc. I. Mech. E.*, 150, 140.

Sims, R. B. (1954). Calculation of roll force and torque in hot rolling mills, *Proc. I. Mech. E.*, 168.

Venter, R. D. and A. A. Abd-Rabbo (1980). Modelling of the rolling process - I and II, *Int. J. Mech. Sci.*, 22, 83-92 and 93-98.

The listings for the programs discussed in this paper are available from the author.

FINITE-ELEMENT ANALYSIS OF PLASTIC DEFORMATION OF POROUS MATERIALS

Y. T. IM and S. KOBAYASHI

Department of Mechanical Engineering, University of California,
Berkeley, USA

ABSTRACT

The finite-element method for the deformation analysis of porous materials is developed here. The yield function, associated flow rules and their finite-element discretization are formulated. These formulations are implemented into a rigid-viscoplastic finite-element program. The program was tested by simulating the compression of a cylinder with an initial relative density of 0.999 and by comparing the solution with that for a solid cylinder. Then the analysis was performed for ring compression and forging for a material of initial relative density of 0.800. The results of the present investigation demonstrate that it becomes possible to study detailed material flow during large deformations in the area of powder metallurgy where the effect of temperature gradient is not significant.

NOTATION

F	yield function
J_1	linear invariant of stress tensor
J_2'	quadratic invariant of deviatoric stress tensor
A, B	functions of relative density
$\sigma_{ij}, \underset{\sim}{\sigma}$	stress component, stress vector
$\dot{\varepsilon}_{ij}, \underset{\sim}{\dot{\varepsilon}}$	strain-rate component, strain-rate vector
$\dot{\varepsilon}_v$	volumetric strain-rate
$u_i, \underset{\sim}{u}$	velocity component, velocity vector
$v_\alpha, \underset{\sim}{v}$	nodal point velocity values, nodal point velocity vector
$T_i, \underset{\sim}{T}$	prescribed traction and its vector representation
U_i	prescribed velocity
$f_i, \underset{\sim}{f}$	frictional stress and its vector representation
S_T, S_u, S_c	surfaces where traction is prescribed, velocity is prescribed, and the die-workpiece contact surface
Y_R	yield stress of porous metals

$\dot{\bar{\varepsilon}}_R$ apparent effective strain-rate

Y_b yield stress of base metal

$\bar{\varepsilon}_b, \dot{\bar{\varepsilon}}_b$ effective strain and strain-rate in base metal

R, R_0 local relative density, current and initial values, respectively

R_{ave} average overall relative density

η a parameter characterizing material property and function of relative density

$\underline{N}, \underline{A}, \underline{B},$ finite-element matrices defined in the appendix
$\underline{P}, \underline{D}$

INTRODUCTION

Since the finite-element method was introduced for analyzing metal-forming processes, the method, particularly in the area of process modelling, has assumed steadily increased importance in metalworking technology. A recent review on the subject (Kobayashi, 1982) indicates that the major accomplishments have been the development of a user-oriented rigid-viscoplastic finite-element code for the analysis of metal-forming processes with arbitrarily shaped dies by Oh (1982) and the coupled analysis of transient viscoplastic deformation and heat transfer by Rebelo (1980).

These accomplishments have made it possible to simulate forming processes with geometrically complex dies not only at room temperatures, but also in the warm and hot regions. Apparently, the applicability of the method can be extended to the area of powdered metal fabrication when the material model includes the behavior of porous metals This paper describes the development of the finite-element method for the deformation analysis of porous metals. As a first step, the working temperature is assumed to be constant and the effect of heat generation during deformation is neglected.

Porous metals are more susceptible to fracture than are solid metals. Thus, flow and fracture are of particular importance in forming these materials. A number of plasticity theories have been proposed with the yield function F of the following form (Kuhn and Downey, 1971; Green, 1972):

$$F = AJ_2' + BJ_1^2 \tag{1}$$

where J_1 is the linear invariant of stress tensor, J_2' is the quadratic invariant of deviatoric stress tensors, and A, B are functions of void ratio or relative density.

Starting with Eq. (1), Oyane and his colleagues derived the plasticity equations. Then, based on this theory, they derived the slip-line field equations and the upper-bound theorem applicable to porous metals (Shima and Oyane, 1976; Tabata and Masaki, 1975; Tabata and Oyane, 1975). Osakada et al. (1982), also used this theory, in developing the finite-element method and applied it to the rigid-plastic analysis of metal forming (Mori, Osakada, and Oda, 1982; Osakada, Nakano, and Mori, 1982). The development described in this paper is based on the same principle, but the formulation is somewhat different.

FINITE-ELEMENT FORMULATIONS

Basic Equations

The governing equations for the boundary value problem associated with plastic deformation of porous metals are as follows:

Equilibrium equation $\sigma_{ij,j} = 0$ (2)

Compatibility $\dot{\varepsilon}_{ij} = \frac{1}{2}(u_{i,j} + u_{j,i})$ (3)

Flow rules $\dot{\varepsilon}_{ij} = \frac{\partial F}{\partial \sigma_{ij}}\dot{\lambda}$ (4)

where σ_{ij} and $\dot{\varepsilon}_{ij}$ are apparent stresses and strain-rates, respectively, considering the porous metal as a continuum. The prescribed boundary conditions are

$$\sigma_{ij}n_j = T_i \text{ on } S_T$$

$$u_i = U_i \text{ on } S_u \qquad (5)$$

$$T_i - (T_j n_j)n_i = f_i \text{ on } S_c$$

where the boundary surface, $S = S_T + S_u + S_c$ and n_j is the unit outward normal to the surface.

The yield surface is defined by

$$AJ_2' + BJ_1^2 = Y_R^2$$

where Y_R is the apparent yield stress of the porous material determined by uniaxial tension or compression tests. Then it can be shown that $B = 1 - \frac{A}{3}$.

The yield function F is now expressed by

$$F = A[\frac{1}{6}\{(\sigma_x - \sigma_y)^2 + (\sigma_y - \sigma_z)^2 + (\sigma_z - \sigma_x)^2 +$$

$$(\tau_{xy}^2 + \tau_{yz}^2 + \tau_{zx}^2)] + (1 - \frac{A}{3})(\sigma_x + \sigma_y + \sigma_z)^2 \qquad (6)$$

With this yield function, the proportionality factor $\dot{\lambda}$ in the flow rule, Eq. (4), is given by

$$\dot{\lambda} = \frac{\dot{\varepsilon}_R}{2Y_R} \qquad (7)$$

where the apparent effective strain-rate $\dot{\varepsilon}_R$ is defined according to $\sigma_{ij}\dot{\varepsilon}_{ij} = Y_R\dot{\varepsilon}_R$ (Hill, 1950; Johnson and Mellor, 1980) and expressed by

$$(\dot{\bar{\varepsilon}}_R)^2 = \frac{1}{A} [\frac{2}{3} \{(\dot{\varepsilon}_x - \dot{\varepsilon}_y)^2 + (\dot{\varepsilon}_y - \dot{\varepsilon}_z)^2 + (\dot{\varepsilon}_z - \dot{\varepsilon}_x)^2\} +$$

$$(\dot{\gamma}_{xy}^2 + \dot{\gamma}_{yz}^2 + \dot{\gamma}_{zx}^2)] + \frac{1}{3(3 - A)} \dot{\varepsilon}_v^2 \tag{8}$$

where $\dot{\varepsilon}_v$ is the volumetric strain rate.

The variational form of the equilibrium equation used as a basis for discretization is expressed as

$$\int Y_R \delta\dot{\bar{\varepsilon}}_R \, dV - \int_{S_t} \underset{\sim}{T}^T \delta\underset{\sim}{u} \, dS - \int_{S_c} \underset{\sim}{f}^T \delta\underset{\sim}{u} \, dS = 0 \tag{9}$$

where δ denotes variation.

Material Properties

The apparent yield stress Y_R in Eq. (9) depends on the properties of the base metal and the relative density R (ratio of the volume of base metal to the volume of porous metal) (Doraivelu et al., 1984) according to

$$F = Y_R^2 = \eta Y_b^2 \tag{10}$$

where Y_b is the yield stress of the base metal and η is a function of the relative density and expressed by

$$\eta = \frac{R^2 - R_c^2}{1 - R_c^2} . \tag{11}$$

In Eq. (11) R_c is the critical density where the compact loses its mechanical strength. The effects of strain, strain rate, and temperature on the yield stress are included in $Y_b = Y_b(\bar{\varepsilon}_b, \dot{\bar{\varepsilon}}_b, T_b)$, where $\bar{\varepsilon}_b$, $\dot{\bar{\varepsilon}}_b$, and T_b are the strain, strain-rate, and temperature of the base metal. The relationship between the apparent strain and strain rate and those of the base metal are given by

$$Y_R \dot{\bar{\varepsilon}}_R = RY_b \dot{\bar{\varepsilon}}_b \text{ or } \dot{\bar{\varepsilon}}_b = \frac{\sqrt{\eta}}{R} \dot{\bar{\varepsilon}}_R$$

and

$$\bar{\varepsilon}_b = \int \frac{\sqrt{\eta}}{R} d\bar{\varepsilon}_R. \tag{12}$$

In order to complete the formulation, A, involved in the yield function, and η must be expressed as functions of the relative density, and determined by experiments.

Discretization

The element used for discretization is an isoparametric quadrilateral

element with a bilinear distribution function. The elemental velocity field $\underset{\sim}{u}$ is approximated by

$$\underset{\sim}{u} = \underline{N}\underset{\sim}{v} \tag{13}$$

where $\underset{\sim}{v}$ is the velocity vector of nodal point value and $\underline{N}$ is the
distribution function matrix. Applying differentiation to Eq. (13),

$$\underset{\sim}{\dot{\varepsilon}} = \underline{A}\underset{\sim}{u} = \underline{B}\underset{\sim}{v} \tag{14}$$

Substitution of Eq. (14) into Eq. (8) leads to

$$(\dot{\bar{\varepsilon}}_R)^2 = \underset{\sim}{\dot{\varepsilon}}^T \underline{D} \underset{\sim}{\dot{\varepsilon}} = \underset{\sim}{v}^T \underline{B}^T \underline{DB}\underset{\sim}{v} \tag{15}$$

$$= \underset{\sim}{v}^T \underline{P}\underset{\sim}{v}$$

and the variation $\delta\dot{\bar{\varepsilon}}_R$ becomes, because of symmetry of the matrix $\underline{P}$,

$$\delta\dot{\bar{\varepsilon}}_R = \frac{1}{\dot{\bar{\varepsilon}}_R} \delta\underset{\sim}{v}^T \underline{P}\underset{\sim}{v} \tag{16}$$

Substituting Eqs. (13) and (16) into Eq. (9) at the elemental level
and assembling the element equations with global constraints, we
obtain

$$\sum_{j=1}^{M} \delta\underset{\sim}{v}^T \left| \int \frac{\bar{Y}_R}{\dot{\bar{\varepsilon}}_R} \underline{P}\underset{\sim}{v} \, dV_j - \int \underline{N}^T \underline{T} \, dS_{tj} - \int \underline{N}^T \underline{f} \, dS_{cj} \right| = \underset{\sim}{0} \tag{17}$$

where M is the total number of elements. Because the variation $\delta\underset{\sim}{v}$
is arbitrary, Eq. (17) results in the stiffness equations. The
matrices in Eq. (17) are defined in the appendix for the case of
axial symmetry.

NUMERICAL PROCEDURES

The stiffness equations resulting from Eq. (17) are of the form
$h(\underset{\sim}{v}) = \underset{\sim}{0}$ and are nonlinear because $\dot{\bar{\varepsilon}}_R$ in the first term of Eq. (17)
is a function of $\underset{\sim}{v}$.

Newton-Raphson Method

The numerical procedure for solving nonlinear equations and adopted
here is the Newton-Raphson method. Writing Eq. (17) as

$$\underline{G}(\underset{\sim}{v}) = \Sigma \, (\pi_{\sim v} - \pi_{\sim t} - \pi_{\sim f}) = \underset{\sim}{0}$$

linearization is accomplished according to

$$\underline{G}(\underset{\sim}{v}) \cong \underline{G}(\underset{\sim}{v}_{n-1}) + \left[\frac{\partial \underline{G}}{\partial \underset{\sim}{v}}\right]_{n-1} \Delta\underset{\sim}{v}_n = 0 \tag{18}$$

where n is the number of the iteration.

In Eq. (18) the first term of $\frac{\partial G}{\delta v}$ is expressed by

$$\int \left| \frac{Y_R}{\dot{\varepsilon}_R} \underline{P} + \left[\frac{1}{\dot{\varepsilon}_R^2} \frac{\partial Y_R}{\partial \dot{\varepsilon}_R} \div \frac{Y_R}{\dot{\varepsilon}_R^3} \right] \underline{P} \underline{v} \underline{v}^T \underline{P} \right| dV \qquad (19)$$

where $\dfrac{\partial Y_R}{\partial \dot{\varepsilon}_R} = \dfrac{\eta}{R} \dfrac{\partial Y_b}{\partial \dot{\varepsilon}_b}$ according to Eq. (12).

Treatment of the Rigid Zone

When $\dot{\bar{\varepsilon}}_R \leq \dot{\bar{\varepsilon}}_0$, with $\dot{\bar{\varepsilon}}_0$ as an offset value which is several orders of magnitude smaller than the strain-rate prevailing in the deforming region, the stress-strain-rate relationship is assumed to be

$$\underline{\sigma} = \left[\frac{Y_R}{\dot{\bar{\varepsilon}}} \right]_0 \underline{D} \dot{\underline{\varepsilon}}$$

Then, for the element which is nearly rigid, Eq. (19) becomes

$$\int \left[\frac{Y_R}{\dot{\bar{\varepsilon}}} \right]_0 \underline{P} \, dv$$

where the subscript 0 denotes the value at $\dot{\bar{\varepsilon}} = \dot{\bar{\varepsilon}}_0$.

Updating of Relative Density

The volumetric strain-rate is related to the density rate according to

$$\dot{\varepsilon}_v = \underline{c}^T \dot{\underline{\varepsilon}} = -\frac{\dot{R}}{R} \qquad (20)$$

where the relative density R is defined by

$$R = \frac{V_b}{V_b + V_v}$$

with V_b being volume of the base metal and V_v as the volume of voids. Integrating Eq. (20), we have

$$R = R_0 \exp(-\int \dot{\varepsilon}_v \, dt) \cong R_0 (1 - \Delta\varepsilon_v) \qquad (21)$$

In Eq. (21), R_0 is the current relative density and $\Delta\varepsilon_v$ is the change of volumetric strain in one step. The average relative density R_{ave} is defined by

$$R_{ave} = \frac{\Sigma R_i V_i}{\Sigma V_i}$$

where R_i and V_i are the relative density and the volume of an element, respectively.

Friction Boundary Condition

The friction representation of a velocity-dependent frictional stress, expressed by

$$f = mk \left\{ (\frac{2}{\pi}) \tan^{-1} \left[\frac{|v_{DW}|}{a} \right] \right\}$$

is used, where k is the shear strength, v_{DW} is the relative velocity of the work-piece material with respect to the die and a is a constant several orders of magnitude smaller than the die velocity and m is the friction factor, ranging from zero to unity.

PROGRAM TESTING

Modifications were made to the rigid-viscoplastic finite-element program ALPID (Oh, 1982) to include compressibility of the material. The program code was tested by analyzing the compression of cylindrical porous materials using the following conditions:

. diameter-to-height ratio, 1.2

. material property, $A = 2 + R^2$, $\eta = 2R^2 - 1$ (with $R_c^2 = \frac{1}{2}$ in

Eq. (11)), and $Y_b = 110(1 + 0.768\bar{\varepsilon}_b) N/mm^2$

(Base metal: aluminium 1100F)

. friction condition, m = 0.1

. initial relative density, $R_0 = 0.999, 0.800$ (uniformly distributed)

. total reduction in height, 50%

The results for the porous material with initial density of 0.999 were compared with those for solid metal in terms of effective strain distribution. It was found that the trend is the same and the magnitudes differ from each other only slightly. After compression of 50% in height, the average relative density increased from 0.999 to 0.99975. For the case of $R_0 = 0.800$, the average relative density was 0.940 at 50% reduction in height. Some detailed results are given for $R_0 = 0.800$. The effective strain distribution and the distribution of relative density at 50% reduction in height are shown in Fig. 1. The apparent strain distribution, when it is compared with that for $R_0 = 0.999$, shows little difference in trend as well as in magnitude. The strain concentration is seen to occur near the edge of the die-workpiece contact surface and larger strains are shown in the central region of the specimen. The relative density distribution shows that densification is effective also near the edge and in the central zone. Although no obvious correlation can be observed between the effective strain and the degree of densification, a more detailed examination of densification revealed that the relative density can be correlated closely to the mean stress. With higher compressive mean stress, more densification results. This is consistent with the practice of compaction for uniform densification.

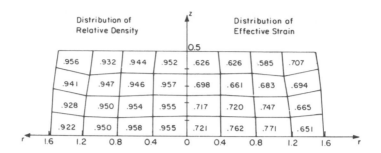

Fig. 1.   Distributions of relative density and
effective strain at 50% reduction in
height for R_0 = 0.800.

It was found during the course of program testing that the following
constraints in numerical procedures were helpful in obtaining well-
behaved convergence for the solution:

(1)   When the relative density R in an element approaches unity
(say R = 0.9999), then convergence behavior during iteration
becomes erratic.   A constraint was incorporated such that an
element R = 0.999 is considered as a fully dense element.

(2)   Usually, the volume integration involved in Eqs. (17) and (19)
uses Gaussian quadrature for the terms involving distortion
energy, but for the terms of volumetric strain rate, reduced
integration is appropriate.   Thus, in Eq. (19), the matrix
$\underline{P}$ is decomposed into two components $\underline{P}_1$ and $\underline{P}_2$ and the reduced
one-point integration scheme was applied to the terms which
involve the volumetric strain-rate.   The matrices $\underline{P}_1$ and $\underline{P}_2$
and their relation to the corresponding matrices for solid
materials are indicated in the appendix.

APPLICATIONS

Two problems were analyzed with the present finite-element program.
One is ring compression and the other is axisymmetric forging.

Ring Compression

Solutions for ring compression were obtained for the following conditions:

. ring specimen geometry: OD:ID:height = 6:3:2
. material property: same as that in cylinder compression
. friction: m = 0.6 and 1.0
. initial relative density: R_0 = 0.800

The results are obtained in terms of a load-displacement curve, average density changes, grid-distortion, and density distributions and

their variations. Figure 2 shows the comparison of the load-displacement curves for fully dense materials and for porous materials with
interface friction of m = 0.6 and 1.0. The load is higher for a
fully dense material than for a porous material and with higher friction and it may be of interest to note that the difference between
the loads for R_0 = 0.999 and 0.800 is of the same amount for various

reductions in height in compression. The average relative density
increases as compression continues, as seen in Fig. 3. Initially,
the increase is linear with respect to reduction in height, but the
rate decreases as the reduction in height increases and eventually
the average density begins to decrease, particularly for a high friction factor of m = 1.0, when the reduction in height becomes large
(say, 70% or 80%).

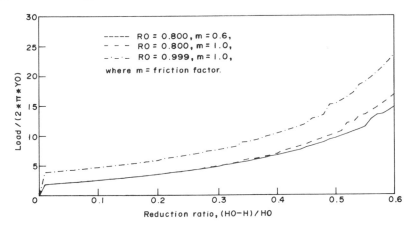

Fig. 2. Load-displacement curves in ring compression
for solid materials (R_0 = 0.999, m = 1.0) and
porous materials (R_0 = 0.800, m = 1.0, 0.6).

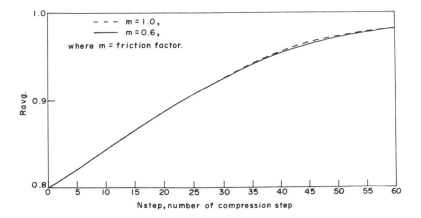

Fig. 3. The average relative density as a function
of reduction in height.

The grid distortions for a porous material are compared for the two friction conditions at several stages of compression in Fig. 4.   At 20% reduction in height, the effect is not seen, but at 40% and 60% reductions in height, the effect of friction becomes obvious from the geometrical changes of the ring specimen.   Greater friction results in corresponding decrease of the inner-diameter.   In interpreting geometrical changes, it should be noted that the relative density varies and thus the apparent volume of the workpiece changes.   This is clearly demonstrated by comparing the grid distortion for solid materials and porous materials, as shown in Fig. 5.   The decrease in apparent volume for a porous material due to densification is obvious from the figure.   The relative density distributions as functions of reduction in height are shown in Fig. 6.   At 20% reduction it can be seen that densification takes place throughout the workpiece.   Increase in density is greatest near the edge of the die-workpiece contact area.   Because of high friction, densification is less in the mid-region under the die.   Also, near the free surface along the equatorial surface, less densification is observed.

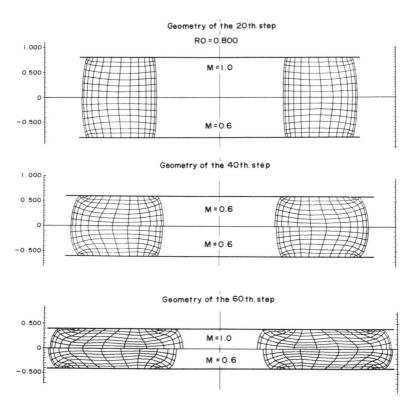

Fig. 4.   Grid distortions for m = 1.0 and 0.6 at 20%, 40%, and 60% reductions in height.

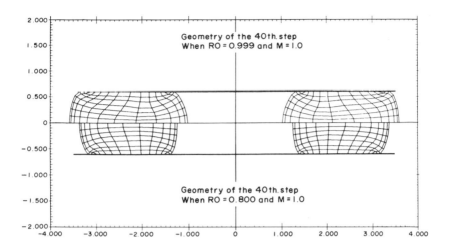

Fig. 5.   Comparison of grid distortions for solid
          and porous materials at 40% reduction in
          height for m = 1.0.

Densification increases as the reduction in height increases, except
near the outer free surface of the equatorial plane. At 60% reduction,
the relative density reaches 0.999 (almost fully dense material).   In
the element near the free surface, however, the relative density
varies from the initial value of 0.800 to 0.829, 0.828, and 0.784 at
reductions in height of 20%, 40% and 60%, respectively.

Axisymmetric Forging

Closed-die forging of two preform shapes (Downey and Kuhn, 1975) were
analyzed. The forging dies and two preform specimens with the mesh
system used for the analysis are shown in Fig. 7.   The base metal is
OFHC copper with the stress and strain property given by

$$Y_b = Y_0 \left[1 + \frac{\bar{\varepsilon}_b}{0.3518} \right]^{0.28} \quad (Y_0 = 290\text{MN/m}^2)$$

and the initial relative density is R_0 = 0.800. The frictional
stress is given by mk and m is assumed to be 0.1.   In Fig. 8, the
grid distortions and the velocity fields are shown at three stages
of forging for the preform I in Fig. 7.

Figure 8 is shown for qualitative observation of metal flow as to the
deforming regions. It is apparent from the figure that the region
beneath the die remains rigid, because grid lines in the region remain undistorted.

This is observed throughout the deformation stages shown in Fig. 8(a)-
(c). The velocity fields shown in the lower half indicate that the
deformation is similar to that in the compression of a solid cylinder
between the two parallel platens with frictional stress equal to the

shear strength of the deforming material. A detail with regard to
the deformation zone and the extent of densification can be seen in
Fig. 9 where the distributions of apparent effective strain and
relative density are shown at three stages of forging. From a glance
at the figure, the pattern of equi-strain and density contours are
remarkably similar. This is because the effective strain contains
not only the term of distortion, but also the volume change which is
associated with the relative density. The effective strains are
largest at the center of the forging and at the edge of the die-
specimen interface, and so are the relative densities.

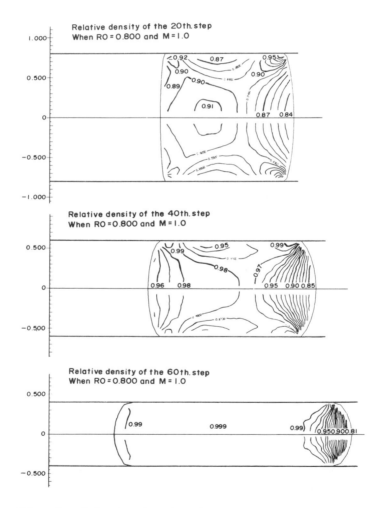

Fig. 6.   Relative density distributions at 20%, 40%,
           and 60% reductions in height for a porous
           material with m = 1.0.

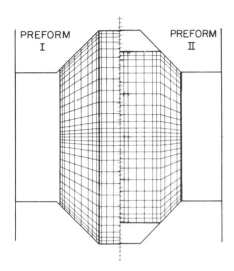

Fig. 7.   Preform shapes and mesh systems in axisymmetric closed-die forging.

A close examination of the distribution at 10% reduction in height
(Fig. 9(a)) reveals that the central region under the die indeed
shows no deformation and no change in relative density from the
initial value (R_0 = 0.800).   However, as the reduction in height
increases, deformation (mostly volume change) extends into this
region, and densification takes place even though the distortion of
grid lines in this region are not seen in Fig. 8.

Another observation of importance is that the relative density near
the bulged free surface increases at first, then begins to decrease
as seen by comparing the densities at 20% reduction and 30% reduction.   This is a site of fracture observed in experiments (Downey
and Kuhn, 1975).

For preform II the grid distortions and velocity fields are shown at
three reductions in height in Fig. 10.   At the smallest reduction in
height, the deformation is similar to that of preform I, and the rigid
zone in the central region near the free surface can be seen.   Although the workpiece is moving toward the center of the forging, its
velocity is slower than the die velocity.   Therefore, the workpiece
is deforming radially as well as axially to fill the cavity relative
to the die motion.   Thus, as the reduction in height increases, the
grid lines show distortions throughout the workpiece.

Figure 11 shows the distributions of effective strain and relative
densities at three reductions of height.   Again, patterns of distributions of the two quantities are similar.   At 10% reduction, the zone
of no deformation and no change in relative density is seen in the
central region near the free surface.   Densification takes place into
this region gradually as the reduction in height increases.   Characteristics of a strong singularity, where the gradients of strain and
density are large, are seen along the central periphery of the die-
workpiece contact surface.   This results in a distinctly different
deformation pattern from that for preform I.

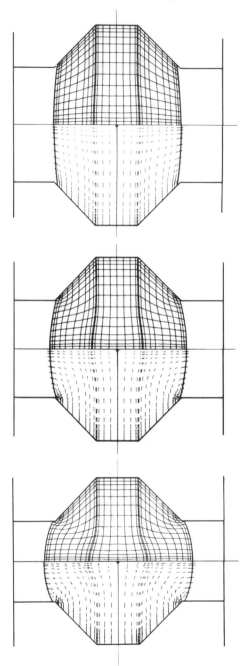

Fig. 8.   Grid distortions and velocity fields at
          (a) 10%, (b) 20%, and (c) 30% reductions
          in height in forging preform I.

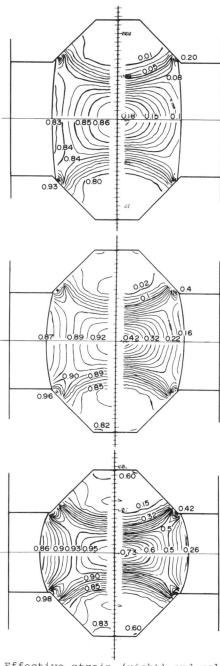

Fig. 9.   Effective strain (right) and relative
          density (left) distributions at (a) 10%,
          (b) 20%, and (c) 30% reductions in height
          in forging preform I.

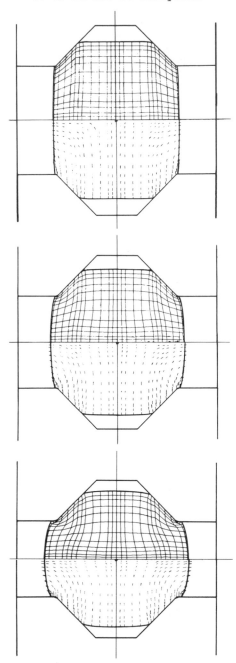

Fig. 10.   Grid distortions and velocity fields at
           (a) 10%, (b) 20%, and (c) 30% reduction
           in height in forging preform II.

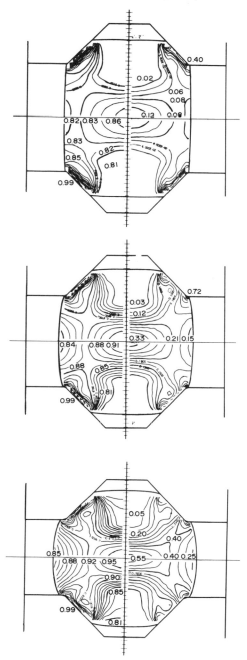

Fig. 11.   Effective strain (right) and relative
           density (left) distributions at (a) 10%,
           (b) 20%, and (c) 30% reduction in height
           in forging preform II.

CONCLUSIONS

It is concluded that the formulation for the deformation of porous
materials and its implementation as a finite element program were
successfully accomplished. During this investigation several difficulties were encountered concerning integration schemes and automatic
initial-guess generation. In determining proper integration schemes,
it was found that one-point integration must be applied to the terms
involving the volumetric strain-rate. It was also noted that friction at the die-workpiece interface must be treated carefully. If
friction is sensitive to the metal flow, then the problem should be
solved with low friction first for easy convergence of the solution.
Although stress fields are not shown in the results, they are obtained as a part of the solution as well as local density variations.
In this program the base metal properties can be strain and strain-
rate dependent. In order to simulate the processes of powdered
metals, the temperature calculation should be coupled with the deformation analysis. This is the next step of development for applications in the field of powder metallurgy.

ACKNOWLEDGEMENTS

The authors wish to thank Dr. T. Altan, Battelle Columbus Laboratories, and Dr. H. Gegel, for the Processing Science Program under
which the present investigation was conducted. They further wish to
thank Dr. W. Spurgeon, National Science Foundation, for grant MEA-
8312062 which concerns preform design in metal forming. They also
wish to thank Ikuko Workman for preparing the manuscript.

REFERENCES

Doraivelu, S. M., H. L. Gegel, J. S. Gunasekera, J. C. Malas, J. T.
    Morgan and J. F. Thomas, Jr. (1984). A new yield function for
    compressible P/M materials. Private communication.
Downey, C. L., Jr. and H. A. Kuhn (1975). Application of a forming
    limit concept to the design of powder preforms for forging. *J.
    Eng. Mat. Tech.*, **97**, 121-125.
Green, R. J. (1972). A plasticity theory for porous solids. *Int.
    J. Mech. Sci.*, **14**, 215-224.
Hill, R. (1950). *The Mathematical Theory of Plasticity*. Oxford at
    Clarendon Press.
Johnson, W. and P. B. Mellor (1980). *Engineering Plasticity*. Van
    Nostrand Reinhold, London.
Kobayashi, S. (1982). A review on the finite-element method and
    metal-forming process modelling. *J. Appl. Metalworking*, **2**, 163-
    169.
Kuhn, H. A. and C. L. Downey, Jr. (1971). Deformation characteristics
    and plasticity theory of sintered powder materials. *Int. J. Powder
    Met.*, **7**, 15-25.
Mori, K., K. Osakada and T. Oda (1982). Simulation of plane-strain
    rolling by the rigid-plastic finite-element method. *Int. J. Mech.
    Sci.*, **24**, 519-527.
Oh, S. I. (1982). Finite-element analysis of metal-forming processes
    with arbitrary shaped dies. *Int. J. Mech. Sci.*, **24**, 479.
Osakada, K., J. Nakano and K. Mori (1982). Finite-element method
    for rigid-plastic analysis of metal-forming-formulation for finite
    deformation. *Int. J. Mech. Sci.*, **24**, 459-468.
Rebelo, N. (1980). Finite-element modelling of metalworking processes
    for thermo-viscoplastic analysis. Ph.D. Dissertation, Department
    of Mechanical Engineering, University of California, Berkeley.

Shima, S. and M. Oyane (1976). Plasticity theory for porous metals.
  Int. J. Mech. Sci., <u>18</u>, 285-291.
Tabata, T. and S. Masaki (1975). Plane-strain extrusion of porous
  materials. Memoirs of the Osaka Institute of Technology, Series
  B., <u>19</u>(2).
Tabata, T. and M. Oyane (1975). The slip-line field theory for a
  porous material. Memoirs of the Osaka Institute of Technology,
  Series B., <u>18</u>(3).

APPENDIX Matrices for the axisymmetric case, with a 4-node
quadrilateral element

The natural coordinates (s, t) and global coordinates (r, z) are related by

$$r(s, t) = \sum_1^4 q_\alpha r_\alpha, \qquad z(s, t) = \sum_1^4 q_\alpha z_\alpha$$

where $q_\alpha = \frac{1}{4}(1 + s_\alpha s)(1 + t_\alpha t)$.

Arranging $\underset{\sim}{u}$ and $\underset{\sim}{v}$, according to

$$\underset{\sim}{u}^T = \{u_r, u_z\} \qquad \text{and} \qquad \underset{\sim}{v}^T = \{v_{r1}, v_{z1}, \ldots, v_{r4}, v_{z4}\},$$

the matrix $\underline{N}$ in Eq. (13) becomes

$$\underline{N} = \begin{bmatrix} q_1 & 0 & q_2 & 0 & q_3 & 0 & q_4 & 0 \\ 0 & q_1 & 0 & q_2 & 0 & q_3 & 0 & q_4 \end{bmatrix} \tag{A.1}$$

The matrix $\underline{B}$ in Eq. (14) is given by

$$\underline{B} = \underline{A}\underline{N},$$

where

$$\underline{A}^T = \begin{bmatrix} \frac{\partial}{\partial r} & 0 & \frac{1}{r} & \frac{\partial}{\partial z} \\ 0 & \frac{\partial}{\partial z} & 0 & \frac{\partial}{\partial r} \end{bmatrix} \tag{A.2}$$

with $\underset{\sim}{\dot{\varepsilon}}^T = \{\dot{\varepsilon}_r \; \dot{\varepsilon}_z \; \dot{\varepsilon}_\theta \; \dot{\gamma}_{rz}\}$,

The matrix $\underline{P}$ in Eq. (15), where $\underline{P} = \underline{B}^T \underline{D} \underline{B}$, is obtained as follows.
The inverse of the flow rules can be expressed by

$$\underset{\sim}{\sigma} = \frac{Y_R}{\dot{\bar{\varepsilon}}} \underline{D} \underset{\sim}{\dot{\varepsilon}},$$

where

$$\underline{D} = \begin{bmatrix} \frac{4-A}{A(3-A)} & \frac{A-2}{A(3-A)} & \frac{A-2}{A(3-A)} & 0 \\ & \frac{4-A}{A(3-A)} & \frac{A-2}{A(3-A)} & 0 \\ & \text{sym} & \frac{4-A}{A(3-A)} & 0 \\ & & & \frac{1}{A} \end{bmatrix} \tag{A.3}$$

Y. T. Im and S. Kobayashi

where $\underset{\sim}{\sigma}^T = \{\sigma_r \ \sigma_z \ \sigma_\theta \ \tau_{rz}\}$. Then, from the requirement that $\underset{\sim}{\sigma}^T\dot{\underset{\sim}{\varepsilon}} = Y_R\dot{\bar{\varepsilon}}$

and

$$\left(\frac{\dot{\bar{\varepsilon}}}{}\right)^2 = \dot{\underset{\sim}{\varepsilon}}^T\underline{D}\dot{\underset{\sim}{\varepsilon}}$$

$$\underline{P} = \underline{B}^T\underline{D}\underline{B}$$

Decomposing the matrix $\underline{D}$ into the two components $\underline{D}_1$ and $\underline{D}_2$, we have

$$\underline{D} = \underline{D}_1 + \underline{D}_2$$

$$\underline{D}_1 = \begin{bmatrix} \dfrac{4}{3A} & -\dfrac{2}{3A} & -\dfrac{2}{3A} & 0 \\ & \dfrac{4}{3A} & -\dfrac{2}{3A} & 0 \\ & \text{sym} & -\dfrac{4}{3A} & 0 \\ & & & \dfrac{1}{A} \end{bmatrix} \tag{A.4}$$

$$\underline{D}_2 = \frac{1}{3(3-A)}\begin{bmatrix} 1 & 1 & 1 & 0 \\ & 1 & 1 & 0 \\ & & 1 & 0 \\ & \text{sym} & & 0 \end{bmatrix} \tag{A.5}$$

$$= \frac{1}{3(3-A)}\underset{\sim}{C}\underset{\sim}{C}^T, \quad \text{with } \underset{\sim}{C}^T = \{1, 1, 1, 0\}.$$

Then, $\underline{P} = \underline{P}_1 + \underline{P}_2 = \underline{B}^T\underline{D}_1\underline{B} + \underline{B}^T\underline{D}_2\underline{B}$. It should be noted that $A = 3$ in Eq. (A.4) and $\dfrac{1}{3(3-A)}$ in Eq. (A.5) is replaced by K, the penalty constant, for a solid material.

FINITE STRAIN DETERMINATION IN SHEET METAL STAMPINGS

R. SOWERBY

Department of Mechanical Engineering, McMaster University,
Hamilton, Ontario L8S 4L7, Canada

ABSTRACT

The article presents a method of strain determination over the deformed surface of a sheet metal stamping. The strains are calculated
from the nodal point measurements of a quadrilateral grid, which has
been previously marked on the undeformed blank. The strain calculations and the plotting of the strain contours are performed with
the aid of a computer. The theory on which the technique is founded
is discussed, and some results for a deep drawn part are provided.

INTRODUCTION

In plasticity any strain determined solely from the initial and final
shape of any observable element cannot be regarded as a state parameter. The material state depends not only on the change in shape,
but also on the path along which the shape evolved. For rate-
independent solids, the assumption of isotropic hardening can lead
to a simple relationship between the current *representative* stress
(a measure of the size of the yield surface) and the integral of the
representative strain increment, $d\bar{\epsilon}$. As is well known $d\bar{\epsilon}$ is a multiple of the second invariant of the plastic strain increment tensor.
This is an Eulerian description of the strain, since the components
are measured with respect to the current configuration. The strain
increment comprises an elastic and a plastic part, and Nemat-Nasser
(1979, 1982) has argued the case for an additive decomposition. The
point is not pursued here.

In sheet metal forming processes, the strain (strain increment) is
typically evaluated from the measurement of a grid marked on the surface of the workpiece. The surface element over which the measurements are made is usually regarded as plane. Furthermore, since the
sheet is thin, it is assumed no quantities vary through the thickness
and the normal to the sheet surface is a principal direction. In
order to obtain a simple measure of strain, the investigator is
anticipating that the straining will occur by *homogeneous deformation*
over the region of a simple grid element. Hence a square grid will
deform into a parallelogram and a circle into an ellipse. The
assumption of linear mapping is also embodied in many theoretical

123

studies of the geometry of deformation, it is merely a matter of
scale as to whether the domain of inspection is considered infinitesimal or finite in extent. This presents no problems from a theoretical point of view, but in practice the grids are finite and inhomogeneous straining may arise within the boundary of a single grid.
Fine grid spacings are likely to provide a better visual representation of non-uniform strain distribution, but detailed measurement
of the grids becomes impracticable. Grid spacings of less than 1 mm
are rarely used; the most commonly employed grid markings on the
surface of test pieces have a spacing of about 5 mm. Uncertainties
associated with strain gradients are generally present regardless of
the grid size. In some cases it may be possible to relate the initial
and final shape of grids within such a field by a simple analytical
function which portrays the strain gradient; but the pragmatic
approach has been to assume that no strain gradient exists within
the boundary of a single grid element, unless the experimental observations render such an assumption totally unreasonable. The well
known grid-circle technique employed in sheet metal forming is based
upon this hypothesis. The circle is assumed to be transformed into
an ellipse, and by measuring the major and minor diameters of the
ellipse the principal surface strains can be determined; implicit
in the technique is that the process is one of pure homogeneous deformation. Not every element deforms in this manner, although over
a large portion of the surface of the workpiece, circles do have the
appearance of ovals when viewed by the naked eye or through a low
powered optical measuring device. There is no need to restrict the
grid markings to circles, and square or irregular quadrilateral grids
can be employed. Recent developments in techniques of automatic
image analysis can be most effectively exploited when using coordinate
or nodal point measures.

If the straining takes place by *pure homogeneous deformation* (pure
stretch), the deformation gradient tensor is symmetric and there
exists within a deforming cell an orthogonal triad which remains
orthogonal throughout the deformation history. Without ambiguity
this triad represents the principal axes, and they remain fixed in
space[1] while all other line elements rotate. From the components of
the deformation gradient tensor, the orientation and magnitude of
the principal stretches is readily determined. The *total representative strain* is then a function of the natural logarithm of the
principal stretches. In pure stretch processes, it is sufficient
to measure the initial and final shape only of a grid element in
order to evaluate the principal stretches. It is for the case of
pure stretch that Hill (1968) proposed the tensor logarithm as a
conjugate strain measure in his work on constitutive inequalities.

When the straining occurs by *homogeneous deformation* the deformation
gradient tensor, say F, is unsymmetric. However, an orthogonal triad
can be identified in the initial configuration (the ground state)
which is also orthogonal in the final (current) configuration. It
must be emphasized that this triad has not remained orthogonal
throughout the deformation history, it undergoes rotation and it is
a moot point whether the name principal should be ascribed to this
triad.

The distinction between *homogeneous* and *pure homogeneous* deformation
is exemplified in Figs. 1 and 2. Figure 1 illustrates the simple

---

[1]The superposition of a rigid body rotation on the deformation by
some other agency is not considered.

shear process where the initially orthogonal line pair, OA and OB, have been deformed to OA' and OB' respectively. The simple shear mode is a homogeneous process and the $\underset{\sim}{F}$ tensor is unsymmetric. The same shape change can be achieved by a pure homogeneous mode and this is demonstrated in Fig. 2. The process depicted in Fig. 1 is shown with greater clarity in Fig. 2(a), where B moves to B' after a shear displacement S and point A remains stationary. Also shown in the same diagram is a circle deformed into an ellipse.

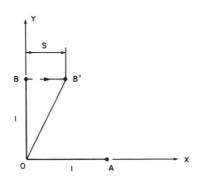

Fig. 1.   The simple shear process.

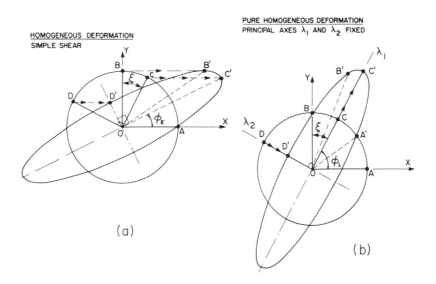

Fig. 2.   A distinction between a) Homogeneous deformation and b) Pure homogeneous deformation.

The pure homogeneous process is represented by Fig. 2(b). The orientation of the principal axes remain fixed and points lying along these axes are constrained to move in the principal directions only. Point B moves to B' and A to A', but point A does not move in a circular arc since OA changes in length during the process. However, it is unity both in the initial position and final position OA'. The deformed ellipses in Figs. 2(a) and 2(b) are identical in shape as are the configurations B'OA in Fig. 2(a) and B'OA' in Fig. 2(b). The *initial* orthogonal line pair, OC and OD are identical in each diagram. They remain orthogonal in Fig. 2(b) since they are aligned with the principal axes. This is not the case in Fig. 2(a), and although they start and finish orthogonal they do not remain orthogonal throughout the motion.

When $\underset{\sim}{F}$ is unsymmetric, this causes some problems in the evaluation of the strain, and recourse is made to techniques to devise a symmetric tensor. The polar decomposition theorem allows $\underset{\sim}{F}$ to be expressed as either $\underset{\sim}{F} = \underset{\sim}{R} . \underset{\sim}{U}$ or $\underset{\sim}{F} = \underset{\sim}{V} . \underset{\sim}{R}$. The tensors $\underset{\sim}{U}$ and $\underset{\sim}{V}$ represent pure deformation and are referred to as the *right* and *left* stretch tensor respectively, while $\underset{\sim}{R}$ provides rigid body rotation and $\underset{\sim}{R}^T = \underset{\sim}{R}^{-1}$. It follows that $\underset{\sim}{U}^2 = \underset{\sim}{F}^T . \underset{\sim}{F}$ and $\underset{\sim}{V}^2 = \underset{\sim}{F} . \underset{\sim}{F}^T$, and it is deemed that the eigenvectors of $\underset{\sim}{V}^2$ define the orientation of an orthogonal triad in the *current* configuration and those of $\underset{\sim}{U}^2$ define the orientation of the same triad in the *initial* configuration. To the associated strain ellipsoids, Hill (1970, 1978) has ascribed the name Eulerian (for current) and Lagrangian (for initial). For the two-dimensional case shown in Fig. 2, the orientation of the Lagrangian ellipse is ϕ_L and that and that of the Eulerian ellipse is ϕ_E. It can be shown that $\phi_L + \theta = \phi_E$, where θ is a measure of the rigid body rotation.

Real deformation processes do not occur, in general, by some combination of pure stretch followed by a rigid body rotation or vice versa. If follows from the polar decomposition theorem that either of the pure stretch tensors $\underset{\sim}{U}$ and $\underset{\sim}{V}$ when acting alone will give the same shape change as $\underset{\sim}{F}$. The same change in shape does not imply identical straining modes, as illustrated in Fig. 2. Sowerby and Chakravarti (1983) have demonstrated that the representative strain in a pure stretch process is always less than the accumulated representative strain in a homogeneous process which produces the same shape.

During the production of a sheet metal stamping it is not possible to view the deformation of a grid element. Only the initial and final configuration are known, and only by invoking the assumption of pure homogeneous deformation can the principal strains be determined simply. In general, this will result in an underestimate of the equivalent strain.

The present article demonstrates how the strain over a finite area of a sheet metal stamping can be evaluated from square or quadrilateral grids based on the assumption of pure homogeneous deformation.

METHOD OF ANALYSIS

Some Theoretical Fundamentals

Homogeneous deformation is a linear process where a vector $\underset{\sim}{X}$ is mapped into $\underline{x}$ according to:

$$\underline{x}(t) = \underline{F}(t) \cdot \underline{X}. \tag{1}$$

The coordinates of the vectors are measured relative to the same basis of the space and $\underline{F}(t)$ is the deformation gradient tensor. At any given time, t, $\underline{F}(t)$ is a 3 x 3 matrix whose determinant is strictly positive and hence admits the polar decomposition theorem

$$\underline{F}(t) = \underline{R}(t) \cdot \underline{U}(t) = \underline{V}(t) \cdot \underline{R}(t). \tag{2}$$

In the above equation $\underline{U}(t)$ and $\underline{V}(t)$ represent pure deformations and are referred to as the *right* and *left stretch* tensors respectively, while $\underline{R}(t)$ is an orthogonal tensor characterizing the rigid body rotation where

$$\underline{R}(t)^T = \underline{R}(t)^{-1} \quad \text{and} \quad \underline{R}(t) \cdot \underline{R}(t)^T = 1 \tag{3}$$

It follows from (3) and (2) that

$$\underline{U} = \underline{R}^T \cdot \underline{V} \cdot \underline{R}, \tag{4}$$

and hereinafter the notation indicating the dependence of these tensors on time is omitted. It also follows that

$$\underline{U}^2 = \underline{F}^T \cdot \underline{F} \qquad \text{(a)}$$

and $\quad \underline{V}^2 = \underline{F} \cdot \underline{F}^T \qquad \text{(b)} \left.\right\} \tag{5}$

Although $\underline{F}$ depends upon the basis selected, $\underline{U}$ and $\underline{V}$ do not. The eigenvalues of the stretch tensors in (5) are identical, and can be readily determined if the components of $\underline{F}$ are known.

As already mentioned, in sheet metal forming, surface grid markings reduce the strain determination to essentially a two-dimensional problem, and the components of F are determined by direct measurement. As a simple illustration~consider the in-plane deformation of an initially orthogonal pair of line elements, OA and OB, in Fig. 3. The deformed line elements are represented by OA' and OB', and for convenience are shown emanating from the original origin 0. The components of the deformation gradient tensor are shown in the diagram.

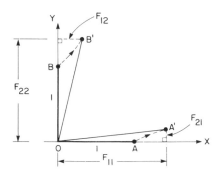

Fig. 3.   Components of the deformation gradient tensor.

In-plane deformation is not usually a general characteristic of sheet
metal forming operations. However, given that the nodal positions
of a grid (square or otherwise) are known in the flat plane (initial
configuration) and that their spatial positions can be measured in
the final configuration, the deformed line lengths can be evaluated.
Note it is still assumed that straight lines deform to straight
lines, and consequently the deformed and initial configurations can
be superimposed. Figure 4 illustrates a triangular element in the
initial and final configurations, and the two configurations can be
superimposed in any convenient but arbitrary fashion.

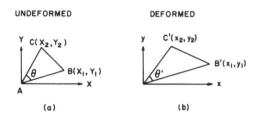

Fig. 4.   Undeformed and deformed triangular elements
          - homogeneous deformation.

The components of a possible deformation gradient tensor can now be
derived as follows:

$$x_1 = F_{11}X_1 + F_{12}Y_1$$

$$y_1 = F_{21}X_1 + F_{22}Y_1$$

$$x_2 = F_{11}X_2 + F_{12}Y_2 \tag{6}$$

$$y_2 = F_{21}X_2 + F_{22}Y_2$$

or in matrix form

$$
\begin{aligned}
x_1 &= \\
y_1 &= \\
x_2 &= \\
y_2 &=
\end{aligned}
\quad
\begin{vmatrix}
X_1 & Y_1 & 0 & 0 \\
X_2 & Y_2 & 0 & 0 \\
0 & 0 & X_1 & Y_1 \\
0 & 0 & X_2 & Y_2
\end{vmatrix}
\quad
\begin{vmatrix}
F_{11} \\
F_{12} \\
F_{21} \\
F_{22}
\end{vmatrix}
\tag{7}
$$

The inversion of the above matrix will yield the values of the four
coefficients

$$
\begin{vmatrix}
F_{11} \\
F_{12} \\
F_{21} \\
F_{22}
\end{vmatrix}
= D
\begin{vmatrix}
Y_2 & -Y_1 & 0 & 0 \\
X_2 & X_1 & 0 & 0 \\
0 & 0 & Y_2 & -Y_1 \\
0 & 0 & -X_2 & X_1
\end{vmatrix}
\begin{vmatrix}
x_1 \\
x_2 \\
y_1 \\
y_2
\end{vmatrix}
\tag{8}
$$

where $D = 1/(X_1 Y_2 - X_2 Y_1)$

The polar decomposition theorem is now invoked, and for convenience (5a) can be expressed as

$$\underset{\sim}{C} = \underset{\sim}{U}^2 = \underset{\sim}{F}^T \cdot \underset{\sim}{F} \tag{9}$$

Upon expanding (9) the components of $\underset{\sim}{C}$ are

$$C_{11} = F_{11}^2 + F_{21}^2$$

$$C_{12} = C_{21} = F_{11} F_{12} + F_{21} F_{22}$$

$$C_{22} = F_{12}^2 + F_{22}^2$$

Consequently the principal values (eigenvalues) of C can be evaluated in the usual manner where,

$$\lambda_{11}^2, \ \lambda_{22}^2 = \frac{C_{11} + C_{22}}{2} + \sqrt{(\frac{C_{11} - C_{22}}{2})^2 + C_{12}^2} \tag{10}$$

The principal logarithmic (natural) surface strains are given by

$$\varepsilon_{11} = \ln \lambda_{11} \quad \text{and} \quad \varepsilon_{22} = \ln \lambda_{22}, \tag{11}$$

while the third principal strain i.e. the thickness strain is furnished by the incompressibility assumption

$$\varepsilon_{11} + \varepsilon_{22} + \varepsilon_{33} = 0 \tag{12}$$

The representative or equivalent strain, $\bar{\varepsilon}$, is

$$\bar{\varepsilon} = \sqrt{\frac{2}{3}(\varepsilon_{11}^2 + \varepsilon_{22}^2 + \varepsilon_{33}^2)} \tag{13}$$

APPLICATION OF THE METHOD

The preceding analysis was applied to the deep drawn component shown in Fig. 5. The component is part of a compressor housing, produced in two stages. The material was a drawing quality steel. Figure 5 shows the partially formed component at the end of the first drawing operation.

An undeformed circular blank had been marked with a square grid pattern of 7.6 mm (0.30 in) side length. After deformation the nodes of the square grid were digitized with the aid of a milling machine (equipped with digital read-out scales), and the x, y, z coordinates of each node recorded. The known nodal positions permit the surface of the deformed part to be displayed graphically. A computer plot of a portion of the component illustrated in Fig. 5, is shown in Fig. 6, about one-quarter of the surface is displayed in the latter diagram. Since the coordinates of the undeformed and deformed nodes are known the strains in each element can be calculated.

Fig. 5. A deep drawn part.

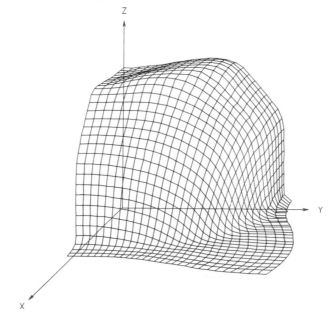

Fig. 6. A computer plot of a portion of the component
         shown in Fig. 5.

To facilitate the strain analysis and the plotting of the strain contours, each original square is divided across a diagonal. After deformation, each original triangle forms a small flat triangular facet on the surface of the stamping - at least, that is the only information the nodal points on the deformed surface would reveal. The principal strains in each triangle were determined using the technique described earlier. It will be recognized that after dividing each square across a diagonal, the nodes of each resulting triangle provide a common vertex to six contiguous triangles, see Fig. 7. Consequently, when assigning a strain measure to a nodal point, an unweighted average of the strain in each of the six adjacent triangles was computed. Once a strain value is assigned to a nodal point, a plotting routine is employed which executes linear interpolation between the nodes to produce strain contours at any desired level. Any strain quantity such as thickness strain, largest surface strain, equivalent strain, etc., can be plotted. Figure 8(a) shows contours of equivalent strain for the deformed portion of the stamping shown in Fig. 6. For convenience and clarity, the contours have been shown plotted in the flat plane, i.e. the undeformed plane. With any moderately sophisticated computer graphics system, it would be easy to encompass any area on the strain contour diagram of Fig, 8(a), and have the corresponding domain delineated on the deformed surface, such as Fig. 6. Contours of thickness strain, over the same region of the part, are shown in Fig 8(b); again the contours are plotted in the flat plane.

UNDEFORMED   DEFORMED

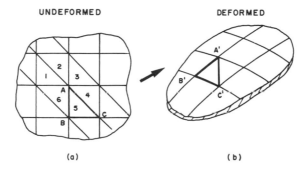

(a)   (b)

Fig. 7. Square and triangular grid elements.

Another way of presenting the strain information is illustrated in Fig. 9. The coordinate axes are the principal natural surface strains, ε_{11} and ε_{22}, and the points shown in the diagram are the calculated principal surface strains for each of the deformed triangles for the portion of the panel shown in Fig. 6. All points which lie on the line with a slope $\varepsilon_{11}/\varepsilon_{22} = -1$, indicate no change in surface area of an individual triangle. Consequently, a diagram like Fig. 9 helps the tool and die designer visualize how well the actual forming process maintains constant surface area of the original blank. Deformation without change in thickness can be regarded as an ideal deformation mode. Fracture due to excessive thinning is eliminated, although buckling failures are still possible. For the component under consideration there was no observable

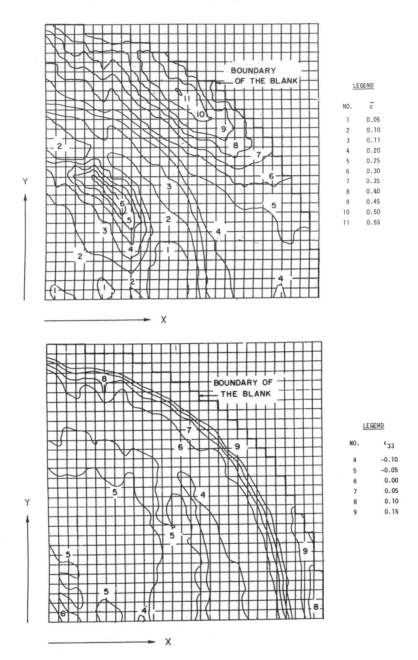

Fig. 8. a) Contours of representative strain for
             the part shown in Fig. 6.
          b) Contours of thickness strain for the
             same part.

evidence that buckling was imminent at any point on the surface, although some of the calculated surface strains i.e. the ε_{22} strains, were highly compressive in certain regions. The component was produced from relatively heavy gauge material, the initial blank thickness was 3.2 mm, and this would assist in resisting the propensity to wrinkle.

MAJOR STRAIN (ε1)

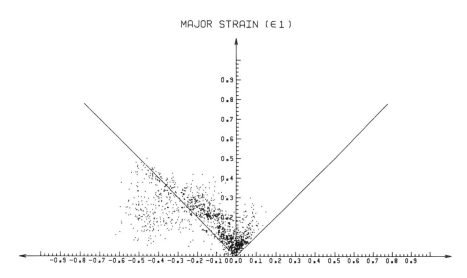

MINØR STRAIN (ε2)

Fig. 9. Principal natural surface strains in each
         of the triangular elements making up the
         surface shown in Fig. 6.

A forming limit diagram (F.L.D.), see Sowerby and Duncan (1971), can be superimposed on Fig. 9. Although, as is well known, a material does not possess a unique F.L.D., some type of a limit strain locus is generally regarded as a press-shop aid. Whether the F.L.D. has been determined theoretically or experimentally, from controlled laboratory tests, a comparison between the anticipated F.L.D. and the measured strains is available to the designer.

CONCLUSIONS

The article has provided the theoretical fundamentals for the evaluation of finite strain, and applied them to the determination of the strain over a large area of an industrial stamping. A case had been made for using a square or rectangular grid pattern, since the nodes of this type of grid are readily digitized. Regardless of the grid pattern employed, grid strain analysis invariably invokes the assumption of pure homogeneous deformation.

In the present work, the nodes of the grid were digitized using a milling machine, a more sophisticated coordinate measuring machine

would reduce the labour involved. A knowledge of the spatial co-
prdinates of the nodal points of the deformed grid permits a graphical display of the surface of the component. In turn, this leads
to an evaluation of the principal surface strains in triangular
elements which make up the surface. The calculated strain data can
be presented in a variety of ways, and some of these have been discussed.

The development of computer software packages is mandatory, in order
to perform the strain calculations and to display the strain contours
and the deformed component. The technique presented is applicable
to any sheet metal stamping and forms the basis for a fully automated
strain analysis package.

ACKNOWLEDGEMENTS

The author would like to thank his colleagues, Dr. J. L. Duncan and
Dr. E. Chu, who have been involved in the work on strain determination
in stampings since its inception at McMaster.

REFERENCES

Hill, R. (1968). *J. Mechs. Phys. Solids*, 16, 229-242.
Hill, R. (1970). *Proc. Roy. Soc.*, A314, 457-472.
Hill, R. (1978). *Advances in Applied Mechs.*, 18, 1-75.
Nemat-Nasser, S. (1979). *Int. J. Solids Structures*, 15, 155-166.
Nemat-Nasser, S. (1982). *Int. J. Solids Structures*, 18, 857-872.
Sowerby, R. and J. L. Duncan (1971). *Int. J. Mech. Sciences*, 13,
    217-229.
Sowerby, R. and P. C. Chakravarti (1983). *J. Strain Analysis*, 18,
    119-123.

DEVELOPMENTS IN THE USE
OF SUPERPLASTIC ALLOYS

G. W. ROWE

*Department of Mechanical Engineering, University of Birmingham,
Birmingham B15 2TT, UK*

ABSTRACT

A brief review is given of the current mechanical and metallurgical
explanations of the phenomena of superplasticity, with examples of
various alloys used in sheet-forming and mould-making applications.
Selection-test procedures are considered and a preliminary analysis
of die filling is presented.

INTRODUCTION

In his Review of Metal Working Plasticity, published in 1972, Johnson
drew attention to the remarkable phenomenon of superplasticity (Johnson, 1972). This had first been studied by Rosenhain in 1920 and
later by Russian workers since 1945 (Beresnev, Vereshchagin, Ryabinin
and Livshits, 1963), but it had only recently been recognised that
it could be put to practical uses (Pearce and Swanson, 1970; Fields
and Stewart, 1971; North, 1970; Lee and Backofen, 1967). Since
then the subject has been extensively studied and at least one Company is devoted exclusively to the manufacture of articles from superplastic sheet (Fig. 1). Many other Companies in U.S.A, U.K., Japan
and elsewhere produce superplastic alloys and components.

The special property that led to the name superplasticity was the
extraordinary ductility in tension that can be obtained. Under
appropriate circumstances a tensile test specimen can be elongated
by 1000% (Johnson, 1972) or even 2000% (Hosford and Caddell, 1983).
This contrasts strongly with more typical values of 50% for conventional annealed alloys.

The reason for this dramatic increase in the possible tensile extension is that the process of neck formation in a normal metallic tensile specimen is inhibited. The underlying cause is that a superplastic alloy exhibits an abnormally high dependence of yield stress
upon strain-rate. If therefore a neck were to form so that the deformation became localised, the local strain rate would be greatly
increased and the material in the neck would become correspondingly
stronger. This would encourage further deformation elsewhere, so
that a local neck does not in fact form.

A somewhat similar phenomenon occurs, for example, though for a quite different reason, in the stretching of a tensile specimen made of nylon. As the polymer extends it becomes oriented and exhibits a greater degree of crystallinity, which increases the strength of the drawn section. Again neck formation is prevented and very large permanent strains can be produced.

Since the superplastic materials also exhibit high ductility in bi-axial and more complex modes of stretching, they can be used to form deep cups of very complex pressings. In addition they are very soft, so it is possible to vacuum-form various shapes as in polymer forming.

It is not really correct to refer to superplastic alloys, though this is often done. Composition is important, since by no means all alloys exhibit this phenomenon, but structure, metallurgical condition and temperature are critical parameters. The grain size must be very small, typically below 10 micrometers, and this is conveniently obtained by using eutectic or eutectoid compositions. To utilise the properties to the full, deformation should be performed at or close to the corresponding temperature. For a representative Zn - 22A1 alloy this is about 275°C. It is also important that the imposed strain rate should be low, usually not exceeding 10^{-2} per sec. At room temperatures the same materials exhibit normal strength and ductility properties.

More recently it has been recognised that another important characteristic is that the hardness changes very rapidly with temperature. This too has found commercial applications, since a steel or other metallic pattern can easily be pressed into an alloy that is in its superplastic condition. A very precise replica is produced, which hardens as it cools so that it can be used as a mould for polymer injection or even for coining soft metals such as aluminium, brass or silver. (Rowe, 1982a).

Various other specialised products, such as a titanium aircraft panel with integral superplastically-formed internal webs, can be made (Hosford and Caddell, 1983).

THEORY OF INHIBITION OF NECK FORMATION IN SUPERPLASTIC ALLOYS

A more detailed explanation of the remarkable elongation without fracture can be produced by considering the influence of strain-rate sensitivity on the inhibition or distribution of local necks in a tensile specimen (Hosford and Caddell, 1983; Edington, Melton and Cutler, 1976).

In a normal tensile test, the flow stress of the specimen increases as the strain is increased, due to strain-hardening. At the same time, the area of cross-section decreases, so the stress applied to the section increases. As is well known from elementary text books, a maximum load is reached in the tensile test, beyond which an instability occurs. The rate of strain hardening becomes insufficient to compensate for the decrease in cross-sectional area, leading to fracture. In analytical terms, at the maximum load P

$$\frac{dP}{d\varepsilon} = 0 = \frac{d(A\sigma)}{d\varepsilon} = A\,\frac{d\sigma}{d\varepsilon} + \sigma\,\frac{dA}{d\varepsilon} \tag{1}$$

But in plastic deformation, volume is conserved so that

$$dV = Adl + ldA = 0 \tag{2}$$

$$\frac{dA}{A} = - \frac{d\ell}{\ell} = -d\varepsilon. \tag{3}$$

Hence from equation (1),

$$\frac{d\sigma}{d\varepsilon} = - \frac{\sigma}{A} \frac{dA}{d\varepsilon} = \sigma \tag{4}$$

When the applied stress increases beyond the current value of $d\sigma/d\varepsilon$ the deformation becomes unstable and a local neck forms.

In a superplastic condition however, strain hardening is a minor factor and the strength of the alloy is determined by the strain rate.

Suppose that a tensile specimen of uniform cross-sectional area A_u is starting to form a neck that has reduced the cross-section to A_n. Since the same tensile force is transmitted by both sections:

$$A_u \sigma_u = A_n \sigma_n \tag{5}$$

where the suffices apply to the relevant stresses. If a further tensile strain is imposed, A_u will decrease to A'_u and A_n to A'_n and the relevant compressive strains are defined as

$$\varepsilon_u = - \ln(A'_u/A_u)$$
$$\varepsilon_n = - \ln(A'_n/A_n) \tag{6}$$

or

$$A_u' = A_u e^{-\varepsilon u}$$
$$A_n' = A_n e^{-\varepsilon n} \tag{7}$$

It is usual to assume a power-law strain-rate dependence

$$\sigma = \sigma \, (\dot{\varepsilon})^m \tag{8}$$

where m is the strain-rate exponent, otherwise known as the strain-rate sensitivity index.

Thus, combining equations 5, 7 and 8

$$A_u e^{-\varepsilon u}(\dot{\varepsilon}_u)^m = A_n e^{-\varepsilon n}(\dot{\varepsilon}_n)^m \tag{9}$$

The neck can be described by an inhomogeneity factor

$$f = A_n/A_u \tag{10}$$

Making this substitution, raising both sides to the power 1/m and integrating with respect to time leads to

$$\int_0^{\varepsilon_u} e^{-\varepsilon_u/m} \, d\varepsilon_u = f^{1/m} \int_0^{\varepsilon_n} e^{-\varepsilon_n/m} \, d\varepsilon_n \tag{11}$$

G. W. Rowe

$$e^{-\varepsilon_u/m} - 1 = f^{1/m} (e^{-\varepsilon_n/m} - 1) \qquad (12)$$

If for example, the inhomogeneity factor f is 0.98 soon after the start of a neck, and m = 0.1 as in a normal material the uniform strain occurring for an increase in neck strain to 0.1 is given by:

$$\varepsilon_u = 0.073 \qquad (13)$$

For a superplastic material with m = 0.6,

$$\varepsilon_u = 0.097 \qquad (14)$$

Even at this small strain the spread of deformation in the super-plastic alloy is evident.

For higher neck strain the difference is much more obvious e.g. at $\varepsilon_n = 1.0$

for   m = 0.1 $\varepsilon_u = 0.17$ $\qquad (15)$

         m = 0.6 $\varepsilon_u = 0.922$ $\qquad (16)$

In the limit if $\varepsilon_n \to \infty$, $\varepsilon_u = 2.04$ and the nominal strain e_t is about 700%.

METALLURGICAL CONDITIONS FOR SUPERPLASTICITY

At high rates of strain or low temperatures the strain-rate sensi-tivity of the superplastic alloys is low, with m ≅ 0.2. This is typical of thermally-activated slip, and the materials behave like normal alloys (Cottrell, 1967). If however the strain rate is low and the temperature is in the correct range, much higher values are found, as stated above, with m ≅ 0.6. Three possible explanations have been suggested.

One is that the process of deformation is essentially diffusional creep, with vacancies migrating from grain boundaries that are normal to the tensile axis and reaching boundaries parallel to that axis. This in effect increases the lengths of the grains in a direction parallel to the tensile axis and decreases their width. The limiting value of m for a process of this type would be 1.0, as for viscous flow. While this mechanism could account for high m values, the evidence is that the grains remain approximately equi-axed during superplastic deformation.

A more widely accepted interpretation is that the process is essen-tially one of grain-boundary sliding. This would not require pre-ferential deformation of the grains, and again would have a limiting value of 1 for viscous sliding.

Both theories require very fine grain size and high temperatures with low strain rates. In practice, because the crystals tend to grow at the operating temperatures, it is necessary to stabilise the grains, usually by addition of a second phase, but some single-phase alloys, such as Sn 1Bi, can show superplastic properties.

A third view is that the superplasticity is due to the transitional nature of the grain structures. This has been found, for example,

in iron cycled through its α - γ transition (Bond, 1979). At or
close to a eutectoid transformation temperature, the structure is
mobile and changes easily from one form to another, thus facilitating
mechanical movement. Opponents of this view suggest that it would
imply a more critical dependence on temperature at the eutectoid
value than is observed, so the explanation cannot be generally valid.

Nevertheless, eutectoid compositions are favoured for superplastic
alloys, but it is argued by the grain-boundary-sliding school of
thought that this is simply for convenience of obtaining the very
fine grain size necessary for grain boundary sliding. Not all superplastic alloys have such composition.

EXAMPLES OF SUPERPLASTIC ALLOYS

Many alloys can be brought into a superplastic condition by suitable
mechanical and thermal treatment. Examples are Sn-40Pb, Al-33Cu,
some low to medium-carbon steels, zirconium and titanium alloys,
stainless steels and even superalloys and tool steel.

Zinc-Aluminium alloys

The most widely publicised of the early alloys was based on the Zinc-
Aluminium system which has a eutectoid composition with about 22% Al
at 275°C.

Early workers (Backofen, Turner and Avery, 1964; Baudelet, 1974;
Belk, 1974; Naziri and Pearce, 1973) studied the strain-rate sensitivity and concluded that an equiaxed fine-grained and stable microstructure could be produced by quenching from 375°C to 0°C, i.e. from
well above the transformation temperature. The physical metallurgy
has been extensively discussed (Alden, 1968). It is found that the
binary alloy has rather poor mechanical properties, especially its
room-temperature creep resistance, which can be markedly improved
by adding a third constituent (Naziri and Pearce, 1970) or, better
still, by forming a quaternary compound 78Zn-21Al-1Cu/Mg, developed
by Belk (1976) and known as ZAM.

These alloys were rolled into sheets for the various deep drawing
and pressing applications for which they are very suitable. The
main deficiency for practical purposes was the tendency to creep,
even under their own weight, at warm room temperatures and their
susceptibility to corrosion. They have however been exploited for
casings, car body construction, etc.

Aluminium-Silicon alloys

Aluminium-silicon alloys have better creep resistance and better
corrosion resistance, but they have not been widely used, partly
because higher temperatures are needed for their superplastic deformation. They are also unsuitable for highly-stressed applications
because they are less ductile than aluminium or duralumin (Grimes,
Stowell and Watts, 1976). They have recently proved useful for hot
forging of moulds, as discussed below.

Aluminium-Bronzes

Certain copper alloys containing 5-10% aluminium and minor additions

G. W. Rowe

of Mn, Fe or Ni are remarkably wear-resistant and can be used as
drawing dies, especially for titanium tubes, or for decorative purposes as cutlery and other household items requiring wear and tarnish
resistance. These alloys can be brought into a superplastic state
which enables them to be formed into intricate shapes.

Superplastic Stainless Steel Sheet

Superplasticity is by no means confined to low-temperature alloys.
A micro-duplex 26Cr-6.5Ni stainless steel stabilised with Ti has the
requisite very fine grain size. (The detailed composition is 26Cr-
6.5Ni-0.06C-0.3Ti-0.4Mn-0.4Si). At the forming temperature (800°C+)
tensile elongations of 500% can be obtained, but at room temperature
the material reverts to a normal condition, showing a proof stress
of 500 N/mm² and UTS of 780 N/mm² with an elongation of 35%. It has
good fatigue resistance and corrosion resistance, comparable with
other stainless steels (Smith, 1976).

It is well known that stainless steel is normally difficult to form
because of problems of lubrication which are avoided to a large extent by thermoforming superplastically. This is of course an advantage additional to the great depths and complexity obtainable by all
superplastic sheet forming techniques.

Superplastic Titanium Alloy Sheet

The grain structure of Ti-6Al-4V is predominantly equiaxed and very
fine, less than 10 microns in average diameter (Budillon, 1977).

This material can be produced by forging in the α-β range and subsequently heat treating in the β domain. It has been shown that
even TIG weldments can deform superplastically (Baudelet, 1974).
During this deformation the grains tend to grow. In the weld region
the acicular grains become more equiaxed and therefore more like the
parent material. (Overall elongations of about 400% have been obtained). This is contrary to the observations with eutectic and
eutectoid alloys which show that a dendritic structure is unfavour-
able for superplasticity. The explanation may be that the α titanium grains slide over their interface with the more ductile
grains.

To obtain the desired structure, the alloy was solution treated at
1020°C, quenched in water and reheated to 730°C. The samples were
then tested at 850-900°C with strain rates from 3×10^{-3} to 3×10^{-1} per
sec. Values of m between 0.45 and 0.6 were found (Lee and Backofen,
1967).

Superplastic Aluminium-based alloys

Grimes, Stowell and Watts (1976) have shown that it is not essential,
as previously thought, to use eutectic or eutectoid compositions to
produce superplasticity in aluminium alloys. In all of these the
volume fraction of the second phase, Cu Al_2 or Si in Al-33Cu, Al-Si,
Al-Cu-Mg and others, is of the order of 50%. This was found necessary
to ensure a sufficiently fine uniform distribution to inhibit the
grain growth of the α Al.

Unfortunately Al-33Cu is not strong enough at room temperature to be
used for sheet-formed products and Al-Si is relatively brittle at
its forming·temperature.

To make a superplastic alloy that is close in properties to ordinary duralumin, an Al-6Cu-0.5Zr alloy was cast with as fine a structure as possible, the Zr acting as a grain refiner during casting and also inhibiting grain growth due to recrystallisation. By suitable casting practice, with a very high cooling rate, followed by appropriate thermo-mechanical treatment, specimens with m = 0.5 could be obtained. Tensile elongations of 1000% or more were recorded at 480°C and $\dot{\varepsilon}$ = 3 x 10^{-3}/sec, with a flow stress of 9 N/mm^2. Other alloys such as Al-5Zn-1Mg-0.5Zr have also been produced with superplastic properties.

Sheets of these materials can be thermoformed and a wide range of superplastically-formed aluminium alloy components is now manufactured commercially (Figure 1).

Fig. 1. A lower engine cover from a Turbo-Prop air-
         craft formed in Superplastic Aluminium
         alloy. (Courtesy of Superform Metals Ltd.)

SUPERPLASTIC ALLOYS FOR PRECISION FORMING

Saller and Duncan (1971) observed that detail of a metal master could be replicated accurately by coining in a superplastic alloy. Katyal, Rowe and Wilson (1971) proposed that this could be utilised for the production of inexpensive superplastic tooling, since the hardness of some superplastic alloys, especially those related to ZAM, increased rapidly as temperature was reduced.

Compression Tests

Tests have been carried out to determine suitable forging schedules for precision forming of a high strength Zn-Al alloy (Balliett, Foster and Duncan, 1978), and for a compression moulding process (Fields & Stewart, 1970) with eutectoid Zn-Al at 275°C, as used by Saller and Duncan (1971) in hot hobbing tests. Excellent die filling was found to be possible in closed-die forging at much lower loads than with conventional alloys (Stewart, 1973).

Balliett, Foster and Duncan, (1978) used HSZ at 260°C for forging at normal ram speeds but with a substantial dwell time under load. Most of the forging occurred in the early part of the cycle but die filling was critically dependent upon dwell time. The alloy used was Zn-22.1 Al-0.95 Cu-0.045 Mg-0.025 Ca, continuously cast into 5in diameter billets, solution treated at 365°C for 48 hours, water quenched and subsequently extruded at 260°C into suitable rods. The average grain size was 3µm. A strain-rate sensitivity value m ≅ 0.4 was obtained and the relationship

$$\sigma = 172 \; \dot{\epsilon}^{0.4} \; \text{MPa}$$

was established. At room temperature, the alloy had a tensile strength of 414 MPa, 19% elongation and a hardness 64RB. It was observed that it was difficult to fill a peripheral lip and that a cavity sometimes appeared at the base of a cup.

These authors introduced a pressure-compensated time parameter as follows. Assuming $\sigma = K\epsilon^m$, the activating mean pressure for compression of a billet of height L is

$$\bar{p} = K\left(\frac{1}{h}\frac{dh}{dt}\right)^m$$

or $$(\bar{p}/K)^{1/m}dt = -\frac{dh}{h}. \tag{17}$$

Thus $$\int_0^t (\bar{p}/K)^{1/m}dt = \int_{h_0}^h -\frac{1}{h} \, dh \quad \text{or} \quad \left(\frac{\bar{p}}{K}\right)^{1/m} t = \ln\frac{h}{h_0} \tag{18}$$

since $\bar{p}$ is constant.

It is then postulated that

$$h = \phi \left\{ t\left(\frac{\bar{p}}{K}\right)^{1/m} \right\} \tag{19}$$

for any other geometrical form where f is a function of the tool geometry.

Flow stress and Activation Energy

The essential property for utilisation of superplasticity in mould making is not the ductility and resistance to neck formation but the steep dependence of flow stress or hardness on temperature. Figure 2 shows some characteristic values for normal and superplastic alloys (Özmen, 1981). In making a mould the flow stress should be as low as possible during the forming stage to allow replication of intricate detail but as high as possible at the subsequent working temperature, to resist deformation and wear.

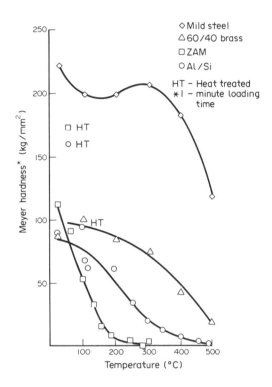

Fig. 2.   The steep fall in hardness of a zinc-based
           superplastic alloy with temperature, com-
           pared with the hardness of brass and mild
           steel.

Since the material is strain-rate sensitive, as above,

$$\sigma = C \, \varepsilon^n \, \dot{\varepsilon}^m. \qquad\qquad (20)$$

The strain-hardening exponent n falls rapidly with increasing temperature while the strain rate exponent m increases.

A commonly accepted quantitative method of considering temperature and time dependence is that proposed by Zener and Holloman (1944). This assumes plastic straining is a rate process with an associated activation energy Q, such that

$$\dot{\varepsilon} = A\ e^{-Q/RT} \tag{21}$$

where R is the gas constant and T the absolute temperature. The constant of proportionality A is both stress and strain dependent. If a constant strain is considered, A is a function of stress alone

$$A = A(\sigma) = \dot{\varepsilon}e^{Q/RT} \tag{22}$$

or $\sigma = f(Z)$

where $Z = \dot{\varepsilon}\ e^{Q/RT}$ is the Zener-Holloman parameter. By plotting the strain rate to produce a given stress at given temperature against the reciprocal of temperature an approximately straight-line plot is obtained (Fig. 3).

$$(\dot{\varepsilon})_{\sigma,T} \propto \frac{1}{T} \tag{23}$$

The slope of this line is Q/R, from which the activation energy can be deduced.

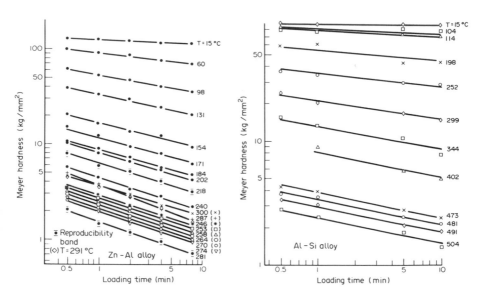

Fig. 3.   Activation energy plots for a zinc-aluminium
          alloy showing the linear relationship between
          indentation pressure and dwell time, on a
          logarithmic basis.

This formulation is found to be valid over small ranges but there is a basic inaccuracy in assuming that the activation barrier is overcome solely by the temperature as in the Arrhenius law applied to other systems. Plastic deformation also involves energy dissipation directly associated with the stress. A more detailed treatment (Jonas, Sellers and Tegart, 1969) suggests that

$$\dot{\varepsilon} = A(\sinh(\alpha\sigma))^{1/m}e^{-Q/RT} \tag{24}$$

gives a better representation, and indeed close correlations have been found over large strain ranges.

Hot Hardness Tests

The deduction of an activation energy from hot tensile tests is arduous and consumes a large number of specimens. Moreover, it is not known that tensile results would apply equally to compression which is encountered in bulk forming of moulds.

Compression tests at high temperatures are notoriously unreliable because of the difficulty of controlling the friction conditions, especially when straining at the very low rates used for superplastic mould making.

Hot hardness testing has proved to be simple (Tabor, 1951; Mulhearn and Tabor, 1960), requiring no special specimen preparation and being more or less independent of the frictional conditions, provided that these are not deliberately changed. Many hardness tests can be made on one small sample of material, and an additional feature is that the temperature can be very accurately controlled in a small oil bath (Özmen, 1981).

To perform such a test, a ball or cone is pressed into the surface under a load W for a known time. The plan diameter d will increase with time in a creep situation and the effective Meyer hardness p at any stage is given by

$$p = 4W/\pi d^2 \tag{25}$$

Using the Arrhenius equation (or the more elaborate stress-dependent version)

$$\dot{\varepsilon} = A \sigma^n e^{-Q/RT} \tag{26}$$

In hardness testing with a ball indenter of diameter D, it has been found (Tabor, 1951) that the strain is a linear function of d/D

$$\varepsilon = K_1 (\frac{d}{D}) \tag{27}$$

This is not strictly accurate. Furthermore the strain is not uniform around a hardness indentation, but the approximation is widely accepted in hardness testing.

Thus $$\varepsilon = \frac{K_1}{D} (\frac{4W}{\pi p})^{\frac{1}{2}} \tag{28}$$

$$\frac{d\varepsilon}{dt} = - \frac{K_1}{2D} (\frac{4W}{\pi})^{\frac{1}{2}} p^{-3/2} \frac{dp}{dt} \tag{29}$$

It is also known that p is a simple multiple K_2, approximately 3, of the flow stress σ. So from equation (26),

$$\frac{d\varepsilon}{dt} = A(\frac{1}{K_2})^n p^n e^{-Q/RT}. \tag{30}$$

Thus using equation (29)

$$p^{-(n+3/2)} \, dp = C_1 \, \frac{D}{W^{\frac{1}{2}}} \, e^{-\Omega/RT} \, dt \tag{31}$$

where $C_1 = \dfrac{A\sqrt{\pi}}{K_1 K_2^{\, n}}$

Integrating at constant temperature, assuming that the hardness number approaches infinity for a loading time equal to zero,

$$\int_{\infty}^{p} p^{-(n+3/2)} \, dp = \int_{0}^{t} C_1 \, \frac{D}{W^{\frac{1}{2}}} \, e^{-\Omega/RT} \, dt \tag{32}$$

or $p^{-(n+1/2)} = C_2 \, \dfrac{D}{W^{\frac{1}{2}}} \, e^{-\Omega/RT} t \tag{33}$

where $C_2 = {}^{-(n+1/2)} C_1 .$

Thus if a test is performed with a ball of known diameter D under a fixed load W at constant temperature T, a plot of ln p against ln t should be linear, with a slope -(n + 0.5), as shown by Mulhearn and Tabor (1960).

By performing experiments at a series of temperatures, they calculated the activation energy from the horizontal intercepts between two lines at selected values of p:

$$\ln t_1 - \ln t_2 = \frac{Q}{R} \left(\frac{1}{T_1} - \frac{1}{T_2} \right) \tag{34}$$

Typical values were 16 and 28 Kcal/deg.gm.mol respectively for indium and lead at about 100°C. Similar measurements by Ozmen (1983) using a ZAM-type superplastic alloy showed 24 kcal/deg.gm.mol at 264°C, near the hardness minimum, rising to about 60 kcal/deg.gm.mol at room temperature.

MOULDING IN SUPERPLASTIC ALLOYS

Zinc-aluminium Alloys

Figure 2 shows the very low hardness of a ZAM type alloy (Zn-21Al-0.5Cu-0.5Mg) at its superplastic temperature compared with copper or mild steel. As the temperature is reduced, especially below 150°C, the hardness increases steeply, until at room temperature it may reach 130HV, or with special treatment as much as 160HV, harder than mild steel.

This is sufficient to stamp replicate medallions in soft silver, gold or even copper. At sub-zero temperatures the hardness of ZAM continues to increase steeply and stronger alloys such as cutlery silver can be stamped. An advantage of these superplastic alloys is that the dynamic hardness increases with the strain rate, even at room temperature. The faster the operation is carried out, the harder the material appears to be.

Such dies are very easy to make from a metal master pattern by a
creep-forming or hot-hobbing process. Usually an isothermal process
is used with the dies being heated to about 265°C, though lower temperatures can be used (Rowe, 1982). The pressure is applied slowly
and held for 10-15 minutes to ensure complete replication of fine
detail (Fig. 4). A remarkable feature is that a mirror-finish can
be obtained by using a highly polished master, without the need to
polish the replica.

Fig. 4.   A 180mm diameter replica plaque made in a
          superplastic zinc-aluminium alloy, showing
          the degree of detail that can be reproduced
          accurately.

It is usually much easier to make and polish a convex or protruberant
shape than a cavity, so the process is well suited to the production
of moulds for polymer forming. An example is shown in Figure 5. A
single metal master is made and polished if necessary, being then
used to form any number of cavities. These can then be incorporated
in a multiple mould for high production rates.

It is also possible to make a positive replica by using two different
superplastic alloys. This is shown in the product (Fig. 6) made in
ZAM which was formed at 260°C in a cavity produced in an aluminium
silicon alloy. This cavity has been pressed at 400°C using an original steel master. The difference is size between the master and

the replica made in two stages is very small, being due only to the
differences in thermal expansion coefficient and the temperature
changes T for the alloys involved.   Thus, taking the above example
and assuming a room temperature of 20°C:

$$\text{Steel master} \quad \alpha = 11 \times 10^{-6}, \quad T = +380, \quad \epsilon = +4.2 \times 10^{-3}$$
$$\text{Al-Si negative} \quad \alpha = 20 \times 10^{-6}, \quad T = -120, \quad \epsilon = -2.4 \times 10^{-3}$$
$$\text{ZAM positive} \quad \alpha = 25 \times 10^{-6}, \quad T = -240, \quad \epsilon = -6.0 \times 10^{-3}$$

The net dimensional change is thus a contraction of about 4×10^{-3}
mm/mm at room temperature.   Because the coefficient of expansion of
brass is 18.9 compared with 11.6 for steel, the contraction is less
if a brass master is used.   Brass has proved entirely satisfactory,
but in principle it would be expected that the wear of a steel master
would be less.   No measurable wear has been found in the trials so
far with several hundreds of pressings.

Fig. 5.   A group of cavities made in zinc-based
          superplastic alloy for making polyethylene
          components of a valve.

Measurements after each 100 mouldings showed no wear, even in the
inlet gate, after 1600 shots.   Subsequent trials have demonstrated
that 200,000 polypropylene hooks could be made in a two-impression
mould without measurable tool wear in the mould cavity or the gate,
though 10^{-4} in (0.002mm) should have been detectable.   All the products were well within the normal tolerances for injection moulding.

Fig. 6.   A group of positive replicas made in a zinc-
          based alloy by pressing at 250°C into a cav-
          ity, itself made by hot-pressing of a steel
          master into an aluminium-silicon alloy at
          500°C.

Superplastic Hobbing of Die Steels

In normal cold hobbing the workpiece strain hardens and very high
loads are required to form even shallow coining impressions. If the
material is heated sufficiently to anneal it and keep it soft, oxi-
dation is likely to occur with loss of detail, surface quality and
accuracy.

Superplasticity offers the possibility of forming dies and other
tools at moderate temperatures, with good replication of the master
pattern and little wear of the hobbing tool.

Pearce and his associates have shown that this is feasible with a
quenched and retempered BS224 No 5 die steel, which is a medium carbon martensitic direct-hardening steel. This is heat-treated to
produce the finest possible grain size which is structurally stable.
At 685°C it shows a maximum strain-rate sensitivity index m of about
0.5.   Certain other steels reach even 0.55 (Naziri and Pearce, 1970).
An unfortunate feature at present is that the hardness at room temperature is only 280-300 HV, but the material can be re-hardened and

tempered to "switch off" the superplasticity. The flow stress in
the superplastic condition, at 700°C with a strain rate ε = 5 x 10^{-4}/
sec, is about 75 MPa. This means that a very strong high temperature
alloy is required as the hobbing master. Nimonic 90 is very suitable
having a yield stress of some 650 MPa under these conditions.

Satisfactory moulds for 0-ring manufacture and some coining dies have
been produced in this way, using a punch speed of 0.25mm/min (Miller
and Pearce, 1981).

The principal limitation at present is the size of billet that can
be uniformly heat-treated to produce the required properties.   30-50
mm blocks can be handled.

A PRELIMINARY ANALYSIS OF DIE-FILLING

One of the major problems in any coining operation is to obtain complete filling and replication of all the low relief cavities on the
surface. If this is not done, the sharpness of the image will be
lost. It is usually necessary to allow some lateral flow of metal
to enhance the filling, and this certainly reduces the pressure required. If however too much flow is permitted the formed protuber-
ance may be sheared or even cut off by the passage of metal beneath
it.

Superplastic alloys are very suitable for the production of intricate
surface detail but because of their exceptionally low creep flow
stress at their forming temperature, they are particularly susceptible to this shearing defect even before the cavity has completely
filled. This can be seen in the periphery of the medallion in Fig.
4.

Özmen (1983) has studied this problem using a simplified plane-strain
model illustrated in Figure 7, so that slip-line field theory can be
introduced. Experiments were first performed using plasticine on
which a square grid had been scribed. A flat die containing one or
more orifices compressed a rectangular billet against a flat platen.
The left-hand side was constrained by a wall, or at a centre-line,
while the billet was able to extend outwards to the right under the
restraint of various friction conditions on the tool surfaces.

From the experimental results, it is possible to propose the slip-
line field shown. The large dead zone in the corner is easily identified but the detailed flow around the orifices was obscured in the
experiments.

This slip-line field has been drawn to maintain orthogonality of the
slip-lines and to satisfy boundary conditions of sticking friction.
A compatible hodograph can be drawn, but it is necessary to make some
assumption about the magnitude of the horizontal velocity or about
the velocity into the orifice. Once either is specified the
other is of course determined by volume constancy but it has not
proved possible to obtain a unique solution. Energy calculations
suggest that either all the metal flows outwards, for zero friction
conditions, or all flows into the orifices when friction is finite.

Upper-bound calculations are also ambiguous, leading to a similar
conclusion. The energy minimum for various velocity ratios is a
flat one.

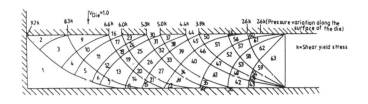

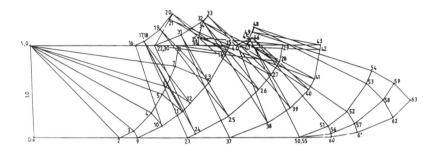

Fig. 7.   A proposed slip-line field and hodograph
for side pressing with a perforated die.
The solution is incomplete because the
outlet velocities 16, 30, 44 are not uniquely defined.

This raises an interesting general problem for situations in which
flow in two orthoginal directions is possible but where there is no
flow divide. A related problem of extrusion-forging with a die containing a hole and a flat platen has been solved by both slip-line
fields (Newnham and Rowe, 1973) and upper bound methods (Kobayashi,
1971), but this differs in having a flow divide or at least a continuous dead zone from top to bottom surfaces.

Very similar problems arise in the flow of water from lateral orifices
fices in a pipe (Keller, 1949) and the flow of gas at both normal
and supersonic speeds across the mouth of an orifice. Although
solutions to these have been published, including an analysis by
characteristics, (Deckker and Yang, 1975), resembling slip-line
field theory, it has been necessary in all instances to invoke an
empirical efflux coefficient. This appears to be an area for further
development of metalworking theory, possible using finite-element
plasticity analysis.

CONCLUSIONS

Although the phenomenon of superplasticity has been known for more
than sixty years, it is only during the past fifteen years that its
practical applications have been appreciated. Superplastic alloys
in sheet form are slowly gaining acceptance in industry, particularly

where superplasticity can be induced in alloys that can also show
normal properties. A few companies are now using superplastic alloys
in block form for mould making and other purposes.

REFERENCES

Alden, T. H. and Schadler, H. W. (1968). The influence of structure
    on the flow stress-strain rate behaviour of Zn-Al alloys. *Trans.
    Met. Soc. AIME*, 242, 825-832.
Backofen, W. A., Turner, I. R. and Avery, D. H. (1964). Superplas-
    ticity in an Al-Zn alloy. *Trans. A.S.M.*, 57, 980-990.
Balliett, R. W., Foster. J. A. and Duncan, J. L. (1978). Precision
    forging of a high-strength superplastic zinc-aluminium alloy.
    Metall. Trans., 9A, 1259-1264.
Baudelet, B. (1974). Temporary superplasticity in conventional
    alloys. *Metals and Matls.*, 8, 117-118.
Belk, J. A. (1974). Indentation tests for superplasticity. *Metals
    and Matls.*, 8, 414-415.
Belk, J. A. (1976). Production of superplastic zinc-aluminium alloys.
    Metals Technol., 3, 161-166.
Beresnev, B. I., Vereschchagin, L. F., Ryabinin, Yu N. and Livshits,
    L. D. (1963). Some problems of large plastic deformation at high
    pressures. *Pergamon Press.*
Bond, M. (1979). Superplastic alloys. *Engineering* (Oct) 1-6.
Budillon, E. and Lechten, J. P. (1977). Aspects fondamentaux de la
    superplasticité. Rapport d'Aerospatiales, Suresnes.
Cottrell, A. H. (1967). An introduction to metallurgy. *Edward
    Arnold Ltd.*, London.
Deckker, B. E. L. and Yang, A. W. (1975). The glancing collision of
    two shock waves in a branched duct. *Proc. I. Mech. E.*, 189, 293-
    303.
Edington, J. W., Melton, K. N. and Cutler, C. P. (1976). Superplas-
    ticity. *Prog. Mat. Sci.*, 21, 63-170.
Fields, D. S. and Stewart, T. J. (1971). Strain effects in the super-
    plastic deformation of 78Zn-22Al. *Int. J. Mech. Sci.*, 13, 63-75.
Grimes, R., Stowell, M. J. and Watts, B. M., (1976). Superplastic
    aluminium-based alloys. *Metals Technol.*, 154-160.
Hosford, W. F. and Caddell, R. M. (1983). Metalforming. *Prentice-
    Hall Inc.*, Engelwood Cliffs.
Jonas, J. J., Sellars, C. M. and Tegart, W. J. McG. (1969). Strength
    and structure under hot working conditions. *Met. Rev.*, 14, 1-24.
Johnson, W. (1972). Review of metalworking plasticity. 2: Hydro-
    static forming and superplasticity, *Metallurgia & Metal Forming*,
    April, 128-129.
Katyal, A., Rowe, G. W. and Wilson, D. V. (1971). The feasibility
    of using extruded superplastic billets for throw-away die and mould
    manufacture, *Proc. Illinois Inst. Tech. Symp.*, "New developments
    in tool manufacture and applications", Illinois U.S.A.
Keller, J. D. (1949). The manifold problem. *J. Appl. Mech. Trans.
    A.S.M.E.*, 16, 77-85.
Kobayashi, S. (1971). Theories and experiments in friction, defor-
    mation and fracture in plastic deformation processes. In *Metal
    forming: interrelationship between theory and plastic* ed. Hoff-
    manner, A. L. Plenum Press, New York.
Lee, D. and Backofen, W. A. (1967). *Trans. Am. Met. Soc. Inst. Mech.
    Engrs.* 239, 1034-1040.
Miller, E. J. W. and Pearce, R. (1981). Superplastic hobbing.
    Metallurgia, 206-210.
Mulhearn, T. O. and Tabor, D. (1960). Creep and hardness of metals:
    a physical study. *J. Inst. Metals.* 89, 7-12.

Naziri, H. and Pearce, R. (1970). The influence of copper additions on the superplastic forming behaviour of the Zn-Al eutectoid. *Int. J. Mech. Sci.*, 12, 513-521.

Naziri, H. and Pearce, R. (1973). Some observations on the behaviour of superplastic zinc-aluminium eutectoid over a wide range of strain rates. *J. Inst. Metals*, 101, 197-202.

Newnham, J. A. and Rowe, G. W. (1973). An analysis of compound flow of metal in a simple extrusion-forging process. *J. Inst. Metals*, 101, 1-9.

North, D. (1970). Superplastic alloy for autobody construction. *Sheet Metal Ind.* 47, 13-16.

Özmen, S. (1981). Forming of tools in superplastic alloys. *M.Sc. Diss.* Birmingham.

Özmen, S. (1983). Moulds and dies in superplastic alloys. *Ph.D. Diss.* Birmingham.

Pearce, R. and Swanson, C. J. (1970). Superplasticity and Metal Forming, *Sheet Metal Ind.* 47, 599-603.

Rowe, G. W. (1982a). Mould making in superplastic alloys. *Metals. Soc., World*, 1, 2-3, 12.

Rowe, G. W. (1982b). Making moulds of superplastic alloy. *U.K. Pat. App. GB 2099743A.*

Saller, R. A. and Duncan, J. L. (1971). Stamping experiments with superplastic alloys. *J. Inst. Metals*, 99, 173-177.

Smith, C. I. (1976). Superplastic deformation of a microduplex stainless steel. *Met. Sci.* May, 1976.

Stewart, M. J. (1973). Superplastic forging of Zn-Al-Cu alloys. Canadian Met. Quarterly, 12(2), 158-169.

Tabor, D. (1951). Hardness of metals, *Oxford Univ. Press.*

Zener, C. and Holloman, H. H. (1944). Effect of strain rate upon plastic flow of steels. *J. App. Phys.*, 15, 22-32.

SHAKEDOWN OF TUBES: A THEORETICAL ANALYSIS AND EXPERIMENTAL INVESTIGATION

O. MAHRENHOLTZ*, K. LEERS**, J. A. KÖNIG***

*Arbeitsbereich Meerestechnik II/Strukturmechanik, TU Hamburg-
Harburg, Federal Republic of Germany
**Institut für Mechanik, Universität Hannover,
Federal Republic of Germany
***Institute of Fundamental Technological Research,
Warsaw, Poland

ABSTRACT

The paper contains an outline of the classical shakedown theory and
an approximate solution for the investigated shell. The experiments,
aimed at modelling the elastic-plastic response of nuclear fuel
cladding shells have been carried out in a completely automatic way
in contrast to other shakedown experiments made hitherto.

INTRODUCTION

Tubular structural elements are frequently used in various structural
systems. A cylindrical tube is an equally good element for carrying
internal pressure, longitudinal forces and bending or twisting moments. Tubes used to transport gas or liquid are often subjected to
a temperature gradient across the wall.

Depending on their manner of production, tubes exhibit some anisotropic properties i.e. different response to circumferential and to
longitudinal stresses to a greater or lesser extent. Metals show
no significant differences in their elastic characteristics i.e. of
Young's modulus or Poisson's ratio. However, more pronounced differences may occur in the yield stress.

Such structures and structural elements should be designed to operate
in the plastic deformation range. However, to evaluate structural
safety in the case of variable repeated loads and temperature field
changes shakedown analysis should be employed rather than classical
plastic limit analysis. This has been recognized by the codes of
practice of several countries. Namely, load/temperature cycles may
result in cycles of plastic strain increments leading to material
breaking due to low-cycle fatigue or to excessive structural deformations (called incremental collapse) due to the accumulation of
permanent strain. Both phenomena may appear below the ultimate load.
Shakedown theory, despite adopting a simple material model, seems to
be able to determine properly the safe load/temperature variation
limits; this guarantees that the plastic strain increments eventually
cease. Such a stabilization of plastic strains is called shakedown.

If a state of shakedown has been attained, a given structure responds in a perfectly elastic way to further load/temperature variations which are within the safe limits.

Results of the previous experimental investigations on simple structures exposed to variable repeated loads have seemed to confirm the validity and the practical applicability of the shakedown approach. However, very rarely have such investigations included thermal effects.

The present investigation of axisymmetric tubes was aimed to check if a simplified shakedown analysis is able to provide safe load and temperature limits. The analysis employed the Tresca yield condition and considered simple incremental collapse mechanisms.

The experimental investigations were made on a specially designed set-up and were performed in a fully automatic way. The switching on and off of each one of the independent load systems was preprogrammed and then executed automatically. The same applied to the temperature changes. In this manner a large number of identical cycles (up to many hundreds) could be performed. Also, the recording of the data was fully automatic.

SHAKEDOWN ANALYSIS OF TUBES

Let us consider a structure made of an elastic, perfectly plastic material and subject to r agents which may vary independently of each other. Let the intensity of each one of them be defined by a certain factor $\beta_k(t)$, $k = 1, \ldots, r$, so that the resulting thermoelastic stress field σ_{ij}^E as well as the temperature field θ are linear functions of the factors β_k:

$$\sigma_{ij}^E (\underline{x},t) = \sum_{k=1}^{r} \beta_k(t) \, \sigma_{ij}^{Ek} (\underline{x}),$$

$$\theta(\underline{x},t) = \sum_{k=1}^{r} \beta_k(t) \, \theta^k(\underline{x}).$$

$$(1)$$

σ_{ij}^{Ek} and θ^k are the individual contributions of a single agent if its intensity is equal to one.

The load and temperature variation limits are prescribed by

$$b_k^- \leq \beta_k(t) \leq b_k^+, \qquad k = 1, \ldots, r; \qquad (2)$$

b_k^-, b_k^+ are given constants.

The temperature dependence of the yield stress is assumed to be linear so that the yield condition takes the form

$$f(\sigma_{ij}) \leq Y_0 (1 - e\theta); \qquad (3)$$

e and Y_0 are material constants. Consequently, the plastic dissipation function depends not only on the plastic strain rate $\dot{\varepsilon}_{ij}^P$ but also on the temperature:

$$\sigma_{ij} \; \dot{\varepsilon}^{P}_{ij} = D(\dot{\varepsilon}^{P}_{ij}, \; \theta) = D_0 (\dot{\varepsilon}^{P}_{ij}) \; (1 - e\theta). \tag{4}$$

D_0 denotes the dissipation in the same plastic flow mode but at a reference temperature.

By a proper rearrangement of the kinematical shakedown theorem (Koiter, 1956) adjusted to take account of thermal effects (Rozenblum, 1965, De Donato 1970), two separate criteria (alternating plasticity; incremental collapse) can be derived (Gokhfeld, 1966, Sawczuk, 1967, König, 1979a, 1979b). The latter can be formulated as follows:

A given structure may suffer incremental collapse associated with a kinematically admissible strain increment field $\Delta\bar{\varepsilon}_{ij}(\underline{x})$ if the following inequality holds:

$$\int_V \sum_{k=1}^{r} a_k(\underline{x}) \; J_k(\underline{x}) \; dV \geq \int_V D_0 \; [\Delta\bar{\varepsilon}_{ij}(\underline{x})] \; dV. \tag{5}$$

Here,

$$J_k(\underline{x}) = \sigma^{Ek}_{ij}(\underline{x}) \; \Delta\bar{\varepsilon}_{ij}(\underline{x}) + e \; \theta^k(\underline{x}) \; D_0 \; [\Delta\bar{\varepsilon}_{ij}(\underline{x})],$$

$$a_k(\underline{x}) = \begin{cases} b_k^{+} & \text{if} \quad J_k(\underline{x}) > 0 \\[2mm] b_k^{-} & \text{if} \quad J_k(\underline{x}) < 0. \end{cases} \tag{6}$$

If there appear, additionally, time independent body forces $F_i^{\,0}(\underline{x})$ and surface tractions $T_i^{\,0}(\underline{x})$ the formula (5) can be written in the following form,

$$\int_V F_i^{\,0}(\underline{x}) \; \bar{u}_i(\underline{x}) \; dV + \int_{S_T} T_i^{\,0}(\underline{x}) \; \bar{u}_i(\underline{x}) dS + \int_V \sum_{k=1}^{r} a_k(\underline{x}) \; J_k(\underline{x}) dV$$

$$\geq \int_V D_0 \; [\Delta\bar{\varepsilon}_{ij}(\underline{x})] dV, \tag{7}$$

where $\bar{u}_i$ is the plastic displacement increment field with which the field $\Delta\bar{\varepsilon}_{ij}$ is associated:

$$\Delta\bar{\varepsilon}_{ij} = \frac{1}{2}(\bar{u}_{i,j} + \bar{u}_{j,i}),$$

$$\bar{u}_i = 0 \qquad \text{on} \qquad S_U;$$

S_T is the part of the structure boundary at which the statical conditions are prescribed. S_U denotes the part with the kinematical ones.

Let us note that the functions $a_k(\underline{x})$ are defined at each point of the structure separately and therefore a formal extension of the formulae

(5) or (7) to bending moments, membrane forces and other generalized
stresses (defined for a cross-section instead of a material point)
may result in considerable errors (König, 1982).

Shakedown of Tubes

In the case of thin-walled, axisymmetric, closed, cylindrical shells
the axial stress σ_x and the circumferential stress σ_φ are the only
non-vanishing stress components. According to the Kirchhoff-Love
hypothesis, the corresponding strains are

$$\varepsilon_x = \lambda_x - z\,\kappa_x \quad \text{and} \quad \varepsilon_\varphi = \lambda_\varphi, \tag{8}$$

where λ_x, λ_φ, κ_x denote the corresponding elongations (stretches) and
the longitudinal curvature increment of the middle surface of the
shell (variation of the circumferential curvature can be neglected).
The latter can be expressed in terms of the longitudinal displacement
u and of the radial displacement w:

$$\lambda_x = \frac{du}{dx}, \quad \lambda_\varphi = \frac{w}{R}, \quad \kappa_x = \frac{d^2 w}{dx^2}; \tag{9}$$

R is the shell radius and z denotes the distance from the middle surface (outwards positive).

If one employs the Tresca yield condition, Fig. 1, it is in general
sufficient to consider only piecewise linear mechanisms of incremental
collapse i.e. mechanisms in which a given shell becomes an assembly
of cylinders and cones with plastic hinges at the joint lines. In
such a case, any kinematically admissible field of plastic strain

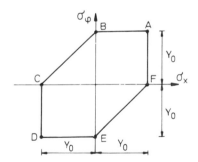

Fig. 1.   The Tresca yield condition at plane stress.

increments is given by the following formulae, see Fig. 2,

$$\Delta\bar{\varepsilon}_\varphi = \frac{1}{R}\,[\bar{w}_\ell + (\bar{w}_r - \bar{w}_\ell)\,\frac{x}{d}],$$

$$\Delta\bar{\varepsilon}_x = (\bar{u}_\ell - \varphi_\ell \cdot z)\,\delta(x) + (\bar{u}_r - \varphi_r \cdot z)\,\delta(x-d).$$

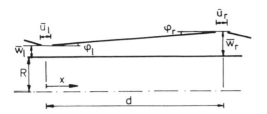

Fig. 2. Segment of a deformed shell.

Here, $\delta(\dot{} \ldots)$ denotes the Dirac impulse function and $\bar{u}_\ell$, $\bar{u}_r$, φ_ℓ, φ_r are concentrated axial stretches and rotation angles (concentrated curvatures) at the left-hand side and right-hand side plastic hinges, respectively.

Moreover, the temperature distribution across the shell wall may be approximated by a linear function:

$$\theta(x,z) = \frac{\theta_e(x) + \theta_i(x)}{2} + [\theta_e(x) - \theta_i(x)] \frac{z}{h}. \tag{10}$$

$\theta_e(x)$, $\theta_i(x)$ denote the external and internal temperature at x, respectively. Therefore, the thermoelastic stresses σ_x^E, σ_φ^E will also be linear with respect to the radial coordinate z. Finally, the integration procedure in formula (5) or (7) may become a problem only along the shell axis. With respect to z, the integrands are linear or quadratic functions. However, one has to keep in mind that the form of the plastic dissipation function as well as the coefficients $a_k(x,z)$ depend on the signs of the corresponding expressions, cf. Table 1.

Example

Let us analyse an infinitely long thin-walled tube, Fig. 3, subjected to a constant ring force Q and to variable repeated internal pressure p and internal temperature θ, the external temperature being zero. Pressure p and temperature θ can vary independently of each other between the limits.

$$0 \leq p \leq \bar{p}, \qquad 0 \leq \theta \leq \bar{\theta}. \tag{11}$$

The unit thermoelastic stresses are

$$\sigma_\varphi^{Ep} = \frac{R}{h}, \qquad\qquad \sigma_x^{Ep} = 0,$$

$$\sigma_\varphi^{E\theta} = \frac{E \alpha z}{(1-\nu) h}, \qquad \sigma_x^{E\theta} = 0; \tag{12}$$

TABLE 1. The Plastic Strains and the Dissipation Function Associated with the Tresca Yield Condition for Plane Stress

Side or vertex	$\dot{\varepsilon}_x^P$	$\dot{\varepsilon}_\varphi^P$	Dissipation function
A	≥ 0	≥ 0	$Y_0(\dot{\varepsilon}_x^P + \dot{\varepsilon}_\varphi^P)$
AB	$= 0$	≥ 0	$Y_0\,\dot{\varepsilon}_\varphi^P$
B	≤ 0	≥ 0	$Y_0\,\dot{\varepsilon}_\varphi^P$
BC	≤ 0	$\dot{\varepsilon}_\varphi^P = -\dot{\varepsilon}_x^P$	$Y_0\,\dot{\varepsilon}_\varphi^P = -Y_0\,\dot{\varepsilon}_x^P$
C	≤ 0	≥ 0	$-Y_0\,\dot{\varepsilon}_x^P$
CD	≤ 0	$= 0$	$-Y_0\,\dot{\varepsilon}_x^P$
D	≤ 0	≤ 0	$-Y_0(\dot{\varepsilon}_x^P + \dot{\varepsilon}_\varphi^P)$
DE	$= 0$	≤ 0	$-Y_0\,\dot{\varepsilon}_\varphi^P$
E	≥ 0	≤ 0	$-Y_0\,\dot{\varepsilon}_\varphi^P$
EF	≥ 0	$\dot{\varepsilon}_\varphi^P = -\dot{\varepsilon}_x^P$	$Y_0\,\dot{\varepsilon}_x^P = -Y_0\,\dot{\varepsilon}_\varphi^P$
F	≥ 0	≤ 0	$Y_0\,\dot{\varepsilon}_x^P$
FA	≥ 0	$= 0$	$Y_0\,\dot{\varepsilon}_x^P$

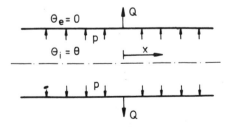

Fig. 3. An infinite tube subject to loads and temperature.

h denotes the shell wall thickness and α is the coefficient of linear thermal expansion. Let us assume the simple mechanism of incremental collapse shown in Fig. 4. The plastic displacement increment field is

$$\bar{w}(x) = \begin{cases} w_0 \ (1 - |x|/a) \ \text{for} \ |x| \leq a \\ \\ 0 \ \text{elsewhere.} \end{cases} \tag{13}$$

Thus, the plastic strain increments are as follows:

$$\Delta\bar{\varepsilon}_\varphi(x,z) = \frac{w_0}{R} \ (1 - \frac{|x|}{a}), \qquad |x| \leq a,$$

$$\Delta\bar{\varepsilon}_x(x,z) = - \frac{z \ w_0}{a} \ [\delta(x+a) + \delta(x-a)] + \frac{2 \ z \ w_0}{a} \ \delta(x). \tag{14}$$

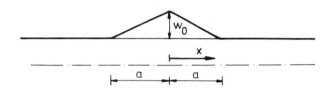

Fig. 4.   The incremental collapse mechanism assumed.

It is easy to see that $\Delta\bar{\varepsilon}_\varphi \geq 0$ everywhere within the shell volume whereas for $|x| = a$:

$$\Delta\bar{\varepsilon}_x \geq 0 \qquad \text{if} \qquad -h/2 \leq z \leq 0,$$

$$\Delta\bar{\varepsilon}_x \leq 0 \qquad \text{if} \qquad 0 \leq z \leq h/2;$$

for $x = 0$:

$$\Delta\bar{\varepsilon}_x \leq 0 \qquad \text{if} \qquad -h/2 \leq z \leq 0,$$

$$\Delta\bar{\varepsilon}_x \geq 0 \qquad \text{if} \qquad 0 \leq z \leq h/2.$$

Now, it is easy to calculate the total energy dissipated at zero temperature in the mechanism (14):

$$2 \ \pi \ R \int_{-a}^{a} \int_{-h/2}^{h/2} Y_0 \ w_0 \ \{\frac{1}{R}(1-|x|/a) + [2\delta(x)+\delta(x-a)+\delta(x+a)]|z|/a\}$$

$$dx \ dz = 2 \ \pi \ Y_0 \ (a \ h \ w_0 + R \ h^2 w_0/a). \tag{15}$$

By substituting the expressions (12) and (14) into (6) one obtains,

$$J_p(x,z) = \frac{w_0}{h} \ (1 - |x|/a),$$

O. Mahrenholtz, K. Leers and J. A. König

$$J_\theta(x,z) = -\frac{E \alpha z w_0}{(1-\nu) Rh} (1 - |x|/a) + \frac{e Y_0 w_0}{2} (1 - \frac{2z}{h})$$

$$\{\frac{1}{R} (1 - |x|/a) + [2\delta(x) + \delta(x-a) + \delta(x+a)] |z|/a\}. \tag{16}$$

Therefore,

$$a_p(x,z) = \bar{p},$$

$$a_\theta(x,z) = \begin{cases} \bar{\theta} & \text{if} \quad z \le 0 \quad \text{and} \quad |x| \le a \\ 0 & \text{elsewhere.} \end{cases} \tag{17}$$

By substituting (17), (16) and (15) into formula (7) one obtains, after some transformations,

$$Q + \bar{p}a + \frac{E h \alpha a \bar{\theta}}{2(1-\nu) R} + \frac{e Y_0 h a \bar{\theta}}{8 R} + \frac{e Y_0 h^2 \bar{\theta}}{12 a} \ge$$

$$\ge \frac{Y_0 h a}{R} + \frac{Y_0 h^2}{a}. \tag{18}$$

To obtain the most stringent incremental collapse condition one has to optimize the relationship with respect to the free parameter a. By differentiating equation (18) with respect to a and setting the derivative to zero, one obtains the optimal value of the distance a:

$$a = 2 Y_0 h^2 (1 - e\bar{\theta}/3)/Q. \tag{19}$$

The relations (18) and (19) define in the three-dimensional $a, \bar{p}, \bar{\theta}$-space the domain within which the incremental collapse mode (14) can occur.

If one introduces the following dimensionless notation

$$q = \frac{Q}{Y_0 h} \sqrt{\frac{2R}{h}}, \quad P = \frac{\bar{p} R}{Y_0 h}, \quad \eta = e \bar{\theta}, \quad \tau = \bar{\theta}/\theta_e, \tag{20}$$

where $\theta_e = 2(1-\nu) Y_0/E\alpha$ is the temperature difference which alone would result in the beginning of yielding of the shell, then the safe domain is defined by the following simple inequality

$$q^2/(4 - \eta/3) + \tau + 2 P \le 2 - \eta/4. \tag{21}$$

Figure 5 gives some examples of the safe domain for various magnitudes of P and η.

EXPERIMENTAL INVESTIGATION OF SHAKEDOWN OF TUBES

The experiments were planned to model the behaviour of nuclear fuel cladding shells. Such shells are exposed to various loads such as the internal pressure of gas emerging from nuclear reactions, local pressure from fuel pellets which grow with time, a high temperature gradient across the shell wall and axial forces. During start-up and shut-down of the reactor, the magnitudes of the above loads increase or decrease. A special set-up was designed for the experiments. Its general layout is given in Fig. 6.

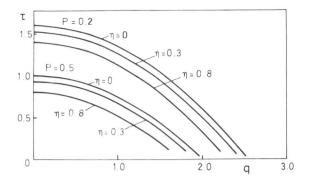

Fig. 5. Safe domains of the tube.

Fig. 6. Experimental plant for shakedown investigations of tubes.

Tubular specimens, Fig. 7, clamped at both ends are subjected to
- internal pressure produced by a hydraulic pump
- a localized radial load applied at some arbitrary position along the tube by means of a ring-force rubber membrane (meridional cross-section shown in Fig. 8), connected to another hydraulic pump
- a longitudinal tensile or compressive force exerted by a hydraulic cylinder
- a radial temperature gradient produced by a heating element along the tube axis.

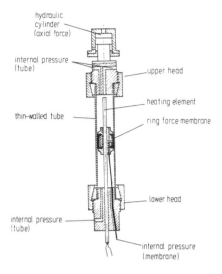

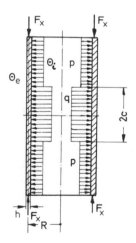

Fig. 7.   Experimental arrangement.

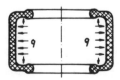

Fig. 8.   The ring-force membrane.

The variations of the loads with time are produced by servo relief valves. The sequencing of the loads (switched on and off) as well as the duration of each loading are arbitrary; they were preprogrammed and then executed by means of an electronic processor.

The same applies to the temperature cycles. To shorten their duration there are cooling nozzles with compressed air directed upon the specimen. They were automatically switched on when the heating ended and switched off when the preprogrammed lower temperature bound was attained. In this way, a very high number of identical load/temperature cycles could be performed. In the case of an unexpected failure (break down of the specimen) the resulting decrease in the liquid pressure of the hydraulic system was used to stop the experiment. This security measure allows the experimental work to continue without anybody supervising it.

The tubular specimens used in our investigations were made of the aluminium alloy Al Mg Si 0.5. The cold drawn tubes had an outside diameter of 50 mm and were 2 mm thick. They were carefully machined to prevent any non-symmetric deformation. Additionally, at the beginning and at the end of every shakedown test, the tube dimensions were measured by means of inductive gauges employed to record the radial displacements. In order to obtain a quantitative comparison of the theoretical predictions with the experimental results, measurements were made to determine the stress-strain relation of the material in both the longitudinal and circumferential directions. In the latter case, specimens used in the shakedown experiments were employed. They were subjected to internal pressure. Since in the central portion of the tube end effects could be neglected, the relation between the radial displacement in the central part of the tube and the pressure provides immediately the σ,ε-relationship. The σ,ε-relationship in the axial direction was determined by standard tensile tests of thin specimens cut longitudinally from the tube. Both stress-strain curves are very similar and therefore no account of anisotropy has been included in the theoretical analysis. On the other hand, these tests have shown a relatively strong strain-hardening effect in the aluminium alloy used. Figure 9 presents some of the σ,ε-curves obtained at various temperatures.

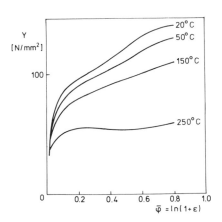

Fig. 9. Stress-strain curves for the aluminium
        alloy. (Meyer-Nolkemper, 1978).

In the course of the shakedown investigations the radial displacement
at a certain central part of the tube was measured by means of three
inductive gauges placed within the same horizontal plane. The mean
value was displayed and, simultaneously, recorded by a plotter versus
time. Examples of the data obtained are given in Fig. 10. In both
the cases the specimens were subjected to cyclic variations of the
internal pressure p̄ and of the localized pressure q̄, cf. Fig. 7.

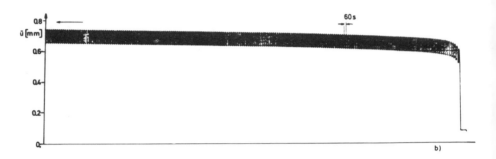

Fig. 10.   Radial displacement drawn against number
           of cycles: (a) shakedown, (b) incremental
           collapse.

Changes of the loads during a single cycle are visualized in Fig. 11.
The lower bound of the internal pressure was p̄ = q̄ = 0,7 N/mm² in
both tests. The overall pressure p̄ remained constant whereas the
upper magnitude of the localized pressure q̄ was 9,5 N/mm² in case
a) and 10,2 N/mm² in case b) shown in Fig. 10.

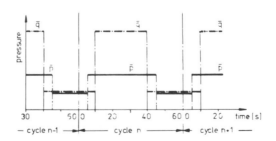

Fig. 11.   Variations of the internal pressure p̄
           and of the local pressure q̄ during a load
           cycle.

In the former case, the displacement eventually stabilized though
only after over a hundred cycles. In the latter case no stabilization could be detected even after several hundred cycles.

In the same way investigations were made for other combinations of
the parameters involved and their maximum values which resulted in
stabilization of the displacement were taken as the experimental
shakedown limits.

It seems worthwhile to mention a practical procedure which was adopted
here. Namely

- if the deflection increment per cycle became less than
  4.10^{-6} mm/cycle, the experiment was terminated and the
  corresponding final deflection $\bar{u}_{max}$ measured;
- if the deflection rate fell below 10^{-5} mm/cycle and then
  became higher again then the experiment was stopped after
  a sufficiently long time or after specimen failure. The
  last measured deflection is taken as $\bar{u}_{max}$.

In all the series of experiments hitherto completed a characteristic
point appeared, marked with W in Fig. 12, in which the curvature of
the $\bar{u}_{max}$-load amplitude curve vanishes. The corresponding magnitude
of the load amplitude, it seems, can be considered as the experimental
shakedown limit.

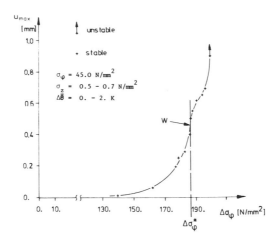

Fig. 12.   Results of a series of experiments.

THEORETICAL ANALYSIS

The theoretical shakedown limits were obtained by means of the
approach outlined in the previous section, cf. (Leers and others,
1983), by assuming the incremental collapse mechanisms as shown in
Fig. 13. They are similar to the mechanism of Fig. 4 but contain,
in addition, some discontinuities in the longitudinal displacement.

The corresponding displacement increments $\bar{w}(x)$ (radial) and $\bar{u}(x)$
(longitudinal) of the middle surface of the shell are

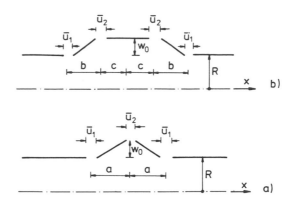

Fig. 13.   Mechanisms of incremental collapse: (a)
          first mechanism, (b) second mechanism.

in the first mechanism (Fig. 13a)

$$\bar{u}(x) = \bar{u}_1 \left[\delta(x-a) + \delta(x+a)\right] + \bar{u}_2\, \delta(x)$$

$$\bar{w}(x) = \begin{cases} w_0\,(1 - |x|/a) & \text{for} \quad |x| \le a \\ 0 & \text{elsewhere,} \end{cases} \qquad (22a)$$

in the second mechanism (Fig. 13b)

$$\bar{u}(x) = \bar{u}_1 \left[\delta(x-b-c) + \delta(x+b+c)\right] + \bar{u}_2\left[\delta(x-c) + \delta(x+c)\right]$$

$$\bar{w}_x = \begin{cases} w_0 & \text{for} \quad |x| \le c \\ w_0\,[1 - (|x|-c)/b] & \text{for } c \le |x| \le b + c \\ 0 & \text{elsewhere.} \end{cases} \qquad (22b)$$

The external agents accounted for in the analysis were, cf. Fig. 7b,
- the internal pressure p allowed to vary between the limits

$$0 \le p \le \bar{p}; \qquad (23a)$$

- the independent internal pressure q distributed within a length 2c
and varying between analogous limits

$$0 \le q \le \bar{q} \qquad (23b)$$

- the longitudinal force F remaining constant
- the internal temperature θ_i varying independently of other agents
within the limits

$$0 \le \theta_i \le \bar{\theta}. \qquad (23c)$$

The strain increments associated with the displacement increments
(22a) or (22b) were calculated, substituted into the formulae (6)

together with the load/temperature limits (23a), (23b), (23c) and
then into eq. (5). After lengthy though simple calculations, one
arrives at the corresponding incremental collapse criteria. To obtain the most stringent structural safety conditions, one has to
optimize these criteria with respect to free parameters: with respect to $\bar{u}_1/w_0$, $\bar{u}_2/w_0$ and a in the case of the first mechanism and
with respect to $\bar{u}_1/w_0$, $\bar{u}_2/w_0$, b in the case of the second one. The
resulting shakedown domains are defined by means of the following
relations:

first mechanism

$$\frac{c}{a}(\bar{q} - \bar{p}) + \bar{p} + \frac{E\alpha\Delta\theta h}{8(1-\nu)R} - \frac{Yh}{R} \leq 0, \tag{24a}$$

where

$$a = \left[\frac{(\bar{q} - \bar{p})c^2 - \dfrac{\bar{F}x^2}{Y} + Yh^2}{-\bar{p} - \dfrac{E\alpha\Delta\theta h}{8(1-\nu)R} + \dfrac{Yh}{R}}\right]^{1/2};$$

second mechanism

$$\bar{p} + \frac{c}{b}\bar{q} + (1 + \frac{c}{b})\left[\frac{E\alpha\Delta\theta h}{8(1-\nu)R} - \frac{Yh}{R} \leq 0, \tag{24b}\right.$$

where

$$b = \left[\frac{-\dfrac{\bar{F}x^2}{Y} + Yh^2}{-\bar{p} - \dfrac{E\alpha\Delta\theta h}{8(1-\nu)R} + \dfrac{Yh}{R}}\right]^{1/2}$$

$$Y = Y_0 - A\frac{\bar{\theta} + \theta_e}{2} - B(\frac{\bar{\theta} + \theta_e}{2})^2.$$

The shakedown domains (24a) and (24b) are presented in Fig. 14 in
the case F = 0. The following notation has been employed

$$\sigma_\varphi = \bar{p}\,R/h, \qquad \Delta\sigma_\varphi = (\bar{q}-\bar{p})\,R/h, \qquad \Delta\theta = \bar{\theta} - \theta_e \tag{25}$$

COMPARISON BETWEEN THE EXPERIMENTAL AND THEORETICAL
SHAKEDOWN LOADS

Figure 15 shows the experimental data (full dots) obtained in a series
performed at F = 0, $\bar{\theta} - \theta_e = 0$, $\bar{\theta} = 20°C$. The theoretical shakedown
limits provided by the first mechanism (solid line) and by the second
mechanism (dashed line) are also given.

According to the kinematical shakedown theorem one could expect that
the experimental points and the actual deformed shell form should
follow the first mechanism rather than the second one. However, the
shakedown loads obtained in these experimental series nearly coincide
with the curve appropriate to the latter. Contrariwise, the deformed
shell form, Fig. 16, seems to be closer to the first mechanism of
incremental collapse.

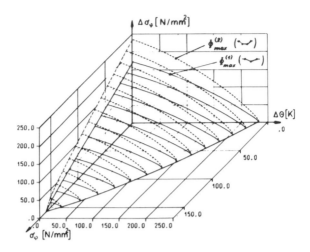

Fig. 14.   Shakedown domains obtained for the tube.

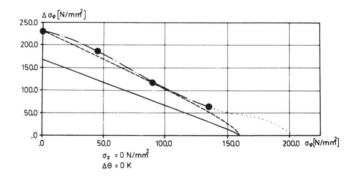

Fig. 15.   The theoretical and experimental shakedown
           limits.

Very similar results have been obtained in other series of the
experiments. It seems reasonable to expect that the material strain-
hardening and, perhaps, some geometrical effects are responsible for
the above discrepancies. Only a proper extension of the shakedown
analysis can answer this question.

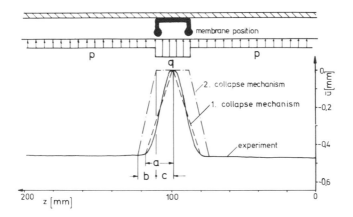

Fig. 16. The deformed shell form.

CONCLUSIONS

1. An example of experimental shakedown investigations in the case of a complex stress/strain state has been presented.
2. The existence of the phenomenon of shakedown has been confirmed. The permanent deformations recorded were clearly plastic deformations and not creep deformations. The latter appeared but were negligibly small in comparison with the former ones.
3. Apart from the above-mentioned inexactness, the simplified analysis provides safe bounds to the shakedown loads.
4. Nevertheless, a more refined approach is needed to enable a more precise evaluation of the shakedown limits of tubes.

REFERENCES

De Donato, D. (1970). Second shakedown theorem allowing for cycles of both loads and temperature. *1st. Lombardo di Scienza e Lettere (A)*, 104, 265-277.
Gokhfeld, D. A. (1966). Some problems on shakedown of plates and shells (in Russian). In *Proc. VI-th Soviet Conf. Plates a. Shells*, Nauka, Moscow. pp. 284-291.
Koiter, W. T. (1956). A new general theorem on shakedown of elastic-plastic structures. *Koninkl. Ned. Ak. Wett. B*, 59, 24-34.
König, J. A. (1979a). On upper bounds to shakedown loads. *Z. Ang. Math. Mech.*, 59, 349-354.
König, J. A. (1979b). On the incremental collapse criterion accounting for temperature dependence of the yield-point stress. *Arch. Mech.*, 31, 317-325.
König, J. A. (1982). On exactness of the kinematical approach in the structural shakedown and limit analysis. *Ing.-Archiv*, 52, 421-426.
Leers, K., J. A. König and O. Mahrenholtz (1981). Anisotropic behaviour of vessel section under distributed load. *Coll. Int. CNRS No. 319*, Villard-le-Lans.
Leers, K., W. Klie, J. A. König and O. Mahrenholtz (1983). Experimental investigations on shakedown of tubes. *Symposium Plasticity Today, CISM*, Udine.

Meyer-Nolkemper, H. (1978). Fließkurven metallischer Werkstoffe.
  HFF-Bericht Nr. 4, Hannover, Vol. 1.
Rozenblum, V. I. (1965). On analysis of uneven heated elastic-
  plastic bodies (in Russian). *Prikl. Mech. Techn. Phys. No. 5*,
  98-101.
Sawczuk, A. (1969). Evaluations of upper bounds to shakedown loads
  for shells. *J. Mech. Phys. Solids*, <u>17</u>, 291-301.

IMPACT
MECHANICS

A REVIEW OF ELEMENTARY APPROXIMATION TECHNIQUES FOR PLASTIC DEFORMATION OF PULSE-LOADED STRUCTURES

P. S. SYMONDS

Division of Engineering, Brown University, Providence, RI 02912, USA

ABSTRACT

The techniques reviewed and compared include single degree of freedom models, the mode approximation technique, and upper and lower bounds on maximum displacement and response time. These are discussed in terms of rigid-perfectly plastic response to general load pulses (Fig. 1). The relations between the techniques and to results for elastic-plastic structures are discussed.

INTRODUCTION

This review describes a family of elementary estimation techniques for determining some features of the plastic deformation of a ductile structure due to pulse loading. We are considering structural response problems, as opposed to such other classes of problems as dynamic metal forming, penetration or perforation, and plastic wave propagation. The kinds of problems to which the methods described are applicable include impact of one body on another, explosive pressure on a building frame, wave impact on a ship or off-shore structure, internal blast pressure on a pipe or protective shell, and ground shock or base acceleration. These can all be treated as short pulse

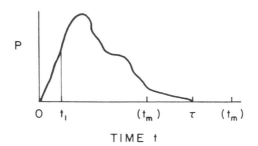

Fig. 1.

loading, if the pulses or ground accelerations are specified. We
shall suppose that this is the case, and hope the results may also
prove helpful in solving the more difficult real problem in which the
force is not specified, but must be inferred from information about
the structure and the velocity and other conditions of impact.

The purpose of the review is to give a resume of the simplest theorems
for obtaining information about the response. Simplicity is gained
at the price of restrictive conditions, and we sketch the derivations
of the theorems so as to make these conditions clear. We then look
at ways of removing or weakening some of the restrictions.

The need for estimation techniques is evident from the nature of
structural responses beyond the elastic range. The first phase of
response is an elastic one, ending when a yield condition is satisfied
at some point. Subsequent to this the typical response involves
plastic deformation in isolated regions separated by elastic regions.
The boundary surfaces between elastic and plastic (or viscoplastic)
regions move during the response, and plastic regions may disappear
and reappear. Eventually all plastic regions vanish, and the subsequent motion consists of wholly elastic vibrations. These are
damped out eventually by the usual mechanisms, leaving the structure
in a state of permanent deflections and an associated field of residual stresses.

The complex intermingling of elastic and plastic behavior is responsible for the absence of general analytical approaches and for much
of the difficulty of numerical solutions. One approach to simplification in treating problems of plastic flow with substantial plastic
deformations is to drop the elastic strains, effectively taking the
elastic moduli as infinite. In this "rigid-plastic" analysis, deformation occurs only in regions where a yield condition is satisfied;
everywhere else only motion as a rigid body occurs.

The assumption of rigid-plastic behavior, together with further
idealizations (small deflections; perfectly plastic behavior without
strain hardening or strain rate sensitivity) allows a substantial
body of theory to be developed. (This body of theory is analogous
to the basic "plastic theory" for statically loaded structures).
The present review collects some of the simplest and most useful
results of this theory. However we are trying also to provide per-
spective, so that the relations between different techniques of rigid-
plastic analysis to each other and to more realistic elastic-plastic
behavior are kept in view.

MASS-SPRING SYSTEM

Background and useful results are provided by considering the response
to a force pulse of a simple mass-spring system, the spring being
supposed to have an elastic-perfectly plastic characteristic, with
yield Y (see Fig. 2). In the initial elastic range the spring force
$Q = ku$, for displacement less than $u_y = Y/k$. For sufficiently large
pulse force P a plastic deformation occurs following time t_1 when
$u(t_1) = u_y$, which continues so long as the velocity $\dot{u} \geq 0$ with $Q = Y$.
At a time t_m the mass is brought to rest and plastic deformation
ceases. Reduction of the spring force takes place elastically, the
spring force being reduced to zero after a negative displacement u_y.
The elastic strain energy $Y^2/2k$ possessed by the spring when it enters

the plastic range is the total energy of the elastic vibration which
is ultimately damped out by the usual mechanisms, the permanent static
displacement being presumably the mean amplitude of the vibration.

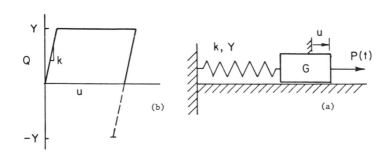

Fig. 2.

The effect of the force in the elastic range is to produce a displacement $u(t_1) = u_y$ and a velocity $\dot{u}(t_1) = \dot{u}_1$. For any pulse these can
be readily computed by the usual formulae. The equation of motion
in the initial elastic range is

$$G\ddot{u} = P(t) - ku \qquad (1)$$

where G is the mass and P(t) is the applied force. We suppose the
displacement and velocity are initially zero. At the end of the
elastic stage at time t_1 the displacement and velocity are given by

$$u(t_1) - u_y = \frac{1}{\omega G} \int_0^{t_1} P(\eta)\ \sin\omega(t_1-\eta)\,d\eta \qquad (2)$$

$$\dot{u}(t_1) = \dot{u}_1 = \frac{1}{G} \int_0^{t_1} P(\eta)\ \cos\omega(t_1 - \eta)\,d\eta \qquad (3)$$

where $\omega = \sqrt{k/G}$ is the radian frequency of the system.

The equation of motion in the subsequent plastic deformation is

$$G\ddot{u} = P(t) - Y \qquad (4)$$

for $\dot{u} \geq 0$. The maximum displacement occurs at time t_m when $\dot{u}(t_m) = 0$.

The equation to be solved for t_m is

$$G\dot{u}(t_m) = G\dot{u}_1 + \int_{t_1}^{t_m} P\,dt - Y\,(t_m - t_1) = 0 \qquad (5)$$

The maximum displacement u_m can then be written as

$$u(t_m) = u_m = u_y + \frac{G\dot{u}_1^2}{2Y} + \frac{1}{G} \int_{t_1}^{t_m} I_1(t)\,dt - \frac{I_{1m}^2}{2GY} \qquad (6)$$

where
$$I_1(t) = \int_{t_1}^{t} P(\eta)d\eta; \quad I_{1m} = I_1(t_m) = \int_{t_1}^{t_m} P(t)dt \tag{7}$$

The final static displacement $u_f = u_m - u_y$. We see that u_f depends on the velocity $\dot{u}_1$ at the end of the elastic stage and on the part of the force pulse after the time t_1.

A convenient alternative form is obtained by introducing the centroid time $\bar{t}_{1m}$ of the pulse area between t_1 and t_m, defined by the equation

$$(\bar{t}_{1m} - t_1) I_{1m} = \int_{t_1}^{t_m} (t-t_1) Pdt \tag{8}$$

Integration by parts of the right-hand side enables the final displacement to be written as

$$u_f = \frac{1}{2GY} [(G\dot{u}_1 + I_{1m})^2 - 2(\bar{t}_{1m} - t_1) YI_{1m}] \tag{9}$$

An "equivalent load" magnitude P_{1m}^e can be defined as the force of a rectangular pulse whose centroid coincides with the centroid of the specified pulse and which has the same total impulse I_{1m}, in both cases considering only the part of the pulse between t_1 and t_m.

$$2(\bar{t}_{1m} - t_1) P_{1m}^e = I_{1m} \tag{10}$$

The permanent displacement then can be expressed as

$$u_f = \frac{1}{2GY} [(G\dot{u}_1 + I_{1m})^2 - \frac{Y}{P_{1m}^e} I_{1m}^2] \tag{11}$$

Rigid-Plastic Case

The simplification attainable if rigid-plastic behavior is assumed can be seen at once. By Eq. (4), motion of the mass starts when $P(t_1) = Y$ at time t_1, but the displacement $u(t_1)$ and velocity $\dot{u}(t_1)$ at the start of plastic deformation are both zero. Equation (11) reduces to

$$u_m = u_f = \frac{I_{1f}^2}{2GY} (1 - \frac{Y}{P_{1f}^e}) \tag{12}$$

In the rigid-plastic system there is no elastic recovery, and the time t_m of reaching maximum displacement is the time t_f when the mass comes to rest in its final position. (Since rigid-plastic behavior plays a large role in this review, for brevity we will speak of a "rigid-plastic structure" when we really mean a structure treated according to the assumptions of rigid-plastic behavior). In dealing with the rigid-plastic response it is appropriate to use t_f instead of t_m, and $I_{1f} = I_{1m}$, etc. The time t_f of the rigid-plastic model can be found from Eq. (5) with $\dot{u}_1 = 0$

$$Y(t_f - t_1) = I_{1f} \equiv \int_{t_1}^{t_f} P dt \qquad (13)$$

The result of Eq. (12) can be expressed equivalently in terms of pulse and response time as

$$u_f = \frac{I_{1f}^2}{2GY} [1 - 2 (\overline{t}_{1f} - t_1) \frac{Y}{I_{1f}}] = \frac{I_{1f}^2}{2GY} [1 - \frac{2(\overline{t}_{1f} - t_1)}{t_f - t_1}]$$

$$(14a,b)$$

Note that all the foregoing results are entirely general as far as the load pulse is concerned, subject only to the proviso that there is only one stage of plastic deformation, from time t_1 to time t_m.

If the pulse is "short" in a certain sense various results are simpler. The definition of short pulse obviously depends on the structure, and in dealing with responses of general structures we will introduce particular definitions where needed. However, we may note that rigid-plastic analysis is particularly appropriate for short pulses, for reasons to be seen shortly.

For the present simple mass-spring model being considered, a "short pulse" may be defined as one whose duration τ is less than the time t_m (or t_f for the rigid-plastic case). This is illustrated in Fig. 3. The impulse that is effective I_{1f} in the rigid-plastic case then is independent of the response, and to indicate this can be written as $I_{1\tau}$. As a simple example consider the triangular pulse of Fig. 3a, with force decreasing linearly from the initial maximum P_0 to zero in time τ. For this pulse we have $t_1 = 0$ and

$$I_{1f} = I_{1\tau} = \frac{1}{2} P_0 \tau; \quad \overline{t}_{1f} = \frac{1}{3} \tau; \quad P_{1f}^e = \frac{3}{4} P_0$$

$$u_f = \frac{I_{1\tau}^2}{2GY} (1 - \frac{4Y}{3P_0}) \qquad (15)$$

This is valid for $P_0/Y \geq 2$, so that $t_f \geq \tau$.

A non-dimensional form is obtained for u_f/u_y in terms of $i_{1\tau} = I_{1\tau}/YT$ (where $T = 2\pi/\omega$ is the natural period), and $\mu = P_0/Y$; namely

$$\frac{u_f}{u_y} = 2\pi^2 i_{1\tau}^2 (1 - \frac{4}{3\mu}) = 2\pi^2 i_{1\tau}^2 (1 - \frac{2}{3i} \frac{\tau}{T}) \qquad (16a,b)$$

where $i = P_0 \tau/2YT$.

The real advantage of writing the formulae for rigid-plastic structures in terms of the equivalent force P_{1f}^e based on the centroid time $\overline{t}_{1f}$ only appears when general structures are considered; Youngdahl (1970) showed from numerical solutions for several classes of structures and different pulse shapes that even though the response differs from a single degree of freedom pattern, the results for all pulse shapes are surprisingly close when the deflections are plotted against

P^e_{1f} rather than against maximum or average force, for example. (This does not hold necessarily for elastic-plastic response, unfortunately)

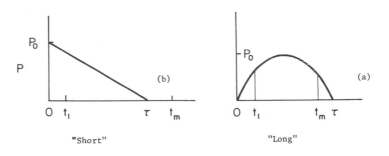

"Short"   "Long"

Fig. 3.

The extreme case of a short pulse is a delta function. The term "impulsive loading" is commonly used for a pulse which imparts a finite impulse I in a vanishingly short time. In the mass-spring model, the mass acquires finite velocity in a time $\tau \to 0$, so the problem can be treated as one of initial velocity specified as $\dot{u}_0 = I/G$ and of zero initial displacement. In the rigid-plastic case $t_1 = 0$, but in the elastic-plastic system t_1 depends on the impulse, and vanishes only for an infinite impulse.

The simplifications obtainable by rigid-plastic analysis are great, but is is important to have as much knowledge as possible about when the neglect of elastic strain components is permissible. The criterion first proposed (Lee and Symonds, 1952) was that a rigid-plastic solution might be anticipated to be valid if the ratio R of plastic work to the capacity of the structure to store elastic strain energy greatly exceeded unity. The mass-spring model we have been discussing here provides a way of examining the conditions for the error from a rigid-plastic solution to be acceptable. The results, of course, are not conclusive for real structures. However it is safe to assume that the error would not be less for a real structure than for the single DOF model.

For a pulse of given shape, such as the decreasing triangular pulse of Fig. 3a, any two of the three non-dimensional quantities impulse, i, force, μ, and time ratio, τ/T, may be used to define the load. Another useful parameter is the energy ratio R, which here means

$$R = \frac{\text{plastic work}}{\text{elastic energy capacity}} = \frac{Yu_f}{\frac{1}{2}Yu_y} = 2\,\frac{u_f}{u_y} \qquad (17)$$

The error e of a rigid-plastic solution may be defined as

$$e = \frac{u_f^{rp} - u_f^{ep}}{u_f^{ep}} \qquad (18)$$

In Fig. 4 are drawn some curves for constant R, with the error e
plotted on the vertical axis and the pulse time ratio τ/T on the
horizontal axis. These curves depend on the pulse shape, and those
shown in Fig. 4 are for a half-sine pulse. This shape is somewhat
more conservative than the triangular pulse, but less so than a
rectangular one. Note that the error is negative for sufficiently
long pulses, but positive for very short ones. These signs are easily
understandable. Consider the limiting case of impulsive loading
$\tau/T \to 0$. In a rigid-plastic analysis all the initial energy goes
into plastic work, whereas in the elastic-plastic system it leads
to an elastic vibration as well, and the plastic deflection must be
smaller. The error of the rigid-plastic final deflection is easily
written for impulsive loading. For the elastic-plastic system Eq.
(11) can be shown to lead to

$$u_f^{ep} = \frac{G\dot{u}_0^2}{2Y} \left(1 - \frac{u_y^2 \omega^2}{\dot{u}_0^2}\right); \quad \frac{u_f^{ep}}{u_y} = 2\pi^2 i^2 - \frac{1}{2} \tag{19}$$

For the rigid plastic systems Eq. (16) gives

$$\frac{u_f^{rp}}{u_y} = 2\pi^2 i^2$$

(In each case $i = I/YT$, $I = G\dot{u}_0$ being the total impulse).

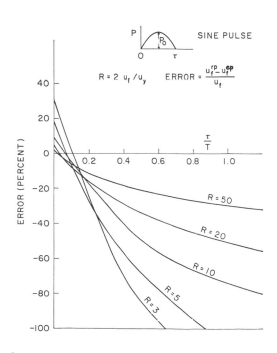

Fig. 4.

Substituting in Eq. (18) and using Eq. (17),

$$e = \frac{2\pi^2 i^2 - (2\pi^2 i^2 - 0.5)}{u_f^{ep}/u_y} = \frac{1}{R} \tag{20}$$

If we suppose a tolerable error of an approximation technique is about 30 percent, the curves of Fig. 4 show that the length of the pulse plays a strong role in determining how large R must be. Evidently with impulsive loading R $\geq$ 3 would be sufficient, but for $\tau/T = 1$, the curve shows that R $\geq$ 50 is demanded for this accuracy. Stated differently, R $\geq$ 20 might be supposed more than adequate since it implies plastic displacement 10 times the maximum elastic one; however Fig. 4 shows that if τ/T exceeds 1, the negative error may be well over 50 percent.

SINGLE DOF TREATMENT OF STRUCTURES

The results for a simple mass-spring model may be used to estimate responses of general structures, if they are treated as moving with a single degree of freedom. The parameters k, Y, G etc. of the mass-spring must be interpreted in terms of properties of the structure.

Notation for Structure and Loading

We may show the general notation for what follows by writing the linear equation of virtual work rate as it will repeatedly be used,

$$\int_s p_i^d \dot{u}_i^c ds - \int_s \bar{\rho} \ddot{u}_i^d \dot{u}_i^c ds = \int_s Q_\alpha^d \dot{q}_\alpha^c ds \tag{21}$$

We write vector force per unit area p_i and displacement u_i, velocity $\dot{u}_i$, and acceleration $\ddot{u}_i$, where i = 1, 2, 3, and the superscripts refer to dynamic and kinematic admissibility; see below. Q_α and $\dot{q}_\alpha$ respectively are generalized stress and conjugate strain rate, where $\alpha = 1, 2 \ldots \gamma$ for a system of stress resultants treated as a vector of γ components. This notation is convenient for the main class of structures of interest, namely one-dimensional (beams, rings, frames, etc.) and two-dimensional structures (plates and shells). It may easily be changed to deal with general bodies by replacing $(Q_\alpha, \dot{q}_\alpha)$ by $(\sigma_{ij}, \dot{\varepsilon}_{ij})$, i.e. by conventional stress and strain rate components. The summation convention for repeated suffixes is used in Eq. (21) to express scalar products, i.e. work rates. The integrations indicated in Eq. (21) should be understood to cover all elements of the structure in question. In a one-dimensional structure (beam, etc.) the position of a particle is defined by a coordinate on a center-line curve, and in a two-dimensional structure (plate, etc.) by a pair of coordinates on the middle surface. The scalar ds is used to mean an element either of a center-line curve or of a middle surface. We use the vector coordinate $\underset{\sim}{s}$ generally to indicate the position of a particle in both cases, $\underset{\sim}{s}$ being either a one-component or a two-component vector. The specific mass $\bar{\rho}$ correspondingly is the mass per unit length or the mass per unit area of a middle surface.

The two systems in Eq. (21), namely $(p_i^d, \bar{\rho}\ddot{u}_i^d, Q_\alpha^d)$ and $(\dot{u}_i^c, \dot{q}_\alpha^c)$, satis equations respectively of *dynamic* equilibrium and of kinematic *compatibility* including boundary constraint conditions of "workless" type. These will be referred to as dynamically and kinematically admissible systems, respectively. When it is necessary to distinguish between boundary surfaces where stresses rather than displacements are prescribed, the integrations on the left hand side of Eq. (21) is designated as over s_p rather than s_u. The total boundary "surface" $s = s_p + s_u$. "Area" of course is interpreted as either points or curves, for one- or two-dimensional structures, respectively.

The load vector $p_i(\underset{\sim}{s}, t)$ is meant to represent applied loads in a generalized sense, i.e. not only distributed or concentrated forces but moments as well. The velocity $\dot{u}_i$ is the conjugate quantity, the product $p_i^{d \cdot c} \dot{u}_i ds$ being a specific work rate in each case. We will frequently suppose that the load system can be written as

$$p_i(\underset{\sim}{s}, t) = \lambda(t)\overline{p}_i(\underset{\sim}{s}) \tag{22}$$

where $\overline{p}_i$ denotes the distribution of loads over the structure with any convenient magnitudes, and the separated form implies that the distribution stays the same during the load pulse. To assume this form is within the spirit of the approximation techniques concerned here.

The case of impulsive loading is interpreted as the limiting case as the pulse duration decreases,

$$\lim_{\tau \to 0} \int_0^\tau p_i dsdt = \lim_{\tau \to 0} \int_0^\tau \lambda(t)\overline{p}_i(\underset{\sim}{s})dsdt = \overline{\rho}\dot{u}_i^0(\underset{\sim}{s})ds \tag{23}$$

This type of loading is equivalent to prescribing an initial momentum field $\overline{\rho}\dot{u}_i^0$ whose distribution over the structure is the same as that of $\overline{p}_i(\underset{\sim}{s})$.

Constitutive Equations

Finally, the constitutive equations of elastic-perfectly plastic flow need to be stated. We suppose that the strain rate components are the sum of elastic and plastic components

$$\dot{q}_\alpha = \dot{q}_\alpha^e + \dot{q}_\alpha^p = C_{\alpha\beta}\dot{Q}_\beta + \dot{q}_\alpha^p \tag{24}$$

where $C_{\alpha\beta}$ are a symmetric tensor of flexibility coefficients.

Stress states Q_α are subject to a yield condition

$$\Phi(Q_\alpha) \leq Q_0 \tag{25}$$

where Q_0 is a constant. The plastic strain rates obey the following "flow rule" relations

$$\dot{q}^P = 0 \begin{cases} \text{if} & \Phi(Q_\alpha) < Q_0 \\ \text{or} & \Phi(Q_\alpha) = Q_0, \ \dot{\Phi} < 0 \end{cases} \tag{26a}$$

$$\dot{q}^P_\alpha = \nu \frac{\partial \Phi}{\partial Q_\alpha}, \ \nu \geq 0 \text{ if } \Phi(Q_\alpha) = Q_0, \ \dot{\Phi} = 0 \tag{26b}$$

where $\quad \dot{\Phi} = \dfrac{\partial \Phi}{\partial Q_\alpha} \dot{Q}_\alpha$

The convexity of the yield surface and the normality of the plastic strain rate vector to it are expressed by

$$(Q_\alpha - Q^s_\alpha)\dot{q}^P \geq 0 \tag{27}$$

where Q^s_α is any stress satisfying Eq. (25) and $\dot{q}^P_\alpha$ is associated with Q_α. For two states of stress and associated plastic strain rates $(Q^a_\alpha, \dot{q}^{pa}_\alpha)$ and $(Q^b_\alpha, \dot{q}^{pb}_\alpha)$, Eq. (27) implies

$$(Q^a_\alpha - Q^b_\alpha) \ (\dot{q}^{pa}_\alpha - \dot{q}^{pb}_\alpha) \geq 0 \tag{28}$$

Beams in bending are an important class of structure, and illustrate the yield condition and flow rules for uniaxial stress states. Here $Q_1 = M = $ bending moment; $q_1 = k = $ curvature rate; shear and axial forces are treated as reactions since the corresponding strain rates are neglected.

$$\dot{k} = \dot{k}^e + \dot{k}^P = \frac{1}{EI} \dot{M} + \dot{k}^P \tag{29}$$

The yield condition is

$$-M'_0 \leq M \leq M_0 \tag{30}$$

where M_0, M'_0 are fully plastic moments. The plastic curvature rates obey the relations

$$(M - M_0) \ \dot{k}^P \geq 0; \qquad (M + M'_0) \ \dot{k}^P \geq 0 \tag{31}$$

$$\dot{M}\dot{k}^P = 0 \tag{32}$$

Plastic curvature rates vanish if

$$-M'_0 < M < M_0 \tag{33a}$$

or if either

$$M = M_0, \ \dot{M} < 0; \qquad M = -M'_0, \ \dot{M} > 0 \tag{33b}$$

Single Degree of Freedom Approximations

With these preliminaries, we note that all the results written for the mass-spring model may be applied to a general structure if its response is assumed to occur with a constant shape. Suppose we write the velocity field as

$$\dot{u}_i^c(\underset{\sim}{s}, t) = \dot{u}^c(t)\phi_i^c(\underset{\sim}{s}) \tag{34}$$

We write the assumed shape function as the vector valued field $\phi_i^c(\underset{\sim}{s})$ and its time dependent scalar amplitude as $\dot{u}^c(t)$; this can have various physical meanings, but we will take it to be the velocity magnitude of a point of interest on the structure. We write it with superscript c because its meaning is related to the shape function $\phi_i^c(\underset{\sim}{s})$.

The velocity field $\dot{u}^c\phi_i^c(\underset{\sim}{s})$ will be assumed to satisfy the boundary constraint conditions and to be compatible with strain rate functions $u^c k_\alpha^c(\underset{\sim}{s})$. This will only exceptionally provide a full solution; although kinematic compatibility and boundary constraint conditions are satisfied, there will usually be violations of yield conditions and of equations of dynamics. However in appropriate circumstances there are single degree of freedom responses that are full solutions, satisfying all the field equations. We will term these "mode form solutions" and designate them as $u^*\phi^*(\underset{\sim}{s})$. These play an important role in certain methods of approximation to be discussed.

To obtain an approximate solution based on the assumed velocity field of Eq. (34), we take the acceleration field to be $\ddot{u}^c\phi_i^c$ and suppose that the internal work rate $Q_\alpha\dot{u}^c k_\alpha^c$ can be determined from constitutive equations. These are substituted in Eq. (21). After deleting $\dot{u}^c$ in each term, and taking the load in the form Eq. (22) we have

$$\ddot{u}^c \int \bar{\rho}\phi_i^c\phi_i^c ds = \lambda \int \bar{P}_i\phi_i^c ds - \int Q_\alpha^c k_\alpha^c ds \tag{35}$$

Comparing with Eq. (1), the forms of G, P, and Q are evident. This can be used for general elastic-plastic response, but is easiest to interpret and use for rigid-plastic behavior. With elastic strain rate components all zero, the strain rate components $k_\alpha^c(\underset{\sim}{s})$ in Eq. (35) are the plastic components, and the stress components Q_α^c are associated with them by the flow rules; thus $Q_\alpha^c k_\alpha^{pc}$ is the specific plastic dissipation rate, uniquely determinable from k_α^{pc}. There is a load factor λ_c for the static plastic collapse such that

$$\lambda_c \int \bar{P}_i\phi_i^c ds = \int Q_\alpha^c k_\alpha^{pc} ds \tag{36}$$

The equation of motion then can be written for $\dot{u}^c \geq 0$

$$\ddot{u}^c \int \bar{\rho}\phi_i^c\phi_i^c ds = [\lambda(t) - \lambda_c] \int \bar{P}_i\phi_i^c ds \tag{37a}$$

or $\qquad G_c\ddot{u}_c = P^c(t) - Y_c \tag{37b}$

where $G_c = \int \bar{\rho} \phi_i^C \phi_i^C ds$; $P^C(t) = \lambda \int \bar{P}_i \phi_i^C ds$; $Y_c = \int Q_\alpha k_\alpha^{PC} ds$.

All the results written above for rigid-plastic response of a mass-spring system can be applied to the single DOF structure model, replacing (G, P(t), Y) by $(G_c, P^C(t), Y_c)$.

The success of such a treatment obviously depends on the choice of the shape function ϕ_i^C. This is arbitrary, apart from requirements of kinematic admissibility. Kaliszky (1970, 1973) who is the main contributor to this type of approximate theory, emphasized using the shape of the mechanism at the plastic limit load. If this shape is ϕ_i^L, the corresponding limit load factor λ_L satisfies an equation like Eq. (36) with ϕ_i^C; we know that $\lambda_c \geq \lambda_L$ by the upper bound theorem of static structural plasticity. The use of the velocity field ϕ_i^L at the limit load would appear to be a good choice for the shape function ϕ_i^C, for loads not greatly exceeding the limit load. On the other hand, if the peak applied loads are not greatly in excess of the limit load, the load duration τ must be relatively long, if plastic deformations are important. This is, unfortunately, the situation where rigid-plastic analysis is most difficult to apply, i.e. where the neglect of elastic deformations may cause serious errors, as discussed earlier.

CONVERGENCE THEOREM AND MODES

A theorem due to Martin (1966) has proved to have special importance for the understanding of dynamic plastic response. This was stated first for rigid-plastic behavior and shows that solutions of rigid-plastic problems are unique at least as far as velocity fields are concerned. A later version included elastic-plastic behavior (Martin and Lee, 1968).

Consider an elastic-perfectly plastic behavior described by Eqs. (24) - (28). Suppose two solutions are designated as $(\dot{u}_i, Q_\alpha)$ and $(\dot{u}_i^*, Q_\alpha^*)$, both systems being full solutions (with specified loads $p_i(\underline{s}, t)$ in the sense that the field equations of dynamics and compatibility, boundary constraint conditions, and constitutive relations are all satisfied everywhere in the structure. All the response quantities are functions of $\underline{s}$ and t, in general. A functional Δ of the "difference" between the two solutions can be defined as

$$\Delta(t) = \frac{1}{2} \int \bar{\rho} (\dot{u}_i - \dot{u}_i^*)(\dot{u}_i - \dot{u}_i^*) ds + \frac{1}{2} \int C_{\alpha\beta} (Q_\alpha - Q_\alpha^*)(Q_\beta - Q_\beta^*) d$$

(38)

The first term is a kinetic energy formed from the velocity difference, while the second is a complementary elastic energy formed from the stress difference. We assume that the time rate of change can be computed as

$$\frac{d\Delta}{dt} = \int \bar{\rho} \ (\ddot{u}_i - \ddot{u}_i^*)(\dot{u}_i - \dot{u}_i^*)ds + \int C_{\alpha\beta}(Q_\alpha - Q_\alpha^*)(\dot{Q}_\beta - \dot{Q}_\beta^*)ds \tag{39}$$

Now each of the two solutions can be used as either the dynamically admissible or the kinematically admissible field in the virtual work equation Eq. (21); and the differences between the various response quantities are fully admissible. For two load systems p_i, p_i^* the virtual work relation can be written as

$$\int (p_i - p_i^*)(\dot{u}_i - \dot{u}_i^*)ds - \int \bar{\rho} \ (\ddot{u}_i - \ddot{u}_i^*)(\dot{u}_i - \dot{u}_i^*)ds$$

$$= \int (Q_\alpha - Q_\alpha^*)(C_{\alpha\beta}\dot{Q}_\beta + \dot{q}_\alpha^p - C_{\alpha\beta}\dot{Q}_\beta^* - \dot{q}_\alpha^{p*})ds \tag{40}$$

where the total strain rate is composed of elastic and plastic components as in Eq. (24). Now if we assume that both solutions satisfy dynamical equations with the same load system, $p_i = p_i^*$, the time derivative is given by

$$\frac{d\Delta}{dt} = - \int (Q_\alpha - Q_\alpha^*)(\dot{q}_\alpha^p - \dot{q}_\alpha^{p*})ds \leq 0 \tag{41}$$

where the inequality results from the inequality (28), which in turn corresponds to the normality-convexity property of the material.

Uniqueness of solutions follows from the inequality (41) and the positive definiteness of Δ. If the two solutions are identical at t = 0, they must be so at all t > 0. In the rigid-plastic case the stresses do not appear in Δ, and may not be uniquely determined by the plastic strain rates.

Convergence of Solutions

The operation of the result that Δ decreases when plastic flow occurs is seen clearly in the case of a rigid-plastic structure subjected to impulsive loading; as pointed out, this is equivalent to specifying a field of velocity over the structure at t = 0. Two solutions $\dot{u}_i$, $\dot{u}_i^*$, that are full solutions satisfy all the field equations, but may correspond to different initial conditions. Any such pair of solutions approach each other in the sense that the "difference" Δ between them continues to decrease so long as plastic deformation occurs in either solution, and they are not identical.

Of special interest are *mode-form* (full) solutions, namely ones with a velocity field which remains constant in shape. In the impulsive loading case the structure is free of external loading for t > 0 and there is a family of mode solutions. The actual rigid-plastic solution starts from some specified initial velocity field $\dot{u}_i^0$ ($\underset{\sim}{s}$), which in general is not proportional to one of the mode-form shapes of the structure. As the structure deforms, obeying equations of rigid-plastic dynamics, the velocity pattern changes. In an initial period there are usually travelling plastic hinges, but these disappear and typically the pattern of velocities becomes smoother. The final phase of motion is apparently always a mode-form solution.

Since the shape of the velocity field is constant in this closing phase, the stresses deduced by rules for rigid-perfectly plastic behavior also are constant. This means that the final mode-form is part of a full solution not only of the rigid-plastic structure but also of the structure treated more realistically as an elastic-plastic one. This helps one to visualize the whole response history when it includes elastic and plastic deformations. After the initial elastic stage and a usually complex stage of mixed elastic and plastic response, the motion becomes smoother, eventually coming to a velocity field which retains its shape while its magnitudes are reduced to zero. The elastic strain energy corresponding to the stresses of the mode shape finally provides the energy of an elastic vibration which continues until absorbed by ordinary damping mechanisms. The final elastic vibration typically is nearly sinusoidal, dominated by the fundamental elastic mode, since it is related to the simple stress field of the closing elastic-plastic mode form pattern.

This description is valid even when large deflections occur in the late stage plastic mode form pattern, although the final elastic vibrations are those of the deformed structure, not its initial configuration. Real materials exhibit strain hardening, strain rate sensitive and history dependent plasticity, and the actual response is somewhat complicated by these effects. In the case of structural steel the main disturbing effect is a strong strain rate sensitivity, but the strain rate dependence is such that the above description remains valid in its main qualitative features.

Mode Approximation Technique

If the pulse is short and the rigid-plastic model is adopted (it may well be highly appropriate in this case), the "mode technique" (Martin and Symonds, 1966) provides a simple solution which is often a good approximation. Suppose the load is actually impulsive, defined by an initial velocity field $\dot{u}_i^0$ ($\underset{\sim}{s}$). The actual response, starting from this field, goes through a sequence of shape changes, settling finally on the single dof shape which it retains to the end of the motion. This mode-form solution has velocity field

$$\dot{u}_i^*(\underset{\sim}{s}, t) = \dot{u}^*(t)\phi_i^*(\underset{\sim}{s}) \tag{42}$$

The functional $\Delta(t)$ can be written for the actual response $\dot{u}_i(\underset{\sim}{s}, t)$ and the above mode. At $t = 0$ its value $\Delta(0) = \Delta_0$ is a function of the initial amplitude $\dot{u}^*(0) = \dot{u}_0^*$ of the mode solution,

$$\Delta_0(\dot{u}_0^*) = \frac{1}{2} \int \bar{\rho}(\dot{u}_i^0 - \dot{u}_0^*\phi_i^*)(\dot{u}_i^0 - \dot{u}_0^*\phi_i^*)ds \tag{43}$$

Now since both fields in Eq. (43) belong to full solutions, $d\Delta/dt \leq 0$; the difference represented by $\Delta(t)$ decreases. The "best" value of $\dot{u}_0^*$ may be taken as the one that minimizes the maximum value of Δ, namely Δ_0, since this can be said to minimize the largest "error" of the approximate solution Eq. (42). This gives

$$\dot{u}_0^* = \frac{\int \bar{\rho}\dot{u}_i^0\phi_i^*ds}{\int \bar{\rho}\phi_i^*\phi_i^*ds} \tag{44}$$

When this choice is made, the approximate and actual solutions are often identical in the closing phase. Their initial kinetic energies are *not* equal; when Eq. (44) is used for $\dot{u}_0^*$, the initial kinetic energy of the mode approximation is always less than that of the specified field. The initial value $\dot{u}_0^*$ might be determined by matching kinetic energies; this will be discussed later. We note here that determining $\dot{u}_0^*$ by the above ("min Δ_0") method implies matching the two fields of momentum $\overline{\rho}\dot{u}_i^0$, $\overline{\rho}\dot{u}_*^0\phi_i^*$, after taking their scalar product with ϕ_i^* and integrating over the structure.

In this method the mode shape function ϕ_i^* plays a critical role. As emphasized already, this is assumed to be part of a full solution. A method must be available for finding the shape of the most suitable mode, and a criterion for preferring one mode over another. Generally the best mode in these problems of impulsive loading is the one with the longest response time. Martin (1981) has given an efficient iterative method that in a few cycles determines the shape of this "primary" mode of the unloaded structure. Criteria for selecting the preferable mode have been discussed by Symonds (1980).

Finite Pulse Duration and the Mode Technique

Impulsive loading is an idealization that is convenient but may be somewhat unrealistic. We know that when the pulse is long (comparable to or longer than the natural elastic period), rigid-plastic analysis may seriously underestimate the final displacement of the actual elastic-plastic structure (even when the permanent displacement is large compared to the maximum elastic one). However we may still wish to take account of the finite length of a pulse that is short enough so that a rigid-plastic treatment is adequate.

The simplest way to account for the finite duration of a pulse in a rigid-plastic treatment is to assume that the correction is the same as that in the single dof case. The expression of Eq. (12) applies to all pulses, regardless of shape and duration. It shows that the correction factor on the deflection for an impulsive load is expressible in terms of the ratio of the equivalent load magnitude P_{1f}^e to the yield force Y. This factor therefore can be written as

$$C_\tau = 1 - \frac{1}{\mu_e} \tag{45}$$

where $\mu_e = P_{1f}^e/Y$

and the notation is a reminder that the factor corrects for finite pulse duration τ. The equivalent load P_{1f}^e and yield force Y must be computed for the structure from an assumed velocity field by formulas such as those of Eqs. (37).

It is interesting that the two assumptions behind the simplest solution of a pulse loading problem involve errors of opposite sign: the neglect of pulse duration in assuming impulsive loading gives an overestimate, while the neglect of elastic deflections involves

an underestimate. Both increase with the pulse duration, and can
have roughly similar magnitude. This may perhaps have something to
do with the agreements reported by several investigators (e.g. Parkes
1955, Bodner and Symonds 1962, Symonds and Jones 1972) between predictions of rigid-plastic solutions and experiments.

Interrelations and Comparisons

Considering rigid-perfectly plastic behavior, theorems for an upper
bound on the final displacement and a lower bound on the response
time were derived by Martin (1964), both for an assumed impulsive
loading. A lower bound on the maximum displacement was given by
Morales and Nevill (1970) for the same conditions and extended to
finite pulses by Morales (1972). Finally, an upper bound on response duration was implicitly provided by Lee (1972). It is of
interest to point out connections between the various bound theorems
and approaches in which the structure is treated as a single degree
of freedom system.

Consider first the upper bound on final displacement. For a structure subjected to impulsive loading producing initial velocities
$\dot{u}_i^0(\underset{\sim}{s})$, a bound on the displacement at any point and in any chosen
direction, is provided by the following

$$u^f_{An} \leq \frac{K_0}{R^L_{An}} \leq \frac{K_0}{R^s_{An}} \tag{46}$$

where R^s_{An} is a force with the location and direction of the wanted
displacement for which an equilibrium stress field can be found
that does not violate the yield condition (25). The mode approximation technique makes use of a full solution in separated variable
form. Writing this as in Eq. (42), then as for any single degree of
freedom model (Eq. 37) gives the final displacement in terms of the
initial velocity $\dot{u}_0$

$$u^*_f = (\dot{u}^*_0)^2 \int \rho\phi^*_i\phi^*_i ds / 2 \int Q^*_\alpha k_\alpha ds = \frac{K^*_0}{\int Q^*_\alpha k^*_\alpha ds} \tag{47}$$

where K^*_0 is the initial kinetic energy of the approximating solution
in mode form. Normally in using the "mode technique" this is computed
from Eq. (44), and differs from the value K_0 of the given velocities.
Now it is implicit in writing Eq. (42) that the mode amplitude $\dot{u}^*$ is
the magnitude of the vector velocity at a chosen point, say $\underset{\sim}{s}_A$, and
in a direction stated by a unit vector n_i determined by the mode
shape vector $\phi^*_i(\underset{\sim}{s}_A)$ at that point. This is the direction also of
the final displacement, whose magnitude is given by Eq. (47). Now
the denominator of Eq. (47) can be identified as an upper bound on
the static limit load magnitude R^L_{An} of a force at point $\underset{\sim}{s}_A$ and in
direction n_i.

$$\int Q_\alpha^* k_\alpha^* ds = R_{An}^* \geq R_{An}^L \tag{48}$$

Furthermore, if we put $K^* = K_0$, the final displacement of Eq. (47) can be expressed as

$$u_f^* = \frac{K_0}{R_{An}^*} \leq \frac{K_0}{R_{An}^L} \tag{49}$$

We recognize the last term as the formula of the bound theorem, Eq. (46). Thus we see that if we compute the final displacement by assuming a kinematically admissible velocity field whose initial kinetic energy is the specified value K_0, this will be an approximation on the unsafe (low) side to the upper bound on this displacement according to Martin's (1964) theorem. It will be a good approximation if the velocity field chosen is close to a mechanism of plastic collapse at the (static) limit load.

The upper displacement bound will not generally coincide with the actual final displacement (of the rigid-plastic structure). For this to happen, the stress field of the static field must always agree with that of the actual response at all points where plastic flow takes place during the response; this is very unlikely.

Finally, we note the relations between response times and displacements predicted by a single degree of freedom treatment based on an assumed velocity shape $\phi_i^c(s)$, and the lower bounds on time of response (Martin, 1964) and on the maximum final deflection (Morales and Nevill, 1970; Morales, 1972). We note first that when the velocity field ϕ_i^c is adopted, deformation starts at time t_1^c such that, with λ_c defined by Eq. (36),

$$\lambda_c = \lambda(t_1^c) = \int Q_\alpha^c k_\alpha^{pc} ds \;/\; \int \overline{p}_i \phi_i^c ds \tag{50}$$

Then Eq. (37) can be integrated, giving

$$\dot{u}^c \int \overline{\rho} \phi_i^c \phi_i^c ds = [\int_{t_1^c}^{t} \lambda dt - (t - t_1^c)\lambda_c] \int \overline{p}_i \phi_i^c ds = R_c(t) \tag{51}$$

Evidently motion in this shape ceases at time t_f^c such that

$$t_f^c = t_1^c + \frac{1}{\lambda_c} \int_{t_1^c}^{t_f^c} \lambda dt \tag{52}$$

A second integration furnishes the displacement at time t_f^c

$$u_f^c \int \overline{\rho} \phi_i^c \phi_i^c ds = [\int_{t_1^c}^{t_f^c} \Lambda_1 dt - \frac{1}{2}(t_f^c - t_1^c)^2 \lambda_c] \int \overline{p}_i \phi_i^c ds \tag{53}$$

where

$$\Lambda_1(t) = \int_{t_1^c}^{t} \lambda \, dt$$

The two lower bound theorems are derived by using an artificial velocity field $\phi_i^c(s)$ in the virtual work rate equation (21) together with the actual system $(\dot{u}_i, Q_\alpha, \dot{q}_\alpha^p)$.

We may write this as

$$\lambda \int \bar{P}_i \phi_i^c ds - \int \bar{\rho} \ddot{u}_i \phi_i^c ds = \int Q_\alpha^c k_\alpha^{pc} ds - \int (Q_\alpha^c - Q_\alpha) k_\alpha^{pc} ds \quad (54)$$

$$\leq \int Q_\alpha^c k_\alpha^{pc} ds \quad (55)$$

since $(Q_\alpha^c - Q_\alpha) k_\alpha^{pc} \geq 0$, in view of Eq. (27). The actual response starts at time t_1 when $\lambda(t_1) = \lambda_L$, the limit load factor. Integrating Eq. (55) from t_1 to t,

$$\int \bar{\rho} \dot{u}_i \phi_i ds \geq [\int_{t_1}^{t} \lambda dt - (t - t_1) \lambda_c] \int \bar{P}_i \phi_i^c ds = R(t) \quad (56)$$

The response is completed at time t_f when $\dot{u}_i(s, t_f) = 0$, such that

$$t_f \geq t_1 + \frac{1}{\lambda_c} \int_{t_1}^{t_f} \lambda dt \quad (57)$$

Comparing this with Eq. (52), we see that

$$t_f \geq t_f^c \quad (58a)$$

provided that

$$\int_{t_1}^{t_1^c} (\lambda - \lambda_c) dt + \int_{t_f^c}^{t_f} \lambda dt \geq 0 \quad (58b)$$

Thus the response time t_f^c of a one-degree of freedom treatment, based on a velocity field $\phi_i^c(s)$ which is kinematically admissible but otherwise arbitrary, furnishes a lower bound on the response time of the actual response time t_f, only when the specified load-time function $\lambda(t)$ satisfies the conditions implied by (58b). For example, if the load decreases monotonically to zero, with $t_1 = t_1^c = 0$, these conditions will be satisfied. This includes impulsive loading, and Martin's (1964) result for this case is recoverable from Eq. (52) when Eqs. (36) and (23) are used, namely

$$t_f \geq t_f^c = \frac{\int_{t_1^c}^{t_f^c} \lambda dt \int \bar{P}_i \phi_i ds}{\int Q_\alpha k_\alpha^{pc} ds} = \frac{\int \bar{\rho} \dot{u}_i^0 \phi_i^c ds}{\int Q_\alpha^c k_\alpha^{pc} ds} \quad (59)$$

The lower bound on maximum final displacement is obtained by integrating Eq. (56) from t_1 to t_f, and invoking a mean value theorem, so that for a final displacement component $u_n^f(s)$ one obtains

$$(u_n^f)_{max} \int \overline{\rho}\phi_n ds \geq \int \overline{\rho}\phi_i^c u_i^f ds \geq \int_{t_1}^{t_f} R dt \qquad (60)$$

where $R(t)$ is the right-hand side of Eq. (56). Under suitable load conditions it can be shown that

$$\int \overline{\rho}\phi_i^c u_i^f ds \geq u_f^c \int \overline{\rho}\phi_i^c \phi_i^c ds = \int_{t_1^c}^{t_f^c} R_c dt \qquad (61)$$

where $R_c(t)$ is the right-hand side of Eq. (52).

Thus for example if $\lambda(t)$ is a function which decreases monotonically to zero (including impulsive loading), the "weighted average" of u_i^f in Eq. (61) is bounded below by the quantity in Eq. (61) furnished by the one-degree of freedom treatment.

In general, the single DOF model furnishes bounds only for special pulse shapes. Its real advantage is in providing a conceptually simple scheme for calculation.

ACKNOWLEDGEMENT

The writer benefited from many discussions with Dr. T. X. Yu, now at Mechanics Department, Peking University, prior to writing this paper.

REFERENCES

Bodner, S. R. and P. S. Symonds (1962). Experimental and Theoretical Investigation of the Plastic Deformation of Cantilever Beams Subjected to Impulsive Loading. *J. Appl. Mech.*, Vol. 29, No. 4, pp. 719-727.

Kaliszky, S. (1970). Approximate Solutions for Impulsively Loaded Inelastic Structures and Continua. *Int. J. Non-Linear Mechanics*, Vol. 5, pp. 143-158.

Kaliszky, S. (1973). Large Deformations of Rigid-Viscoplastic Structures under Impulsive and Pressure Loading. *J. Struct. Mech.*, Vol. 1, pp. 295-317.

Lee, L. S. S. (1972). Mode Responses of Dynamically Loaded Structures. *J. Appl. Mech.*, Vol. 39, pp. 904-910.

Lee, L. S. S. and J. B. Martin (1970). Approximate Solutions of Impulsively Loaded Structures of a Rate Sensitive Material. *J. Appl. Math. Physics (ZAMP)*, Vol. 21, pp. 1011-1032.

Lee, E. H. and P. S. Symonds (1952). Large Plastic Deformations of Beams under Transverse Impact. *J. Appl. Mech.*, Vol. 19, No. 3, pp. 308-314.

Martin, J. B. (1964). Impulsive Loading Theorems for Rigid-Plastic Continua. *Proc. ASCE Jour. Eng. Mech. Div.*, Vol. 90, No. EM5, pp. 27-42.

Martin, J. B. (1966). A Note on Uniqueness of Solutions for Dynamically Loaded Rigid-Plastic and Rigid-Viscoplastic Continua. *J. Appl. Mech.*, Vol. 33, pp. 207-209.

Martin, J. B. (1967). Time and Displacement Bound Theorems for Viscous and Rigid-Viscoplastic Continua Subjected to Impulsive Loading. *Developments in Theoretical and Applied Mechanics*, Pergamon Press, Vol. 3, pp. 1-22.

Martin, J. B. (1972). Extremum Principles for a Class of Dynamic Rigid-Plastic Problems. *Int. J. Solids Structures*, Vol. 8, pp. 1185-1204.

Martin, J. B. (1981). The Determination of Mode Shapes for Dynamically Loaded Rigid-Plastic Structures. *Meccanica*, Vol. 16, pp. 42-46.

Martin, J. B. and L. S. S. Lee (1968). Approximate Solutions for Impulsively Loaded Elastic-Plastic Beams. *J. Appl. Mech.*, Vol. 35, pp. 803-809.

Martin, J. B. and P. S. Symonds (1966). Mode Approximations for Impulsively Loaded Rigid-Plastic Structures. *J. Eng. Mech. Div.*, *Proc. ASCE*, Vol. 92, No. EM5, pp. 43-66.

Morales, W. J. (1972). Displacement Bounds for Blast Loaded Structures. *Proc. ASCE, Jour. Eng. Mech. Div.*, Vol. 98, No. EM4, pp. 965-974.

Morales, W. J. and G. E. Nevill, Jr. (1970). Lower Bounds on Deformations of Dynamically Loaded Rigid-Plastic Continua. *AIAA J.*, Vol. 8, No. 22, pp. 2043-2046.

Parkes, E. W. (1955). The Permanent Deformation of a Cantilever Struck Transversely at its Tip. *Proc. Roy. Soc.*, A, Vol. 228, pp. 462-476.

Robinson, D. N. (1970). A Displacement Bound Principle for Elastic-Plastic Structures Subjected to Blast Loading. *J. Mech. Phys. of Solids*, Vol. 18, pp. 65-80.

Symonds, P. S. (1980). The Optimal Mode in the Mode Approximation Technique. *Mechanics Research Communications*, Vol. 7 (1), pp. 1-6.

Symonds, P. S. and C. T. Chon (1975). Bounds for Finite Deflections of Impulsively Loaded Structures with Time Dependent Plastic Behavior. *Int. J. Solids & Structures*, Vol. 11, pp. 403-423.

Symonds, P. S. and N. Jones (1972). Impulsive Loading of Fully Clamped Beams with Finite Deflections and Strain-Rate Sensitivity. *Int. J. Mech. Sci.*, Vol. 14, pp. 49-69.

P. S. Symonds and T. Wierzbicki (1975). On an Extremum Principle for Mode Form Solutions in Plastic Structural Dynamics. *J. Appl. Mech.*, Vol. 42, pp. 630-640.

Youngdahl, C. K. (1970). Correlation Parameters for Eliminating the Effect of Pulse Shape on Dynamic Plastic Deformation. *J. Appl. Mech.*, Vol. 37, pp. 744-752.

ON THE LARGE PLASTIC DEFORMATION ON CLAMPED ANNULAR DIAPHRAGMS UNDER IMPULSIVE LOADING

S. K. GHOSH* and F. W. TRAVIS**

*Department of Mechanical and Computer-Aided Engineering,
North Staffordshire Polytechnic, Beaconside,
Staffordshire ST18 0AD, UK
**Department of Mechanical Engineering,
Sunderland Polytechnic, UK

ABSTRACT

A simple theoretical study, together with experimental vindication, is presented for the inertial forming of annular diaphragms, clamped at their outer edges and subjected to impulsive loading. Starting from the von Mises yield condition and the associated flow rule, a governing equation is deduced for the title problem which can be reduced to a single second order partial differential equation which is solved numerically. Theoretical values of the major parameters over the entire deformation history are compared with corresponding experimental results obtained from tests on such diaphragms using an inertial forming machine. Results of high-speed photographic recordings are presented showing instantaneous profiles, velocities and hoop strain distributions.

NOTATION

a	outer radius of diaphragm
T_0, T	initial and momentary thickness of the diaphragm, respectively
R	initial hole radius
δr	width of annular element
p	internal pressure
δs	meridional arc length of element
r, ψ, z	cylindrical radial, circumferential and axial co-ordinates, respectively
σ, σ_ψ, σ_z	meridional, circumferential and thickness stress, respectively
$\bar{\sigma}$	representative stress
ε, ε_ψ, ε_z	meridional, circumferential and thickness strains, respectively
$\bar{\varepsilon}$, $\dot{\bar{\varepsilon}}$	representative strain and strain rate, respectively
θ	angle between meridional tangent and the horizontal
ρ	density of diaphragm material
t	time
u	radial displacement
V_0	initially uniform transverse velocity
$\dot{u}$, $V = \partial z / \partial t$	radial and axial velocities of a particle, respectively

195

INTRODUCTION

The quasi-static behaviour of an annular plate, where the plate is subjected to a static load and the edges - outer or inner - are either
clamped or simply supported has been treated by Johnson and Mellor
(1973). The method of analysis employed took account of the 'finite
plastic deflection' experienced by such a structural member under a
prescribed load. The problem of the dynamic or inertial loading of
an annular plate, where the inner edge of the plate is rigidly clamped,
has also been treated: an expression for the peripheral outer-edge
deflection in this situation is derived in terms of the plastic bending
moment. Johnson and Mellor (1973) state that "for dynamic investigations, it should be kept in mind that the value of M_p (fully plastic
bending moment) will depend on the rate of straining of the material".

Similar dynamic loading problems encountering *finite deflections only*,
as described above, have also been treated by Shapiro (1959), Jones
(1968 a and b), Perrone (1967) and Florence (1965), and Aggarwal and
Ablow (1971) have studied the plastic bending of an annular plate by
a uniform impulse. However, the extreme impulsive overload of a thin
annular diaphragm clamped along its boundary, has not been examined
in the literature - either experimentally or theoretically. Further,
most of the work on plate problems reported to date has been concerned
with whole circular diaphragms; the practical difficulties of uniformly loading an annular diaphragm using explosives would be formid-
able. It may be noted that Al-Hassani, Duncan and Johnson (1968)
employed the electromagnetic forming process with some degree of
success in a somewhat similar situation to explore the capping of a
glass jar by means of the dynamic deformation of the unsupported
periphery of a whole circular diaphragm placed coaxially on the open
mouth of the jar. In this situation, as in the explosive loading
situation, the uniformity of the velocity imparted would, however,
be very difficult, if not impossible, to achieve. To the best of
the authors' knowledge, the present work is the first attempt to explore the large deformation of annular diaphragms where a controlled,
uniform and known velocity is imparted to the workpiece.

The objectives of the present study were to formulate in simple terms
an approximate governing equation of motion for the title problem,
to obtain a straightforward numerical solution and to compare experimental results such as the permanent deformed shape, the peak deflection, the strain distribution etc., of an inertially formed annular
diaphragm with theoretical predictions. The theory takes account of
membrane stresses only and makes a number of simplifying assumptions
that have been found to be reasonably substantiated by subsequent
experimental findings.

The experiments on annular diaphragms were carried out using an inertial forming machine, the advantages and details of which are described
for example by Ghosh, Balendra and Travis (1978) and Ghosh and Travis
(1978).

ANALYSIS

The isotropic annular diaphragm, shown in Fig. 1(a), is supported at
$r = a$. Since bending moments are neglected in the membrane theory
used below, it does not matter whether the support is considered
rigid (clamped) or pin-jointed. The diaphragm is assumed to be flat
at $t = 0$ (i.e. $z(r,0) = 0$) of uniform thickness T_0, and to receive

an instantaneous impact over its entire surface, i.e. an initially uniform transverse velocity distribution $V(r,t)$ which satisfies

$$V_0 = V(r,0) < 0 \qquad (R \leq r < a) \tag{1}$$

The diaphragm deforms freely thereafter under the influence of its own inertia forces. The impact, as such, is not considered, so that the initial acceleration need not be prescribed. There is, further, no external pressure p acting on the diaphragm surface during deformation and therefore for $t > 0$,

$$p \equiv 0 \tag{2}$$

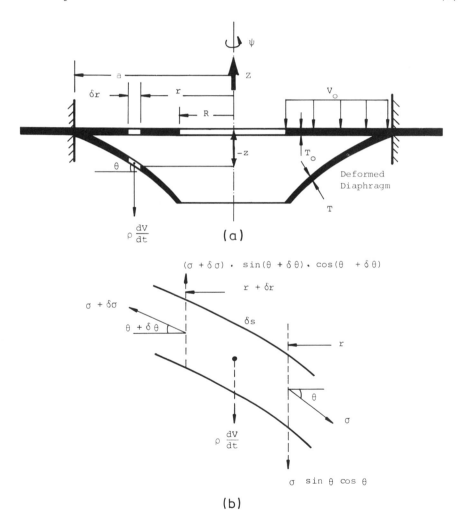

Fig. 1.   Showing (a) the theoretical model and (b) the equilibrium of the annular element considered in the analysis.

S. K. Ghosh and F. W. Travis

Consider the annular element from Fig. 1(a), shown in detail in Fig. 1(b). At radius r, there is a longitudinal stress σ parallel to the surface (in the meridional direction) and the diaphragm has deformed through an angle θ from its original flat position. Over the small radial increment δr, the diaphragm has rotated through a small angle δθ, and thus the total angle of rotation at r + δr is θ + δθ.

The inertial loading is considered to produce axially-symmetric motion of the diaphragm. The equation of motion can be written using only the radial components of strain ε and stress σ along the tangent to the meridional line of the diaphragm. An assumption is made here to simplify the problem: the radial acceleration is neglected. This assumption was made in the particular case of whole circular diaphragms by Hudson (1951), Boyd (1966) and Frederick (1959). In addition to other arguments, it was stated by Frederick (1959) that this assumption has been substantiated experimentally, where radial motions were found to be quite small. Excessive computing times were reported by Witmer et al (1963) when allowing for radial inertia and, in fact, they were so great that the authors suggested that alternative, simpler approaches should be sought. Lippmann (1974) made the same assumption and considered that, for not-too-extensive deformations resulting from a uniform vertical impulse, all the particles of a whole diaphragm could be considered to move vertically downwards only. As a result of this asssuption, however, there have been discrepancies noted in relation to strain distributions obtained experimentally (Travis and Lippmann, 1974), and in the theoretical-experimental correlation of Ghosh and Weber (1976). This problem has been considered further recently by Ghosh, Reid and Johnson (1984). Notwithstanding this, in the present work, to enable the development of a simple, approximate theory for an annular diaphragm, the assumption is made that *the material moves vertically downwards only* i.e.

$$V = V(r,t) = \dot{z}(r,t) < 0 \qquad (R \leq r < a)$$

and $\qquad u = u(r,t) \equiv 0 \qquad\qquad (R \leq r < a)$

Thus there is no radial acceleration i.e. $\frac{\partial u}{\partial t} = 0$ and the motion is governed by just one equation of motion (in the z-direction) due to the symmetry of the problem. The annular element is acted upon by the following forces in the z-direction at r

$$2\pi \, rT \, \sigma \, \sin\theta \, \cos\theta$$

and at r + δr

$$2 \pi(r + \delta r)T \, (\sigma + \delta\sigma) \, \sin(\theta + \delta\theta) \, \cos(\theta + \delta\theta)$$

Since the vertically downward force on this element is $-2\pi \, r \, \delta r \, \rho$ $T\frac{\partial V}{\partial t}$, equilibrium of forces in the z-direction gives

$$-2\pi \, r \, \delta r \, \rho \, T \, \frac{dV}{dt} = 2\pi \, r \, T \, \sigma \, \sin\theta \, \cos\theta - 2\pi(r + \delta r)T \, (\sigma + \delta\sigma)$$
$$\sin(\theta + \delta\theta) \, \cos(\theta + \delta\theta)$$

Thus in the limit this reduces to

$$2\rho \, r \, \frac{\partial V}{\partial t} - \frac{\partial}{\partial r} \, (r \, \sigma \, \sin 2\theta) = 0 \qquad\qquad (3)$$

This equation can be expressed as a second-order non-linear partial differential equation for $z(r,t)$, as shown below. σ and $\sin 2\theta$ are unknown quantities which can be found from the geometry of the deformed diaphragm and the constitutive equation for the material. It is to be noted that the above governing equation of motion is basically the same as that proposed by Lippmann (1974), Munday and Newitt (1963) and Cristescu (1967), for whole circular diaphragms under an initially uniform transverse velocity. From the geometry of the deformed diaphragm,

$$\tan \theta = \frac{\partial z}{\partial r} \tag{4}$$

and therefore,

$$\sin 2\theta = 2 \sin \theta \cos \theta$$

$$= \frac{2 \cdot \partial z/\partial r}{[1 + (\partial z/\partial r)^2]^{\frac{1}{2}}} \tag{5}$$

In thin-diaphragm deformation problems the stress in the thickness direction is assumed to be zero, i.e.,

$$\sigma_z = 0 \tag{6}$$

Available experimental data on the impact forming of thin annular diaphragms - see Ghosh, Balendra and Travis (1978) and Ghosh and Travis (1978) - demonstrate especially for annular diaphragms deformed not-too-extensively, that to a first approximation the circumferential strain may be taken as zero. It is appreciated that this assumption is strictly incompatible with the earlier assumption that radial displacements are zero, and, indeed, examination of subsequent figures in this work will demonstrate that the latter are *not* insignificant. For the purposes of the present *approximate* analysis however, both of these assumptions will be assumed to hold - this will be discussed later in the chapter.

Considering the basic Levy-Mises equation of flow

$$\delta \varepsilon_{ij}^p = \sigma'_{ij} \; \delta \lambda \tag{7}$$

then gives the following relationship between the circumferential and meridional stresses;

$$\sigma = 2 \; \sigma_\psi. \tag{8}$$

σ_ψ is clearly the intermediate stress. The condition of volume constancy gives,

$$\varepsilon = -\varepsilon_z, \tag{9}$$

since $\varepsilon_\psi = 0$.

In the aforementioned situation, the representative stress

$$\bar{\sigma} = \frac{1}{\sqrt{2}} \; [(\sigma_\psi - \sigma)^2 + (\sigma - \sigma_z)^2 + (\sigma_z - \sigma_\psi)^2]^{\frac{1}{2}} \tag{10}$$

can be expressed, substituting (8) into (10), as

$$\bar{\sigma} = \frac{\sqrt{3}}{2} \sigma \tag{11}$$

where $\bar{\sigma}$ depends on $\bar{\varepsilon}$ and $\dot{\bar{\varepsilon}}$ through an appropriate constitutive equation.

Similarly, the representative strain,

$$\bar{\varepsilon} = \frac{\sqrt{2}}{3} [(\varepsilon_\psi - \varepsilon)^2 + (\varepsilon - \varepsilon_z)^2 + (\varepsilon_z - \varepsilon_\psi)^2]^{\frac{1}{2}} \tag{12}$$

may be defined in terms of ε_z,

$$\bar{\varepsilon} = \frac{2}{\sqrt{3}} \varepsilon_z = \frac{2}{\sqrt{3}} \ln \frac{T_0}{T} \tag{13}$$

To enable a solution of the governing equation of motion (3) to be found, the quantity ε_z must be determined theoretically to evaluate $\bar{\varepsilon}$ and $\dot{\bar{\varepsilon}}$, and hence $\bar{\sigma}$. This may be done by using the condition of incompressibility of the annular element before and after deformation. Thus

$$T_0 \cdot \delta r = T \cdot \delta s$$

Since from Fig. 1(b) $\delta s = \delta r/\cos \theta$,

$$T = T_0 \cos \theta = T_0/[1 + (\partial z/\partial r)^2]^{\frac{1}{2}}. \tag{14}$$

Equation (14) predicts that in a deformed diaphragm the final thickness decreases from the clamped periphery to the inner edge of the diaphragm, as θ likewise increases. This particular prediction is confirmed experimentally, as will be seen later from the profiles and strain distributions of typical deformed diaphragms.

Using relation (13), the representative strain rate may be found as follows

$$\dot{\bar{\varepsilon}} = \frac{2}{\sqrt{3}} \frac{\partial}{\partial t} (\ln \frac{1}{\cos \theta}) = \frac{2}{\sqrt{3}} \tan \theta \frac{\partial \theta}{\partial t} = \frac{2 \tan \theta}{\sqrt{3}(1 + \tan^2 \theta)} \cdot \frac{\partial V}{\partial r} \tag{15}$$

since $\frac{\partial V}{\partial r} = \frac{\partial}{\partial t} (\frac{\partial z}{\partial r}) = \frac{1}{\cos^2 \theta} \cdot \frac{\partial \theta}{\partial t}$

All the unknowns in the governing equation have now been defined. Hence, the function $z = z(r,t)$, which defines the shape of the deformed sheet, may be obtained by numerical integration of equation (3).

NUMERICAL SOLUTION

The axial-displacement equation (3) is a non-linear second-order partial differential equation which can be solved in the manner described below.

Boundary Conditions

The boundary conditions which are consistent with axially-symmetric displacement and with a rigidly-clamped outer edge, are as follows

$$z(r,t) = 0 \text{ for } t \geq 0 \text{ and } r = a, \tag{16}$$

while at the inner boundary,

$$\sigma = 0 \text{ for } t \geq 0 \text{ and } r = R. \tag{17}$$

Initial Conditions

The diaphragm is assumed to be initially flat. Therefore

$$z(r,0) = 0 \text{ for } (R \leq r < a) \text{ and } t = 0 \tag{18}$$

The initial velocity is directed axially and is uniform over the diaphragm.

Hence

$$\frac{\partial}{\partial t} z(r,t) = V(r,t) = V_0 \text{ for } (R \leq r < a) \text{ and } t = 0. \tag{19}$$

Solution

To determine the shape of the diaphragm after time Δt, the following numerical scheme was adopted

$$z(r,t + \Delta t) \approx z(r,t) + V(r,t) \cdot \Delta t \tag{20}$$

The new velocity distribution over the diaphragm, according to (3) is

$$V(r,t + \Delta t) = V(r,t) + \frac{1}{2\rho r} \frac{\partial}{\partial r} (r \sigma \sin 2\theta) \cdot \Delta t \tag{21}$$

To complete (21), the value of σ and $\sin 2\theta$ have to be found from (5) and (11), substituting in relation (11) the stress-strain-strain-rate characteristic of the material is given by (22).

The computational cycle goes on until $V(r,t + n \cdot \Delta t) = 0$ which physically represents the total dissipation of the kinetic energy initially delivered over the diaphragm. Therefore, with conventional numerical techniques employing finite difference quotients, the governing equation of motion can be solved for $z(r,t)$, thus giving the transient and final values of deflection, velocity and strain.

EXPERIMENTAL INVESTIGATION

Commerically pure lead (ρ = 11.3 g/cm³) of 0.61 mm initial thickness was used as the test material. It was found, using a servo-hydraulic testing machine, that the representative stress-strain-strain-rate characteristics for this material may be approximated by the following relationship, within a strain-rate range of 0.001 - 10 s⁻¹

$$\bar{\sigma} = 46.4 \, \varepsilon^{-0.365} \, \dot{\varepsilon}^{0.03} \text{ MPa} \tag{22}$$

The results of tests using annular diaphragms of 120 mm outside dia-meter at different velocity levels employing a specially designed forming machine, have already been reported by Ghosh, Balendra and

Travis (1978) and Ghosh and Travis (1978). Of the range of initial hole diameters employed in the investigation reported in Ghosh and Travis (1978), a diaphragm having an initial hole diameter of 40 mm was chosen to examine the theoretical-experimental correlation. This represents the situation where the ratio of exposed radius (a) to initial hole radius (R) is 3 : 1. For a detailed description of the forming machine and the test procedure, the reader is referred to the above references. However, a simple description of the machine - shown in Fig. 2(a) - is that it consists of a spring-loaded arm to which the diaphragm and clamping rings are attached. The arm is strained to different angular locations - thus providing a range of impact velocities - and upon release of the arm the diaphragm and clamping rings are rapidly accelerated to achieve a high velocity. Upon the virtually instantaneous arrest of the clamping rings by a rigid hollow anvil, the diaphragm deforms entirely under its own inertia forces. The final profile measurements were done whilst the specimens were still in the clamping rings.

High-Speed Photographic Recordings

High-speed photographic recordings were made using a 'Hyspeed' camera at a framing speed of about 8000 frames/second, employing an annular specimen tested at an impact velocity of 13.5 m/s. These films show that the deformation propagates form the clamped periphery in the form of a plastic wave front which separates the part of the annulus which has undergone rotation from that which has not. This wave has been described as a 'hinge wave' (Frederick, 1959), a 'plastic bending wave' (Hudson, 1951) and 'plastic strain waves and inertia waves' (Munday and Newitt, 1963). The term 'plastic bending wave' will be used herein as it is rotation of the membrane which is the most observable feature of the deformation. From the above recordings, transient profiles, velocity distributions and the speed of propagation of the plastic bending wave were determined.

THEORETICAL-EXPERIMENTAL CORRELATION

Figure 3 shows the profiles of annular diaphragms with an initial hole diameter (= 2R) of 40 mm, deformed at different initial impact velocities. Profiles computed according to equation (20) are also shown in this figure. Comparison shows that the present analysis provides a reasonably accurate approximation to the true behaviour. However, from examination of the results it is noted that the small differences between the experimental and theoretical profiles on opposite sides of a particular diaphragm - at equal radial distances - are not the same; this situation may be attributed to the fact that the experimental deflections are not *exactly* axially-symmetric. This arises from two causes. First, there is a difference of 6% in the velocity distribution across the diaphragm between the nearer and the outer points of its clamped periphery - measured from the axis of rotation of the spring-loaded arm of the forming machine. Secondly, at the higher velocities, there is 'curling' of the material surrounding the free inner edge of the diaphragm, with irregularly disposed lip fractures, which cannot be incorporated into the present analysis, as the latter does not take into account any assumption at the free inner edge other than the prescribed boundary condition (17).

It must also be reiterated, that in formulating the present simple theoretical model, the radial displacement was taken to be zero everywhere and that the material was assumed to move vertically

Fig. 2.   Showing (a) the overall layout of the forming
          machine, and (b) an annular diaphragm tested
          at an impact velocity of 13.5 m/s.

downwards only. However, the radial strain and the radial displace-
ment are related by $\varepsilon_r = \partial u/\partial r + \frac{1}{2}(\partial w/\partial r)^2$, so that non-zero radial

displacements are required if $\varepsilon_r \neq 0$. The final measured permanent

strain distributions of impulsively loaded annular diaphragms reported
in Ghosh, Balendra and Travis (1978) reveal that equation (9) is not
satisfied exactly, and further, the radial displacement in the present
work - as can be seen, say, in the profiles of diaphragms in Fig. 3
- are not insignificant and cannot be neglected in formulating an
exact solution. It may be noted from the present results that for
very heavily deformed annular diaphragms showing no curling and lip
fracture (of the material surrounding the hole) ε_r is of the order
of 0.4 - 0.5.

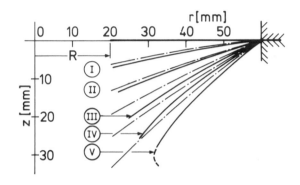

Fig. 3.   Comparison of experimental deflections at
          different velocities with results computed
          according to eqn. (20). Chain-dotted lines
          refer to theoretical predictions and full
          lines to experimental observations, where
          I to V refer to impact velocities of 3.5,
          8.5, 13.5, 17.44 and 23.5 m/s respectively.

Transient velocity distributions and corresponding diaphragm deflections computed using equations (20) and (21) for an initial velocity
at 13.5 m/s, are presented in Figs. 4 and 5 respectively. It can be
seen from Fig. 4 that up to about 0.5 ms after impact the velocity
in the neighbourhood of the inner edge of the annulus is maintained
at almost the initial magnitude, whereas that towards the clamped
periphery decreases in a near parabolic fashion. Further, after this
period has elapsed, the reduction in velocity is faster towards the
clamped periphery that it is towards the centre of the diaphragm.
The same feature is manifested in the transient deflection profiles
of Fig. 5, where the deflection of the material surrounding the hole
is entirely vertically downwards up to 0.5 ms approximately after
the start of the deformation. This behaviour is in agreement with
the behaviour of a whole circular diaphragm under uniform velocity
distribution as presented by Hudson (1951). As an illustration,
the theoretical deformation profile corresponding to t = 1 ms of
Fig. 5 may be compared with the diaphragm - shown in Fig. 2(b) -
which was tested at the same velocity level using the inertial forming
machine.

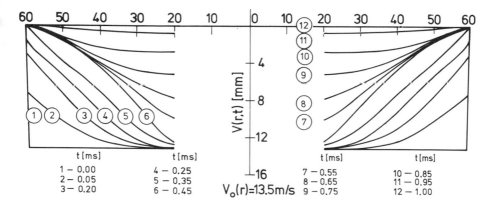

Fig. 4.   Showing the transient velocity distributions
           computed using eqn. (21):   also presented
           are the times after impact.

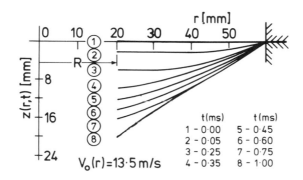

Fig. 5.   As for Fig. 4, but showing corresponding
           diaphragm deflections.

Figure 6 shows the comparison between the thickness strain distributions computed from the experimentally recorded hoop and radial
strains and the values computed using equation (14).   It may be noted
that the theoretical analysis inevitably assumes a state of no-
thinning at the clamped periphery of the diaphragm, whereas all the
experimentally-tested diaphragms show a small degree of thinning at
their clamped boundaries, the amount of which increases, in general,
with increasing initial velocity.   This is possibly due to the radius
(~2 mm) provided on the edge of the clamping ring to avoid peripheral
tearing of the diaphragm upon impact.   As a result of the bend-radius,
the effective clamping is not exactly at a radius of 60 mm, as assumed
theoretically, but somewhat further away from the centre which leads
to the afore-mentioned discrepancy.

The comments made earlier concerning the validity of the assumption $\varepsilon_\psi = 0$ and the neglect of radial displacements in the formulation of the present theory should be recalled. These considerations give rise to theoretical strain distributions that are different from those of the actual specimens. However, it is clearly seen from Fig. 6 that, if all the experimental thickness-strain values at the clamped periphery of the diaphragms are made to coincide with the no-thinning line at r = 60 mm, there is negligible deviation between the overall strain distribution and the theoretical strain profiles.

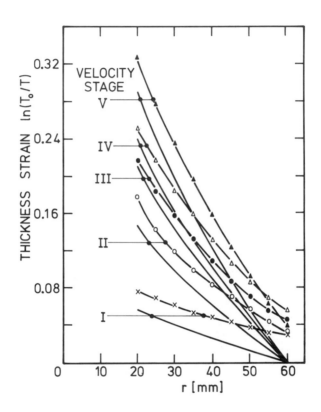

Fig. 6.   Comparison of computed and experimental
           thickness strain distributions. Impact
           velocities are as for Fig. 2.

HIGH-SPEED PHOTOGRAPHIC MEASUREMENTS

Figure 7 shows the photographs of the course of deformation of an annular diaphragm which was deformed at a velocity of 13.5 m/s, as presented in Fig. 2(b). Due to the arrangement employed to arrest the diaphragm, instantaneous plan views of the deforming specimen could not be obtained. The high-speed photographic recordings were carried out therefore by placing the camera and its objective lens at an angle α (in this case, 70°) to the vertical plane passing through the axis of the diaphragm. Thus, if the apparent deformation of a point on the diaphragm is Δ, the true deflection is then $\Delta/\cos \alpha$. The transient deflections of the diaphragm were calculated from measurements taken from individual frames of the set of photographs in Fig. 7 and the results are presented in Fig. 8.

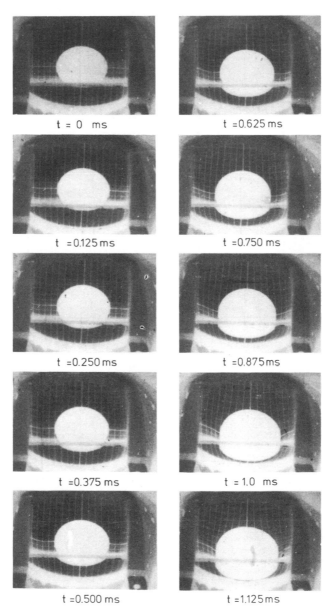

Fig. 7. High-speed photographs showing the process of deformation of the annular diaphragm presented in Fig. 2(b). The time interval between frames is approx. 0.125 ms.

It can be seen from the profiles of Fig. 8 that up to about 0.5 ms
after impact, the material surrounding the hole remains flat and moves
vertically downwards: it is then tilted by a bending wave which
propagates radially inwards generated by the action of peripheral
edge restraint. It may be noted that until a time of 0.5 ms, enlargement of the central hole is minimal. Thereafter, further
deformation towards the clamped periphery reduces considerably,
whilst that towards the free inner edge is enhanced.

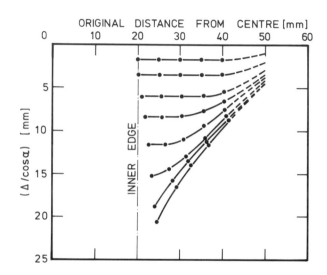

ORIGINAL DISTANCE FROM CENTRE [mm]

Fig. 8.   Transient displacement of the annular
           diaphragm obtained from Fig. 7. (Time
           interval between successive curves =
           0.125 ms).

Since the framing speed of the film is known, the transient velocity
profiles may be easily calculated from the curves of Fig. 8. Figure
9 shows the instantaneous velocity distributions over the diaphragm,
from which it will be noted that the results of the high-speed films
confirm the corresponding theoretical results shown in Figs. 4 and 5.

From the instantaneous plan views of the specimen, the positions of
the bending wave fronts were noted and these positions were plotted
against time after impact, as shown in Fig. 10. Hudson (1951) proposed a model showing the idealised mode of deformation in an explosively loaded blank. In his model for a whole circular specimen,
it is assumed that the diaphragm has some initial impulsive velocity
normal to its plane. A plastic bending wave travels with a speed of
$\sqrt{Y/\rho}$ (where Y is the static yield stress of the material and ρ its
mass density) from the clamped periphery towards the centre of the
specimen, material in the central flat portion of the diaphragm being
tilted as the wave passes through it. The central flat portion of
the blank possesses a radially outward speed, produced by waves
which are propagating radially inwards because of the clamped boundary. Kiyota (1962) has presented high-speed photographic recordings
showing the course of deformation in lead diaphragms (including

diaphragms with a central hole) loaded explosively in air.   The results confirm, in general, Hudson's assumed model.   According to
Hudson, the final theoretical shape of the diaphragm is conical, and
the effect of strain and strain-rates is said to be small, primarily
serving to round-off the apex of the cone.

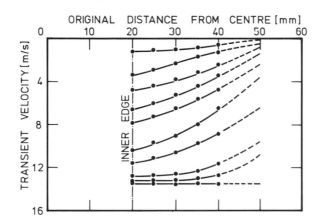

Fig. 9.   Transient velocity distributions over the
          deforming diaphragm, evaluated from Figs.
          7 and 8.   (Time interval between successive
          curves = 0.125 ms).

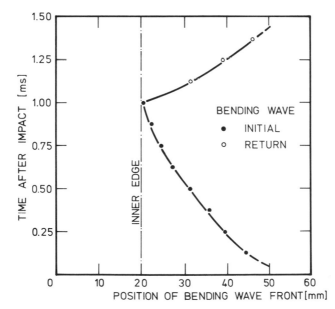

Fig. 10.   Presenting the positions of both the initial
           bending wave front and the return bending
           wave front at different times after impact.

It would seem reasonable to assume that the presence of a central hole in a diaphragm eases the radially outward flow of material during the advancement of the bending wave front towards the hole boundary. The velocity of the bending wave front in the present case is 40 m/s when calculated according to Hudson's equation - using a value of 22.5 MPa for Y which corresponds to a strain rate of 1 s^{-1} for the material used — whilst from the experimental data of Fig. 10 the average velocity of the bending wave is found to be approximately 37 m/s; there is thus quite close agreement.

Once the initial bending wave has reached the hole periphery, in contrast to the case for whole circular diaphragms, a reflected bending wave now proceeds radially outwards towards the periphery. The positions of the returning bending wave front obtained from the high speed film are also plotted in Fig. 10. The velocity of the return bending wave is found to be about 73.2 m/s, which is almost twice the magnitude of that of the initial bending wave: it is difficult, however, to deduce why the velocity of the return bending wave should be any greater than that of the initial bending wave. This feature, presumably, can be related to the state of the material after the initial bending wave has reached the hole periphery. The value of Y (or more presisely $\bar{\sigma}$) in the bending wave speed relation, $\sqrt{Y/\rho}$, is greater since the deformed material is now work-hardened according to $\bar{\sigma} = \bar{\sigma}(\bar{\epsilon}, \dot{\bar{\epsilon}})$: an accurate account of the current value of stress would thus raise, to some extent, the value of the return bending wave speed.

From the grid printed on the diaphragm, transient hoop-strain values were obtained at the hole periphery and at radii of 30 and 40 mm; results are presented in Fig. 11. In general, and considering only the in-plane motion of the diaphragm, straining starts apparently at all locations immediately after impact takes place: this is true whether or not the bending wave has passed through the point in question. It is clear that even if the bending wave has not progressed to a particular point, the effect of the constraint at the periphery of the diaphragm is to also generate a tensile plastic wave which propagates radially inward, thus producing measurable hoop strains. Alter and Curtis (1956) have shown that the speed of a tensile plastic wave travelling through lead is approximately 200 m/s, which is much greater than that of the bending wave. This statement is supported by a comparison of Figs. 10 and 11, and also by the observation of the points in the neighbourhood of the hole periphery: as seen in the first frame following impact (Fig. 7) there is straining in the hoop direction.

Finally, it is necessary to mention that the results of Figs. 8 to 11 have some small error as a consequence of not recording the course of deformation of the diaphragm as plan views, but at an angle with respect to the impact axis, since the deforming diaphragm advances towards the camera. However, the calculated maximum error amounts to about 2%. The effect of this small error is to overestimate hoop strains. It is not significant in relation to the recorded values, and does not affect the foregoing discussion. The greatest error occurs at the end of the process whilst the period of interest is towards the beginning of the operation, when transverse deflections are small.

CONCLUSIONS

(i) The approximate theory presented herein can be used to predict, with reasonable accuracy, the instantaneous and final profiles, and velocities and the thickness strain distributions of inertially-loaded annular diaphragms. The general neglect of radial strain and radial displacements, however, results in some discrepancy.

(ii)   High-speed photographic recordings carried out on a typical annulus have afforded a simple and reliable means of confirming theoretical predictions of some in-process behaviour, whilst the post-process measurement of the formed specimen enabled confirmation of terminal predictions.

(iii) The forming machine employed for the experimental tests reported herein affords an accurate and repeatable means of providing the impact and ensuring conditions required to fulfil the assumptions made in the formulation of the present theory.

ACKNOWLEDGEMENTS

The forming machine used for the tests reported herein was designed by the first author during his stay at the University of Karlsruhe. He wishes to thank the University of Karlsruhe for generously giving the machine to Sunderland Polytechnic, to enable him to continue his research.

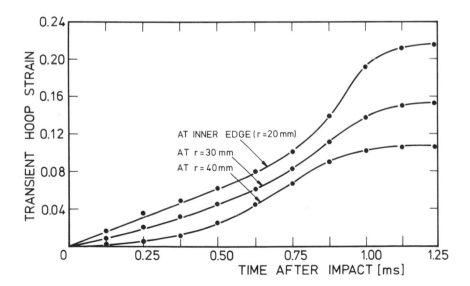

Fig. 11.   Showing the calculated hoop strain at
           different times after impact for 3
           different locations on the deforming
           diaphragm.

REFERENCES

Aggarwal, H. R. and C. M. Ablow (1971). *Int. J. Nonlinear Mech.*, $\underline{6}$, 69.

Al-Hassani, S. T. S., J. L. Duncan and W. Johnson (1968). *Proc. 8th Int. MTDR Conf. (Manchester), 1967*, p. 1333, Pergamon, Oxford.

Alter, B. E. K. and C. W. Curtis (1956). *J. Appl. Phys.*, $\underline{27}$, 9.

Boyd, D. E. (1966). *J. Eng. Mech. Div. ASCE*, $\underline{92}$, 1.

Cristescu, N. (1967). *Dynamic Plasticity*. North Holland Publ. Co., Amsterdam, pp. 307-316.

Florence, A. L. (1965). *AIAA J.*, $\underline{3}$, 1726.

Frederick, D. (1959). *Proc. 4th Annual Conf. Solid Mech.* University of Texas, Austin, p. 18.

Ghosh, S. K., R. Balendra and F. W. Travis (1978). *Proc. 18th Int. MTDR Conf. (London), 1977*, p. 845, Macmillan, London.

Ghosh, S. K., S. R. Reid and W. Johnson (1984). *Proc. Int. Conf. on Structural Impact and Crashworthiness*, July 16-20, 1984, Imperial College, London, pp. 471-481.

Ghosh, S. K. and F. W. Travis (1978). *Int. J. Mach. Tool Des. Res.*, $\underline{18}$, 169.

Ghosh, S. K. and H. Weber (1976). *Mech. Res. Comm.*, $\underline{3}$, 423.

Hudson, G. E. (1951). *J. Appl. Phys.*, $\underline{22}$, 1.

Johnson, W. and P. B. Mellor (1973). *Engineering Plasticity*, Van Nostrand Rheinhold, London, pp. 521-525.

Jones, N. (1968a). *J. Appl. Mech.*, $\underline{35}$, 349.

Jones, N. (1968b). *Int. J. Solids Struct.*, $\underline{4}$, 593.

Kiyota, K. (1962). *Mems. Fac. Eng.* Kumamoto University IX, No. 2.

Lippmann, H. (1974). *Int. J. Mech. Sci.*, $\underline{16}$, 297.

Munday, G and D. M. Newitt (1963). *Phil. Trans. Roy. Soc. (London)* A 251, 1.

Perrone, N. (1967). *J. Appl. Mech.*, $\underline{89}$, 380.

Shapiro, G. S. (1959). *Prik. Mat. Mekh.*, $\underline{23}$, 172.

Travis, F. W. and H. Lippmann (1974). *Int. J. Mech. Sci.*, $\underline{16}$, 945.

Witmer, E. A., H. A. Balmer, J. W. Leech and T. H. H. Pian (1963). *AIAA J.*, $\underline{1}$, 1848.

AN ENERGY CRITERION FOR THE DYNAMIC PLASTIC BUCKLING OF CIRCULAR CYLINDERS UNDER IMPULSIVE LOADING

R. WANG and C. Q. RU

Department of Mechanics, Peking University, Beijing,
People's Republic of China

ABSTRACT

A modified energy criterion is given for the dynamic plastic buckling of a circular cylindrical shell under axial impulsive loading.
It yields some simple analytic formulae for determining the buckling
wave number and the threshold velocity. The method assumes that when
the impulsive velocity is below the threshold value the work done by
the loading is insufficient to meet the energy required for the development of buckling. Numerical examples show good agreement with
experiments. Simple formulae are derived for a stringer stiffened
cylindrical shell and a rectangular plate. Numerical results agree
well with those obtained after tedious computation.

INTRODUCTION

When considering the dynamic plastic buckling of rectangular plates
and cylindrical shells under in-plane or axial impulsive loading,
the current practice is to follow the method used by Goodier (1968)
and Florence and Goodier (1968), who introduced the concept of an
amplification function. Accordingly, the wave number that yields
the largest amplification corresponds to the buckling mode and it
is obtained after complicated numerical calculations. However, the
wave numbers obtained for displacement perturbation and for velocity
perturbation cannot be proved to be close to each other.

Although the critical loading condition is a problem of fundamental
importance in the theory of buckling, there is still a lack of theoretical understanding for the case under consideration. A threshold
velocity is usually defined as a velocity which, if not exceeded,
the amplification will be below a certain arbitrarily prescribed
value. For example, Vaughan (1969), on the basis of his experimental
results, defined this velocity as that corresponding to $(\varepsilon_\theta)_{max}$ =
h/4a for cylindrical shells, where h and a are the wall thickness
and radius of the shell respectively. The difficulty in providing
an objective definition for the critical loading condition lies in
the fact that the commonly accepted stability criteria are no longer
applicable. Liaponoff's definition of stability is also not suitable
here, since stability will depend entirely on the tendency of

initial perturbations to grow at sufficiently large t. It in turn
depends on the asymptotic behavior of the dominant motion at that
time, as the equations of motion for the perturbed motion is determined by the dominant motion. In the case of impulsive loading, to
analyse the behavior at sufficiently large t is very difficult and
impractical if not impossible. However, since buckling takes place
within a very short time t_f, it is assumed that stability of the
dominant motion depends on the behavior of the dominant motion within
$[0, t_f]$ and investigation of the motion after t_f is not necessary.
Here t_f is the response duration of the motion.

In this chapter, we shall reconsider this type of problem from a more
fundamental point of view, make use of the idea of a classical energy
criterion, and take into consideration the characteristics of
stability under impulsive loading. A modified energy criterion is
suggested together with an analytical method which is used to treat
the dynamic plastic buckling of circular cylindrical shells, stringer
stiffened cylindrical shells and a rectangular plate under impulsive
loading.

BASIC IDEA OF ENERGY CRITERION

The classical energy criterion states that if any admissible perturbation is imposed on the deformed state for the system under a given
loading and leads to an increase in the total energy of the system
including the potential energy of the load, then, this state of the
system is stable. Note especially, the system here is conservative
and the state is an equilibrium one. In calculating the energy, the
initial disturbing energy that causes the perturbation is neglected.

In accordance with the above notion, the basic idea of the proposed
modified energy criterion is: for a system under a certain impulsive
loading, when any kinematically admissible perturbation is imposed
on its dominant motion, if the energy absorbed or dissipated by the
system due to the perturbation is always larger than the work done
by the load on the perturbation, then, the dominant motion of the
system is a stable one. Suppose that the kinematic equation for the
perturbed motion w(x,t) is represented by

$$f(\lambda, w) = 0 \tag{1}$$

where f is a linear differential operator in which the coefficient
for the acceleration $\ddot{w}$ is positive, $0 \le x \le L$, and λ is the load
parameter. Then the fundamental criterion can be stated as follows:
the dominant motion is stable if for any admissible perturbation
w(x,t),

$$F(\lambda, w) \equiv \int_0^L \int_0^{t_f} f(\lambda, w) \, \dot{w} \, dt \, dx > 0. \tag{2}$$

The response duration t_f can be treated as a function of the loading
parameter and calculated from other considerations. The dot denotes
time derivative.

Note especially that the perturbation w(x,t) in addition to satisfying the geometric constraints, for the present problem should also
possess the following characteristics:

A. The energy of the initial perturbation determined by $w(x,t)$
    is small, and can be neglected while computing $F(\lambda,w)$.
B. Buckling does not occur at the initial instant, so that the
    initial values of w and its derivatives with respect to x
    (needed to describe the buckled state) as represented by their
    averages can be neglected compared with their final values at
    $t = t_f$.

These characterize the perturbation at the very beginning of buckling. It is this initial value of the perturbation that is important
in dealing with buckling problems within a finite time. We believe
that this value can best be taken care of by an energy method. These
characteristics are obtained on the basis of heuristic, physical
arguments. In the following we shall show how to express them in
terms of an equivalent mathematical formulation related to specific
problems.

We shall denote the entirety of all admissible perturbations that
possess the two above characteristics by a set B. Then a sufficiency
condition for stability may be written as

$$\underset{w \in B}{F(\lambda,w)} > 0 \tag{3}$$

In practical calculations, one of the main difficulties to deal with
is the presence of non-conservative terms.

CIRCULAR CYLINDER UNDER AXIAL IMPULSIVE LOADING

We shall make use of the two following inequalities:
A. If $u(0) = u(L) = u''(0) = u''(L) = 0$ and $u(x)$ is not identically
    zero, with $c > 0$, we have

$$\frac{\int_0^L u''^2 \, dx + c \int_0^L u^2 \, dx}{\int_0^L u'^2 \, dx} \geq 2\sqrt{c} \tag{4}$$

The equality holds if and only if $u(x)$ differs from $\sin(n\pi x/L)$
by a constant factor, where the natural number n
equals approximately

$$n^4 = cL^4/\pi^4 \tag{5}$$

The prime here denotes derivative with respect to x.

B. If $v(t)$ renders the following integral meaningful, then the
    inequality

$$\int_0^a \frac{\dot{v}(t)^2}{(1-t/a)} \, dt \geq \frac{2}{a} [v(a) - v(0)]^2 \tag{6}$$

where $a > 0$, is always true, and the equality can be attained.

The above inequalities can be proved by changing the problem into
an eigenvalue formulation.

In the plastic buckling problem of a circular cylindrical shell under axial impulsive loading, the lateral perturbation $w(x,t)$ satisfies the following equation of motion according to Vaughan (1969) with a slight change of notation:

$$f(\lambda,w) = \frac{1}{1-t/t_f} \left(\frac{h^3\sigma^\circ L}{36V} \dot{w}^{IV} + \frac{4L\sigma^\circ h}{3a^2 V} \dot{w} + \frac{h^3\sigma^\circ L}{9a^2 V} \dot{w}'' \right) - \frac{h^3 E_h}{18a^2} w'' +$$

$$+ \frac{h^3}{12} E_h w'^V + \sigma^\circ h w'' + \rho h \ddot{w} = 0 \tag{7}$$

Where $w(x,t)$ satisfies the simple support boundary conditions, $t_f = \rho V \dot{L} (1+M/m)/\sigma^\circ$, V is the impulsive velocity, h the shell thickness, L the shell length, σ° the average stress, E_h the coefficient of linear hardening, a the shell radius, ρ the density and M and m are the masses of the impact weight and of the shell respectively.

Since $\sigma^\circ h \gg h^3 E_h/18a^2$, in general, the 4th term is small compared with the 6th term and can be neglected. Besides, since w satisfies the simply supported boundary condition, and usually $1 \gg h/2\sqrt{3}\,a$, it can be shown by using (4) that

$$\int_0^L \frac{h^3\sigma^\circ L}{36V} \dot{w}''^2 \, dx + \int_0^L \frac{4L\sigma^\circ h}{3a^2 V} \dot{w}^2 \, dx \gg \int_0^L \frac{h^3\sigma^\circ L}{9a^2 V} \dot{w}'^2 \, dx \tag{8}$$

the r. h. s. can then be neglected as compared to the l. h. s.

Noting the above two arguments and that the loading parameter λ is now the impulsive velocity V, we have

$$\mathbf{F}(\lambda,w) = \dot{F}(V,w) = \frac{h^3\sigma^\circ L}{36V} \int_0^L \left(\int_0^{t_f} \frac{\dot{w}''^2}{(1-t/t_f)} \, dt \right) dx + \frac{4L\sigma^\circ h}{3a^2 V}$$

$$\int_0^L \left(\int_0^{t_f} \frac{\dot{w}^2}{1-t/t_f} \, dt \right) dx + \int_0^L \left[\frac{h^3 E_h}{12} \frac{w''^2}{2} - \sigma^\circ h \frac{w'^2}{2} + \frac{\rho h \dot{w}^2}{2} \right]_0^{t_f} dx \tag{9}$$

It should be noted that the omission of the two terms above is equivalent to using the following simple expression in the geometric relation

$$\dot{\varepsilon}_\theta = -(v/a)(1-t/t_f) \tag{10}$$

instead of the more accurate formula

$$\dot{\varepsilon}_\theta = -(v/a)(1-t/t_f)(1-z/a) \tag{11}$$

used by Vaughan (1969). This is also equivalent to saying that the use of a simple geometric relation does not effectively influence the accuracy. We shall make use of this simplification to avoid complications in the following.

Making use of the two characteristics of the perturbation ($v(0) = V$ and $v(t_f) = 0$), one gets with the aid of inequality (6):

$$F(V,w) \geq \frac{h^3 \sigma^\circ L}{36V} \int_0^L \frac{2}{t_f} w''^2 dx + \frac{4L\sigma^\circ h}{3a^2 V} \int_0^L \frac{2}{t_f} w^2 dx + \int_0^L \frac{h^3 E_h}{24} w''^2 dx$$

$$- \sigma^\circ h \int_0^L \frac{w'^2}{2} dx \qquad (12)$$

where w and its derivatives are taken at $t = t_f$. No special notation is used here in order to simplify the writing.

The sufficiency condition for stability $F(V,w) > 0$ can be written equivalently as:

$$\frac{\int_0^L \left(\frac{h^2 L^2}{18} w''^2 + \frac{h^2 V t_f L E_h}{24\sigma^\circ} w''^2 + \frac{8L^2}{3a^2} w^2 \right) dx}{\frac{V t_f L}{2} \int_0^L w'^2 dx} > 1 \qquad (13)$$

where w satisfies the simply supported boundary condition.

The critical velocity V_{cr} can be obtained by setting the minimum of the l. h. s. (this exists according to (4)) equal to 1. After some reduction, we get:

$$\frac{\rho V_{cr}^2 (1+M/m)}{\sigma^\circ} = \frac{8h}{3\sqrt{3}a} \left(\frac{\sqrt{3}hE_h}{3a\sigma^\circ} + \sqrt{1 + \left(\frac{\sqrt{3}hE_h}{3a\sigma^\circ} \right)^2} \right) \qquad (14)$$

The function corresponding to the minimum value is the buckling mode, viz. $\sin(n\pi x/L)$, where n is approximately equal to

$$n^4 = \frac{48L^4}{\pi^4 a^3 h^2} \frac{1}{(1 + 3V t_f E_h /4L\sigma^\circ)} \qquad (15)$$

This is a rather neat expression for the half wave number. Equation (14) for the critical velocity is quite close to the empirical expression given by Vaughan (1969), the main difference is that the former contains E_h, while the latter does not.

The half wave number given by Eq. (15) can be compared with the experimental data given in Vaughan (1969), in which $a = 0.45$ in., $h = 0.1$ in. and $E_h/\sigma^\circ \approx 2.5$. We have taken V to be the approximate critical velocity, since n is not sensitive to the magnitude of the velocity, this approximation will not make much difference. The results are compared in the following table.

In view of the approximation made, the agreement is fairly good. The method used here is a much simpler one and has a distinctive physical basis.

From the above analysis, it can be seen that the two physical properties for the elements in the set B can be defined mathematically as $w(x,0) = \dot{w}(x,0) = 0$. Since the energy expression is a continuous functional, its dependence on the initial value of w is also continuous. An infinitesimal initial value will then be equivalent to a zero initial value. This is probably one reason for the success of the energy method in dealing with the present problem.

TABLE 1 Comparison of Eq. (15) to Experimental Result

Tube No.	L (in.)	Experiment n	Theoretical n Vaughan 1969	n by Eq. (15)
1,2,3	3	8	9	10
4	3	8	9	10
10	4	12	11	13
13,14,15	4	12	12	13
20	4	12	12	13
22	6	14	16	20
23,24	6	15	16	20

STRINGER STIFFENED CYLINDRICAL SHELLS UNDER AXIAL IMPULSIVE LOAD

For stringer stiffened cylindrical shells, the equation of motion is given by Jones and Papageorgiou (1982) as:

$$\frac{1}{1 - t/t_f} \left[\frac{-t_f A_1 \overset{\bullet}{w}^{IV}}{a} + \frac{A_5(-t_f) \overset{\bullet}{w}'^V}{a} - \frac{t_f A_7 \overset{\bullet}{w}}{a} \right] + \frac{A_2 \overset{\bullet}{w}^{IV}}{a}$$

$$+ (A_3 + A_4) \frac{w''}{a} + A_6 \frac{w}{a} + t_f^2 \frac{\ddot{w}}{a} = 0 \tag{16}$$

Note that the prime here denotes differentiation with respect to x/L and that the coefficients are expressed as follows:

$$A_1 = -\frac{\alpha^2 \bar{\sigma}}{18\bar{h}} + \frac{2\bar{\sigma}(A^* e^* - 2I^*)}{3\bar{h}} - \frac{\alpha^2 \beta^2 (1 - \sigma_y/\sigma^\circ)}{72} - \frac{\beta^2 I^* (1 - \sigma_y/\sigma^\circ)}{6}$$

$$A_2 = \frac{\beta^2 \bar{E}(I^* + \alpha^2/12)}{\bar{h}}$$

$$A_3 = \bar{E}\{1 + A^* + A^* e^*/3 - (1 - \sigma_y/\sigma^\circ) \bar{h}\beta^2 A^* e^*/4\bar{\sigma}\}/\lambda\bar{h}$$

$$A_4 = \bar{E}\{A^* e^* \beta^2 \bar{h}/2\bar{\sigma} - \alpha^2/18 - 2I^*/3\}/\bar{h}$$

$$A_5 = -\bar{\sigma} A_2/3\bar{E}$$

$$A_6 = \beta^2 \bar{E} \bar{h}/4\bar{\sigma}^2$$

$$A_7 = -4 \bar{\sigma}/3\beta^2 \bar{h}$$

where $\alpha = h/a$, $\bar{\sigma} = 1 + M/m$, $\bar{h} = 1 + h'/h$, h' is the additional shell thickness due to smearing of the stiffeners, $\beta = a/L$, $\bar{E} = E_h t_f^2/\rho L^2$, E_h is the linear hardening coefficient, $\lambda = E_h/\sigma^\circ$, σ_y is the yield stress, $e^* = e/a$, $e = \int_{h/2}^{h/2+h'} z \frac{dA}{A}$, $A^* = A/bh$, A is the cross-sectiona|

area of a stiffener, b is the circumferential spacing on mid-surface of cylindrical shell between adjacent stiffeners. $I^* = I/a^2bh$, $I = \int_{h/2}^{h/2+h'} z^2 dA$, $t_f = \rho VL(1+M/m)/\sigma^\circ$.

Note that for the special case of no stiffener ($e^* = 0$) the above equation does not reduce to that of equation (7). This is because the authors had used expressions of different accuracies in relating stress and generalized force.

Utilizing the method used in the last section and noting in general that

$$\frac{64 \; L^4 \; (1+M/m)^2 \sigma^\circ L}{3a^4 E'} \frac{}{t_f V} \gg 1, \tag{17}$$

we can prove that the two equations are equivalent to within second order small quantities. The inequality (17) is used to show the smallness of A_6. In the following computation A_6 is taken to be zero.

Considering the definition of set B, we obtain

$$F(V,w) = \int_0^1 (A_2 \frac{w''^2}{2} + (A_3+A_4)\frac{(-w'^2)}{2}) dx + \int_0^1 \int_0^{t_f} \frac{1}{1 - t/t_f}$$

$$(A_1 t_f \; \dot{w}'^2 - A_5 \; t_f \; \dot{w}''^2 - A_7 \; t_f \; \dot{w}^2) dt \; dx \tag{18}$$

Where the kinetic energy term evaluated at $t = t_f$ should have been zero and is removed here. The factor l/a common to all terms is also eliminated. x is non-dimensionalized by L.

In the subsequent derivation in order to obtain an inequality we have substituted the time interval in the second term on the r. h. s. of Eq. (18) by a smaller one by virtue of inequality (6). In this regard, we have to assume that the value of the parenthesis is positive. This happens to be a basic requirement representing an essential property for this type of buckling phenomenon. It is equivalent to requiring certain quantities to be positive in the analysis of Jones and Papageorgiou (1982). Although this assumption is actually satisfied for the buckling phenomena, we shall give its prerequisites later on; it is also the postulate that has to be satisfied in using the present theory.

Under this assumption, using the inequality (6), we have,

$$F(V,w) \geq \int_0^1 [\frac{A_2}{2} w''^2 - 2A_5 \; w''^2 - 2A_7 \; w^2 - (\frac{A_3+A_4}{2} - 2A_1)w'^2] dx$$

$$\tag{19}$$

In a manner similar to that used in the last section, we obtain the following relation for the critical quantity

$$\bar{E}_{cr} = 4 \frac{\left(A_1 \frac{A_3+A_4}{\bar{E}} - 2A_7 \frac{A_2}{\bar{E}}\right) + \sqrt{\left(A_1(A_3+A_4)/\bar{E} - 2A_7 A_2/\bar{E}\right)^2 - \left(\frac{A_3+A_4}{\bar{E}}\right)^2 (A_1^2 - 4A_5 A_7)}}{\left(\frac{A_3+A_4}{\bar{E}}\right)^2} \quad (20)$$

And the expression for the buckling half wave number is

$$n^4 = n_0^4/(1+12 \ I* \ a^2/h^2) \quad (21)$$

where n_0 is the buckling half wave number for the unstiffened cylindrical shell under the same loading. This is also a rather simple and neat formula for n.

It is easy to show by Eq. (20) that the critical load increases with stiffening and that it is more efficient to place the stiffeners on the outer shell surface than on the inner surface. These agree with the conclusions obtained by Jones and Papageorgiou (1982).

For example, for Fig. 8 in Jones and Papageorgiou (1982) with e* = 0.2, h/a = 0.10, we have

n_0/n A*	0.5	1.5	1.5	2.0
J. & P.(1982)	2.5	2.77	3.2	3.47
Eq. (21)	2.33	2.76	3.05	3.27

For Fig. 9 in Jones and Papageorgiou (1982) with e* = 0.2, h/a.= 0.15, we have

n_0/n A*	0.5	1.0	1.5
J. & P.(1982)	1.85	2.174	2.35
Eq. (21)	1.9	2.236	2.45

In the above calculation we have used the formula

$$I* = A* (4e*^2 - e*\alpha + \alpha^2/4)/3$$

given in Jones and Papageorgiou (1982). It is obvious that the results agree very well. The simple expression here is able to gather the numerical results expressed previously in graphical form into one analytic formula.

As has been pointed out above, the present theory relies on the assumption that the value of the parenthesis in Eq. (18) is positive. This condition can be written in the following form after deleting the common factor:

$$\int_0^1 [A_1(w')^2 - A_5(w'')^2 - A_7 w^2] \ dx > 0 \quad (22)$$

If $A_1 > 0$, this condition is obviously satisfied since both A_5 and A_7 are negative. However, often A_1 is negative and then it can be proved by using the inequality (4) that this condition can also be satisfied if

$$(A_1{}^2 - 4A_5A_7) < 0$$

In the general case, we may first assume the condition to be satisfied and go on with the calculation. After the buckling mode and half wave number are obtained, one can check again if the original assumption is correct. For the two examples stated above, it can easily be verified that the above inequality holds.

RECTANGULAR PLATE UNDER IN-PLANE IMPULSIVE LOADING

According to Goodier (1968), for buckling independent of the transverse coordinate, the equation of motion can be written as:

$$\frac{h^3}{36V} \sigma^\circ L \; \dot{w}^{IV} + \frac{h^3 E_h}{12} w^{IV} + \sigma^\circ h w'' + \rho h \ddot{w} = 0 \tag{23}$$

where $w(x,t)$ still satisfies the simple support boundary condition. Goodier (1968) had treated the case of sustained plastic flow, i.e. initial velocity V is retained after impact, so the response duration $t_f = \rho VL/2\sigma^\circ$, other notations are the same as in Eq. (7). Since the experimental results presented by Goodier deal with a plate projected end-on against a rigid target with velocity V, we shall discuss this case only.

For this case, inequality (6) is replaced by

$$\int_0^a \dot{v}(t)^2 dt > \frac{1}{a} [v(a) - v(0)]^2 \tag{24}$$

we have

$$F(V,w) \geq \int_0^L \left[\left(\frac{h^3 \sigma^\circ L}{36Vt_f} + \frac{h^3 E_h}{24} \right) w''^2 - \frac{\sigma^\circ h}{2} w'^2 + \frac{\rho h \dot{w}^2}{2} \right] dx \tag{25}$$

Note that the kinetic energy term can no longer be neglected in this case.

Since (after separation of variables and factoring out the spatial function), the time function satisfies a second order equation with constant coefficients, it is easy to visualize that the buckling process takes an exponential form in t. Suppose it is exp (pt/t_f), then $w(t_f) = pw(t_f)$. Substituting this relation into the r. h. s. of the inequality (25), the sufficiency condition for stability is then:

$$\frac{\int_0^L \left[\left(\frac{h^3 \sigma^\circ L}{36Vt_f} + \frac{h^3 E_h}{24} \right) w''^2 + \frac{p^2 \rho h}{2t_f{}^2} w^2 \right] dx}{\frac{\sigma^\circ h}{2} \int_0^L w'^2 dx} > 1 \tag{26}$$

From this one gets the expression for the critical velocity (containing an unknown p):

$$\frac{\rho V^2}{2\sigma^\circ} = \frac{hp}{3L}\left(\frac{E_h h}{\sigma^\circ L}p + \sqrt{4 + \left(\frac{E_h h}{\sigma^\circ L}p\right)^2}\right) \tag{27}$$

and the buckling half number will be given by

$$\left(\frac{n\pi}{L}\right)^4 = \frac{p^2\rho}{h^2\left(\frac{\sigma^\circ L}{18Vt_f} + \frac{E_h}{12}\right)t_f^2} = \frac{36p^2}{L^2 h^2\left(1 + 3E_h Vt_f/2\sigma^\circ L\right)} \tag{28}$$

Strictly speaking this formula applies only to the critical state.

Since the buckling function $\exp(pt/t_f)\sin(n\pi x/L)$ satisfies the equation of motion, we have:

$$\left(\frac{h^3\sigma^\circ L}{36Vt_f}p + \frac{h^3 E_h}{12}\right)\left(\frac{n\pi}{L}\right)^4 - \sigma^\circ h\left(\frac{n\pi}{L}\right)^2 + \frac{\rho h p^2}{t_f^2} = 0 \tag{29}$$

Substituting the expression for n, Eq. (28), into Eq. (29), one gets a second degree equation for p which will usually be a function of V. Substituting it back into Eq. (27), one can solve for the critical velocity and into Eq. (28) one gets the corresponding n. These equations provide the complete analytic solution for the problem.

We shall choose an example given by Goodier (1968), solve it by the present method and compare the result with the corresponding data. The example chosen is denoted by "4CSC-3" in that paper. It is a square tube slit along all four corners, each side can be regarded as a rectangular plate independent of the others. Obviously the most relevant case for the present theory is the critical state. There L = 5 in., h = 1/16 in. and $E_h/\sigma^\circ \sim 10$. Since $2 \gg (E_h h/\sigma^\circ L)p$, for the sake of simplicity, we may take $\rho V^2/2\sigma^\circ = 2hp/3L$, and $(n\pi/L)^4 = 36p^2/L^2 h^2$ approximately. Substituting these values into Eq. (29) and solving for p, one gets p = 2. The critical quantity, the total axial strain, is then

$$\varepsilon_{cr} = \frac{\rho V_{cr}^2}{2\sigma^\circ} = 0.0167.$$

The corresponding half wave number satisfies

$$(L/n) \approx 0.5 \text{ (in.)}$$

From the upper photograph of plate 3 in Goodier (1968), the impact velocity 59 ft/sec is close to the critical state. The corresponding experimental data (accurate to two significant figures) are: $\varepsilon = 0.01$, L/n = 0.61 (in.) and so the agreement with our theoretical value is fairly good.

For the post-critical case (V > V_{cr}) but not far from it, we may expect that the formula for the half wave number is still more or less applicable. According to the data given for this case, $E/\sigma^\circ \sim 5$, so again we take approximately $(n\pi/L)^4 = 36p^2/L^2 h^2$. Substituting these into Eq. (29) and solving for p, one gets the expression:

$$p = \sqrt{1 + 6\varepsilon L/h} - 1$$

Where ε is the total axial strain. From these, one gets the following
relation between n and ε

$$(\frac{n\pi}{L})^2 = \frac{6}{Lh}(\sqrt{1+6\varepsilon L/h} - 1)$$

The experimental value given in Goodier (1968) is $\varepsilon = 0.03$, corresponding to $(L/n) = 0.51$, while according to the above formula for
$\varepsilon = 0.03$, the corresponding $(L/n) = 0.42$ (in.), the agreement is
reasonable. Compared to the complicated computation used in that
paper, the present scheme is a very simple one.

CONCLUSION

From the discussions above for three particular cases, we may con-
clude that the proposed energy criterion is a suitable method for
dealing with these types of buckling problems, especially for the
critical state. They agree fairly well with experimental data or
previous theoretical results obtained by numerical computation. The
formulae are rather simple. It may be emphasized that it provides
a good basis for estimating the critical velocity under which the
structure will buckle. This is an important problem not discussed
theoretically before.

Finally it should also be noted that the energy criterion provides
only a sufficient condition for stability, the critical load so
determined is less than the actual one, i.e. on the conservative
side. However, sometimes it may be so much smaller than the actual
load that it is meaningless. This is because a buckling process
which is possible according to energy considerations (i.e. does not
violate the energy principle), is not always an actually realizable
process.

REFERENCES

Florence, A. and J. N. Goodier (1968). Dynamic plastic buckling of
    cylindrical shells in sustained axial compressive flow. *J. Appl.
    Mech.*, 35, 80-86.
Goodier, J. N. (1968). Dynamic buckling of rectangular plates in
    sustained plastic compressive flow in J. Heyman and F. A. Leckie
    (Ed.), *Engineering Plasticity*, Cambridge University Press, Cam-
    bridge, 183-200.
Jones, N. and E. A. Papageorgiou (1982). Dynamic axial plastic
    buckling of stringer stiffened cylindrical shells. *Int. J. Mech.
    Sci.*, 24, 1-20.
Vaughan, H. (1969). The response of a plastic cylindrical shell to
    axial impact. *Z.A.M.P.* 20, 321-328.

STATIC AND DYNAMIC AXIAL CRUSHING OF CIRCULAR AND SQUARE TUBES

N. JONES* and W. ABRAMOWICZ**

*Department of Mechanical Engineering, The University of
Liverpool, UK
**Institute of Fundamental Technological Research, Warsaw, Poland

ABSTRACT

A comparison is made between the experimental results, approximate
theoretical predictions and empirical relations for the static and
dynamic progressive buckling of thin-walled tubes having circular or
square cross-sections. Satisfactory agreement is achieved between
theoretical predictions and experimental results when the effective
crushing distance is considered and provided that the influence of
material strain rate sensitivity is retained in the dynamic crushing
case.

NOTATION

c	length of side of a square cross-section
h	wall thickness
p	Cowper-Symonds exponent in equation (23)
A	$2\pi Rh$ and $4ch$ for circular and square tubes, respectively
A_1	πR^2 and c^2 for circular and square tubes, respectively
D	Cowper-Symonds coefficient in equation (23)
$2H$	Initial distance between plastic hinges at top and bottom of a basic folding element
L	initial axial length of a tube
M	striking mass
M_0	$\sigma_0 h^2/4$ or $(2/\sqrt{3})(\sigma_0 h^2/4)$
P_m	theoretical prediction for mean static crushing load
$P_m^{\ d}$	average value of dynamic force in experimental tests
$P_m^{\ s}$	average value of static force in experimental tests
$\bar{P}_m$	theoretical prediction for mean static crushing load corrected for effective crushing distance
$\bar{P}_m^{\ d}$	theoretical predictions for mean dynamic crushing load corrected for effective crushing distance and material strain rate sensitivity
R	mean radius of a cylindrical shell

V   impact velocity of striking mass
δ_e effective crushing distance
$\dot{\varepsilon}$ strain rate
η $P_m/A\sigma_u$ or $\overline{P}_m/A\sigma_u$
ρ density of material
σ_a average stress
σ_0 uniaxial yield stress
σ_0^d dynamic uniaxial yield stress
σ_u uniaxial ultimate tensile stress
σ_u^d dynamic uniaxial ultimate tensile stress
ϕ A/A_1

INTRODUCTION

There has been considerable activity on the dynamic plastic response
of structures during the past decade and a significant part of this
effort has been concerned with energy absorbing systems and the
structural crashworthiness of vehicles (Johnson, 1972; Rawlings,
1974; Johnson and Mamalis, 1978; Johnson and Reid, 1978; Jones
and Wierzbicki, 1983; Davies, 1984). The behaviour of thin-walled
tubes with circular and rectangular cross-sections when subjected
to axial loads has been of particular interest since the pioneering
work of Pugsley (1960) on the impact of idealised railway coaches.
The initial buckling response of these members is less important
from the viewpoint of energy absorption than the subsequent post-
buckling behaviour with large strains and deflections. This behaviour is often studied using rigid plastic methods of analysis since
the energy absorbed elastically is usually not significant.

Many articles have been published on the static and dynamic axial
crushing of circular tubes which are known to be efficient absorbers
of energy per unit weight of material (Johnson and Reid, 1978; Ezra
and Fay, 1972; Thornton, Mahmood and Magee, 1983). Thin-walled
members with closed hat and rectangular cross-sections are relevant
to the energy absorption characteristics of automobile body structures (Postlethwaite and Mills, 1970) as well as trains (Pugsley,
1960; Macaulay and Redwood, 1964) and buses (Lowe, Al-Hassani and
Johnson, 1972).

Theoretical studies on axially crushed tubes with circular or rectangular cross-sections usually neglect dynamic (inertia) effects
and treat the problem as quasi-static, which is probably reasonable
for low impact velocities which do not activate dynamic plastic
buckling phenomena (Jones, 1984; Jones and dos Reis, 1980). Several
authors (Wimmer, 1975; Ohkubo, Akamatsu and Shirasawa, 1974; Wierzbicki and Akerstrom, 1977; Wierzbicki and Abramowicz, 1981), however, have examined material strain rate sensitive effects which can
be important even when inertia effects are not (Jones, 1983). Optimal
energy absorption for quasi-static loadings is achieved through progressive buckling which avoids overall (Euler) buckling (Thornton,
Mahmood and Magee, 1983; Mahmood and Paluszny, 1981).

A simplified kinematical method of analysis which is capable of
estimating the crushing characteristics of thin-walled metal structures has been developed recently by Abramowicz and Wierzbicki
(1979), Hayduk and Wierzbicki (1982), Wierzbicki and Abramowicz

(1984) and Wierzbicki (1983a). The particular studies of Hayduk and
Wierzbicki (1982) and Wierzbicki and Abramowicz (1984) focus on the
behaviour of two basic folding or collapse elements. The importance
of this theoretical procedure lies in its rigorous kinematical approach and the retention of extensional as well as bending contributions to the energy dissipation. The basic collapse elements were
used by Hayduk and Wierzbicki (1982) to predict the axial collapse
behaviour of a cruciform member; the behaviour of a honeycomb by
Wierzbicki (1983b) and one kind of collapse (symmetric) of rectangular and square box columns by Wierzbicki and Abramowicz (1984). It
transpires that the extensional deformations, though highly localised,
dissipate at least one-third of the total energy.

The theoretical studies on basic collapse elements reported by Hayduk
and Wierzbicki (1982) and Wierzbicki and Abramowicz (1984) for the
crushing behaviour of thin-walled sections have been used by Abramowicz and Jones (1984a) to derive various progressive crushing modes
for square tubes loaded axially. This theory predicts four deformation modes which govern the behaviour for different ranges of the
parameter c/h. New asymmetric deformation modes were predicted
theoretically and confirmed in the experimental tests. These asymmetric modes cause an inclination of a column which could lead to
collapse in the sense of Euler even for relatively short columns.

Recently, it was shown by Abramowicz (1983) that the effective
crushing distance is seventy per cent of the initial length for
axially crushed box columns undergoing a symmetric deformation mode.
This value was obtained by idealising the corner collapse of a box
column and assuming that the deformations are the result of bending
about two orthogonal axes. The approximate theoretical analysis
derived by Abramowicz and Jones (1984a) for the axial crushing of
square box columns considers the effective crushing distance together with the influence of material strain rate sensitivity, which
is important for steel even when dynamic loadings may be treated as
quasi-static. The simple equations developed by Abramowicz and Jones
(1984a) gave reasonable agreement with the corresponding experimental
results on axially crushed square box columns.

A modified version of Alexander's (1960) theoretical analysis for
axisymmetric, or concertina, deformations of axially crushed cylindrical shells is developed by Abramowicz and Jones (1984b) and gives
good agreement with experimental results when the effective crushing
distance is considered and provided that the influence of material
strain rate sensitivity is retained in the dynamic crushing case.
The experimental results from the static and dynamic axial crushing
tests on circular and square steel columns reported by Abramowicz
and Jones (1984a,b) are discussed herein and compared with previously
published experimental results and with various empirical relations
and approximate theoretical predictions.

THEORETICAL PREDICTIONS FOR STATIC AXIAL CRUSHING OF A
CIRCULAR TUBE

Alexander (1960) developed a theoretical analysis for the axisymmetric crushing of a thin-walled cylindrical shell subjected to a
static axial load and predicted the mean crushing load

$$P_m/M_0 = 20.73 \ (2R/h)^{\frac{1}{2}} + 6.283,$$ (1)

where

$$M_0 = (2\sigma_0/\sqrt{3})(h^2/4)$$

and

$$H/R = 1.905 (h/2R)^{\frac{1}{2}},\tag{2}$$

when H is defined in Fig. 1. Alexander (1960) used the von Mises yield condition to generate the fully plastic bending moment M_0 for the plane strain case. However, $M_0 = \sigma_0 h^2/4$ for the Tresca yield criterion.

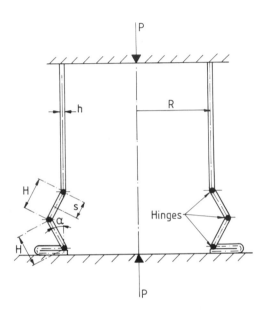

Fig. 1.   Idealised axisymmetric, or concentric, collapse mode for an axially compressed cylindrical shell.

A slight modification is made by Abramowicz and Jones (1984b) to remove an assumption in Alexander's analysis. This yields a mean static crushing load

$$P_m/M_0 = 20.79 (2R/h)^{\frac{1}{2}} + 11.90\tag{3}$$

with

$$H/R = 1.76 (h/2R)^{\frac{1}{2}}.\tag{4}$$

Several authors have examined the non-axisymmetric or diamond crushing mode of cylindrical shells.

Pugsley (1960) and Pugsley and Macaulay (1960) found that the mean axial static force required to crush a cylindrical shell is

$$P_m/M_0 = 0.326 (2R/h) + 217.7, \quad R/h \geq 50,\tag{5}$$

where the coefficients were selected to agree with some experimental

test results on stainless steel and soft aluminium cylindrical shells.
More recently, Pugsley (1979) has suggested the equation

$$P_m/M_0 = 198 \ \sigma_a/\sigma_0, \tag{6}$$

where σ_a is an average stress which is selected to take account of
strain hardening. Wierzbicki (1983c) has derived the approximate
expression

$$P_m/M_0 = 62.88 \ (2R/h)^{1/3}, \tag{7}$$

and obtained good agreement with some experimental results.

A simple theoretical analysis was developed by Abramowicz (1983) to
determine the effective crushing distance of thin-walled open columns
which was taken as 2H to obtain equations (1) to (7). The effective
crushing distance for the symmetric, or concertina, deformations indicated in Fig. 2 was found by Abramowicz and Jones (1984b) to equal

$$\delta_e/2H = 0.86 - 0.568 \ (h/2R)^{1/2}. \tag{8}$$

It was also shown that

$$\delta_e/2H = 0.73 \tag{9}$$

for the static non-symmetric crushing of cylindrical shells.

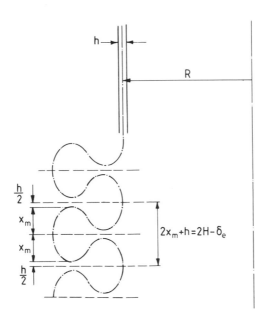

Fig. 2.   Effective crushing distance.

The total plastic energy (E_t) absorbed during the development of one lobe was divided by 2H in order to predict the mean crushing loads given by equations (1), (3), (5)-(7). A more realistic estimate of the mean crushing load $\overline{P}_m$ is given by E_t/δ_e, or

$$\frac{\overline{P}_m}{M_0} = \frac{20.79 \ (2R/h)^{1/2} + 11.90}{0.86-0.568 \ (h/2R)^{1/2}} \tag{10}$$

according to equations (3) and (8) for symmetric, or concertina crushing. Similarly, equations (7) and (9) predict

$$\overline{P}_m/M_0 = 86.14 \ (2R/h)^{1/3} \tag{11}$$

for the non-symmetric, or diamond, mode of crushing.

THEORETICAL PREDICTIONS FOR STATIC AXIAL CRUSHING OF A SQUARE TUBE

A theoretical analysis using the basic collapse elements developed by Hayduk and Wierzbicki (1982) and Wierzbicki and Abramowicz (1984) is reported by Abramowicz and Jones (1984a) for the progressive buckling of thin-walled square box columns subjected to axial loads. It transpires that four different deformation modes govern the behaviour for different ranges of the dimensionless parameter c/h.

Symmetric Collapse Mode

The idealised symmetric[1] collapse mode in Fig. 3 for a square tube with wall thickness h was examined by Wierzbicki and Abramowicz (1984) The external work was equated to internal work then minimised to predict the mean static crushing load (P_m)

$$P_m/M_0 = 38.12 \ (c/h)^{1/3} \tag{12}$$

with
$$H/h = 0.99 \ (c/h)^{2/3}, \tag{13}$$

where $M_0 = \sigma_0 h^2/4$.[2] 2H is the initial axial separation between the locations of the plastic hinges at the top and bottom edges of a basic folding element.

Asymmetric Mixed Collapse Mode A

It was observed during the experimental programme reported by Abramowicz and Jones (1984a) that some of the square tubes developed the asymmetric[3] mixed deformation mode A which is indicated in Fig. 4.

---

[1]The symmetric collapse mode is a quasi-inextensional mode.

[2]The von Mises yield criterion is used for M_0 in equation (1) and in all equations for circular tubes, while the Tresca yield criterion is used in equation (12) and in all subsequent expressions for square tubes. This difference is historical and artificial but is maintained here to remain consistent with earlier studies.

[3]The asymmetric mixed collapse modes are quasi-extensional modes.

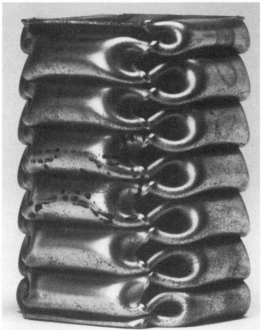

Fig. 3.   Symmetric collapse mode of square box columns
          (a)   paper model
          (b)   specimen number I31

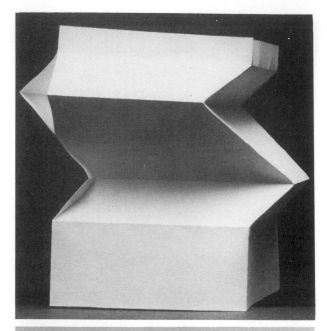

Fig. 4.   Asymmetric mixed collapse mode A.
          (a) paper model
          (b) specimen number 6.

Repeated patterns of pairs of lobes developed in this case and, as far as the authors are aware, this type of behaviour has been neither reported previously nor studied theoretically. However, the pattern of deformation indicated by the paper model in Fig. 4(a) may be constructed from combinations of the two basic folding elements I and II studied by Hayduk and Wierzbicki (1982) and Wierzbicki and Abramowicz (1984). It was shown by Abramowicz and Jones (1984a) that the mean static crushing load P_m is

$$P_m/M_0 = 33.58 \ (c/h)^{1/3} + 2.92 \ (c/h)^{2/3} + 2. \qquad (14)$$

Asymmetric Mixed Collapse Mode B

As far as the authors are aware, the asymmetric[3] mixed deformation mode B shown in Fig. 5 has not been studied previously. However,

Fig. 5.   (a) paper model.

[3]The asymmetric mixed collapse modes are quasi-extensional modes.

it was shown by Abramowicz and Jones (1984a) that the mean crushing load is

$$P_m/M_0 \cong 35.54 \ (c/h)^{1/3} + 1.65 \ (c/h)^{2/3} + 1. \qquad (15)$$

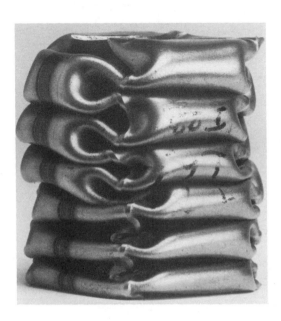

Fig. 5.   Asymmetric mixed collapse mode B
(b)   specimen number I22.

Extensional Collapse Mode

This particular collapse mode was not observed in any of the tests reported by Abramowicz and Jones (1984a), but it is reported here for completeness. The mean static crushing load is

$$P_m/M_0 = 16\sqrt{\pi}(c/h)^{1/2} + 8. \qquad (16)$$

Effective Crushing Distance

The theoretical analysis of Abramowicz (1983) is extended by Abramowicz and Jones (1984a) to predict the effective crushing distances of the four deformation modes. The effective crushing distance is

$$\delta_e/2H = 0.73 \qquad (17)$$

for the symmetric deformation mode which compares favourably with the average value $\delta_e/2H = 0.75$ for four test specimens examined by Abramowicz and Jones (1984a) one of which is shown in Fig. 6.

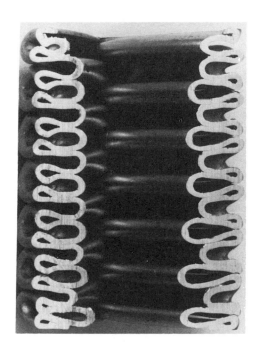

Fig. 6.   Section across diagonal plane of specimen
          number I31 in Fig. 3.

Moreover,

$$\delta_e/2H = 0.77 \tag{18}$$

for the asymmetric mixed collapse modes A and B and the extensional
deformation mode. Thus, equations (12) and (14)-(16) predict dimensionless mean crushing loads,

$$\overline{P}_m/M_0 = 52.22 \ (c/h)^{1/3}, \tag{19}$$

$$\overline{P}_m/M_0 = 43.61 \ (c/h)^{1/3} + 3.79 \ (c/h)^{2/3} + 2.6, \tag{20}$$

$$\overline{P}_m/M_0 = 46.16 \ (c/h)^{1/3} + 2.14 \ (c/h)^{2/3} + 1.3, \tag{21}$$

and

$$\overline{P}_m/M_0 = 36.83 \ (c/h)^{1/2} + 10.39, \tag{22}$$

respectively.

DYNAMIC AXIAL CRUSHING OF A CIRCULAR TUBE

This article is concerned only with impact loadings which are sufficiently small to cause quasi-static behaviour and not excite a

dynamic buckling response (Jones, 1984). Thus, it is assumed that the theoretical work in an earlier section remains valid when one end is struck by a mass M travelling with an initial velocity V. However, even though inertia effects are small, it is still important to retain the influence of material strain rate sensitivity, particularly for tubes made from mild steel (Jones, 1983).

The empirical Cowper-Symonds uniaxial constitutive equation

$$\sigma_0^d/\sigma_0 = 1 + (\dot{\varepsilon}/D)^{1/p} \tag{23}$$

is widely used to assess material strain rate effects in structures. Equation (23) with the customary coefficients $D = 40.4$ sec^{-1} and $p = 5$ for mild steel, is obtained from experimental tests on specimens having relatively small strains in the neighbourhood of the yield stress (Jones, 1983).

The average strain associated with the development of axisymmetric, or concertina, crushing of circular tubes is estimated by Abramowicz and Jones (1984b). It transpires that the stresses associated with the average strains present in axially crushed tubes with $10 < R/h < 100$, approximately, are similar to the ultimate tensile stress (σ_u) rather than the yield stress (σ_0). However, it was shown by Abramowicz and Jones (1984a) that the equation

$$\sigma_u^d/\sigma_u = 1 + (\dot{\varepsilon}/6844)^{1/3.91} \tag{24}$$

fits the experimental data for the ultimate tensile stresses of the steel specimens examined by Campbell and Cooper (1966).

The average strain rate ($\dot{\varepsilon}$) was estimated by Abramowicz and Jones (1984b) as

$$\dot{\varepsilon} = 0.250 \, V/[R\{0.86 - 0.568 \, (h/2R)^{1/2}\}] \tag{25}$$

for the symmetric, or concertina, crushing mode and

$$\dot{\varepsilon} = 0.370 \, V/R \tag{26}$$

for the non-symmetric, or diamond, crushing mode.

Equations (1), (3), (5)-(7) are valid for quasi-static behaviour with σ_0 replaced by σ_u for a strain rate insensitive material, while σ_u is replaced by σ_u^d from equation (24) when the material is strain rate sensitive. Thus, equations (10), (11) and (24)-(26) give the mean dynamic crushing loads

$$\frac{\bar{P}_m^d}{M_0} = \left\{ \frac{20.79 \, (2R/h)^{1/2} + 11.90}{0.86 - 0.568 \, (h/2R)^{1/2}} \right\} \left[1 + \left\{ \frac{0.250V}{6844R\{0.86 - 0.568 \, (h/2R)^{1/2}\}} \right\}^{1/3.91} \right] \tag{27}$$

and

$$\bar{P}_m^d/M_0 = 86.14 \, (2R/h)^{1/3} \left\{ 1 + (0.370V/6844R)^{1/3.91} \right\} \tag{28}$$

for symmetric and non-symmetric crushing modes, respectively.

DYNAMIC AXIAL CRUSHING OF SQUARE TUBES

The procedure outlined earlier for the dynamic crushing of circular tubes is repeated by Abramowicz and Jones (1984a) for square tubes. Equations (12), (14), (15) and (24) give the mean dynamic crushing loads

$$\bar{P}_m^{\ d}/M_0 = 52.22 \ \{1 + (0.33V/cD)^{1/p}\}(c/h)^{1/3} \tag{29}$$

$$\bar{P}_m^{\ d}/M_0 = \{1 + (0.49V/cD)^{1/p}\}\{43.61(c/h)^{1/3} +$$
$$+ 3.79(c/h)^{2/3} + 2.6\} \tag{30}$$

and

$$\bar{P}_m^{\ d}/M_0 = \{1 + (0.41V/cD)^{1/p}\}\{46.16(c/h)^{1/3} +$$
$$+ 2.14(c/h)^{2/3} + 1.30\} \tag{31}$$

for the symmetric, asymmetric mixed collapse mode A and asymmetric mixed collapse mode B, respectively, and where $D = 6844 \ sec^{-1}$ and $p = 3.91$ for the present tests according to equations (23) and (24).

DISCUSSION

Circular Tubes

The predictions of equations (1), (3), (5) to (7), and (10) and (11) for the static mean axial crushing loads of circular cylindrical shells are compared in Figs. 7(a) and 7(b) with various experimental results and empirical relations. The dimensionless parameter $\eta = P_m/A\sigma_u$, or $\eta = \bar{P}_m/A\sigma_u$, is known as the structural effectiveness and $\phi = A/A_1$ is the relative density, or solidity ratio, where $A = 2\pi Rh$ and $A_1 = \pi R^2$. The empirical relation

$$\eta = 2\phi^{0.7} \tag{32}$$

was proposed by Thornton, Mahmood and Magee (1983) and is also shown in Fig. 7. Mamalis and Johnson (1983) also proposed an empirical relation which can be cast into the form

$$\eta = 7\phi/(4 + \phi) + 0.07 \tag{33}$$

when taking $\sigma_u = \sigma_0$ which is a reasonable assumption according to the static tensile stress-strain curve in Fig. 1 of Mamalis and Johnson (1983). Equation (10), which is a corrected version of Alexander's analysis for symmetric, or concertina, behaviour, gives fair agreement with the empirical equation (32) and the mean of all the experimental results. Equations (5) and (10) predict that symmetric deformations control the behaviour when $\phi \geq 0.053$, approximately, or $R/h \leq 37.8$, while the diamond deformation pattern governs for $R/h \geq 37.8$. However, the mean static crushing loads are sufficiently close for $R/h \geq 25$, approximately, that both diamond and concertina modes of deformation would be encountered in tests over a wide range of R/h values according to equations (5) and (10). Indeed, specimen number 7 tested by Abramowicz and Jones (1984b) initially deformed with symmetric, or concertina, deformations, which were then followed by the development of non-symmetric, or

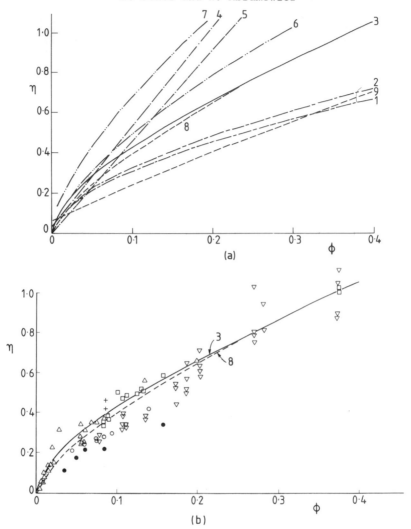

Fig. 7. Static axial crushing of cylindrical shells.
(a) Comparison of theoretical predictions
     (with $\sigma_0 = \sigma_u$) and empirical relations.
(b) Comparison of equations (10) and (32)
     with experimental results.
— – — 1: equation (1); — — — 2: equation (3);
——— 3:   equation (10); —·— 4: equation (5);
—·— 5: equation (6); —··— 6: equation (7);
—··— 7: equation (11); - - - 8: equation (32);
- - - 9:   equation (33).
Experimental results. +: Abramowicz and Jones
(1984b)(mild steel); ●: Mamalis and Johnson (1983)
(aluminium 6061 T6); o: Alexander (1960) (mild
steel); Δ: Macaulay and Redwood (1964); □ ,∇:
Taken from Fig. 4.4 of Thornton, Mahmood and Magee
(1983).

diamond, deformation modes. It is interesting to note that the recent static axial crushing tests by Mamalis and Johnson (1983) on
aluminium alloy 6061 T6 circular cylinders exhibited a similar phenomenon. Andrews, England and Ghani (1983) also observed a transition
from symmetric to non-symmetric crushing in some HT 30 aluminium
circular tubes. The results of this Reference cannot be plotted in
Fig. 7(b) because the static crushing loads were not presented.

The mean axial crushing loads for the dynamically loaded cylindrical
shells tested by Abramowicz and Jones (1984b) are normalised in Fig.
8 with respect to the corresponding static collapse loads and compared
with equations (27) and (28). It is evident that smaller mean dynamic
loads are predicted for symmetric crushing but the differences between
equations (27) and (28) are less than the scatter in the experimental
results. The horizontal line in Fig. 8 is associated with the response of a cylindrical shell which is made from a strain rate insensitive material.

Equation (30b) of Vaughan (1969) predicts a critical velocity $V_{cr} \cong$
$h \, (2\pi L \sigma_u / M)^{\frac{1}{2}}$ for the dynamic axial plastic buckling of a cylindrical
shell when $M \gg 2\pi R L h \rho$ and, for simplicity, σ_u is taken as the mean
flow stress. It was observed by Abramowicz and Jones (1984b) that
V_{cr} was smaller than any of the impact velocities in their experimental tests. Nevertheless, the type of buckling with wrinkling

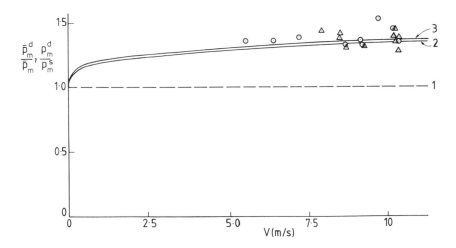

Fig. 8.   Ratio of dynamic axial crushing loads to
          static axial crushing loads of cylindrical
          shells.

          - - - 1:   equations (27) and (28) for a
                     strain rate insensitive material;
          ———— 2:   equation (27);
          ———— 3:   equation (28).

          Experimental results.(R = 28.032 mm, h =
          1.2 mm) (Abramowicz and Jones (1984b);
          o: symmetric;   Δ: non-symmetric.

along the whole length of a tube, which is a characteristic of the
test specimens examined by Vaughan (1969), was not observed in any
of their tests.   Thus, Vaughan's theoretical predictions are not
valid below a certain magnitude of velocity which is larger than
V_{cr} for the test specimens examined by Abramowicz and Jones (1984b).

Vaughan's theoretical analysis uses a perturbed motion superposed on
a fully plastic axial response which, therefore, requires $\eta \geq \sigma_0/\sigma_u$.
Thus, it appears that the progressive axial plastic buckling examined
in this article develops when $\eta \leq \sigma_0/\sigma_u$.   Indeed, if one defines
$\eta = P_m^d/A\sigma_u$, then with two exceptions, the experimental results of
Abramowicz and Jones (1984b) have $0.54 \leq \eta \leq 0.63$ which should be
compared with $\sigma_0/\sigma_u = 0.66$.   It is suggested, therefore, that when
neglecting material strain rate sensitivity effects, the structural
effectiveness $\eta = \sigma_0/\sigma_u$ might provide a transition between quasi-
static progressive buckling examined in this article and dynamic
plastic buckling studied by Vaughan (1969).

Square Tubes

The approximate theoretical predictions of equations (19) to (22)
for axially crushed square box columns are shown in Fig. 9 for the
symmetric, asymmetric mixed collapse mode A, asymmetric mixed collapse mode B and extensional modes, respectively.   Extensional collapse modes govern the static progressive behaviour of thick square

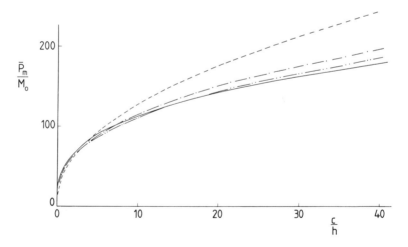

Fig. 9.   Variation of dimensionless mean static
          crushing loads of square box columns with
          c/h.
          ————:   equation (19);
          —·——·:   equation (20);
          —··—··:   equation (21);
          - - -:   equation (22).

columns with c/h $\leq$ 2.2, approximately, except for c/h $\cong$ 0. However, the mean collapse loads for extensional behaviour are considerably larger than those associated with the other three collapse modes when c/h $\geq$ 10, approximately.

It is evident that the asymmetric mixed collapse mode A does not control the progressive crushing of square tubes within the range of c/h in Fig. 9. The asymmetric mixed collapse mode B has the smallest associated mean crushing load when 2.2 < c/h < 17.5, approximately, while the symmetric mode governs the behaviour for thinner square tubes with c/h $\geq$ 17.5, approximately. However, the mean static crushing loads associated with the symmetric mode and the two asymmetric mixed modes are similar over the entire range of c/h in Fig. 9. Experimental data on the static axial crushing of box columns is collected in Fig. 3.8 of Wierzbicki (1983a) and Fig. 4.4 of Thornton, Mahmood and Magee (1983).

This data is plotted in Fig. 10 together with the relations

$$\eta = 1.42 \, (\sigma_0/\sigma_u)\phi^{2/3} \tag{34}$$

and

$$\eta = 1.4\phi^{0.8} \tag{35}$$

from Wierzbicki (1983a) and Thornton, Mahmood and Magee (1983), respectively, where σ_0 is the average flow stress, $\eta = \overline{P}_m/4ch\sigma_u$ is known as structural effectiveness and $\phi = 4 \, hc/c^2$ is the relative density or solidity ratio. Equation (34) was derived theoretically, while equation (35) is an empirical fit to the experimental data.

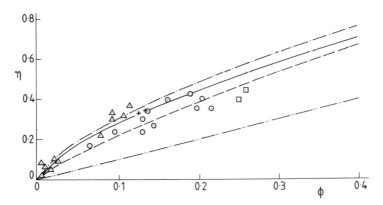

Fig. 10. Static axial crushing of square box columns.
o, $\triangle$, $\square$: experimental data from three test programmes taken from Fig. 3.8 in Wierzbicki (1983a)

+: experimental results of Abramowicz and Jones (1984a).
—·—·—·: equation (34) with σ_0/σ_u = 1
— — — : equation (35)
————: equation (36) with σ_0/σ_u = 1
—··—··: equation (37) with σ_0/σ_u = 1

Equation (19) for the static symmetric crushing of square box columns may be cast into the dimensionless form

$$\eta = 1.30 \ (\sigma_0/\sigma_u) \phi^{2/3} \tag{36}$$

and gives good agreement with the experimental results in Fig. 10.

Meng, Al-Hassani and Soden (1983) have recently examined the static axial crushing of square box columns and developed an inextensional analysis which predicts that

$$\eta = (\sigma_0/\sigma_u) \phi. \tag{37}$$

Equation (37) is linear in contradistinction to equations (34) to (36) and gives an equation for P_m/M_0 which is independent of the ratio c/h unlike equations (19) to (22).

The theoretical predictions associated with the symmetric mode (equation (29)) and asymmetric mixed mode B (equation (31)) are normalised with respect to the corresponding static crushing loads and presented in Fig. 11 together with the experimental test results on 37.07 mm square tubes which were loaded dynamically by Abramowicz and Jones (1984a). These theoretical results (i.e., equations (41) and (43)) are corrected for both the effective crushing distance and material strain rate sensitivity as discussed in a previous section.

Wierzbicki, Molnar and Matolscy (1978) reported some axial impact tests on 40 mm x 40 mm square steel box columns having a 2 mm wall thickness. These test specimens have different dimensions to those examined by Abramowicz and Jones (1984a) but are nevertheless added to Fig. 11 because of the paucity of experimental results. Ohkubo, Akamatsu and Shirasawa (1974) studied the dynamic axial crushing of columns with closed hat box sections, of which square tubes are a special case, and found the empirical relation

$$\overline{P}_m^{\,d}/\overline{P}_m = 1 + 0.0668 \ V, \tag{38}$$

with V in m/s.

Equation (38) is compared in Fig. 12 with the empirical equation

$$\overline{P}_m^{\,d}/\overline{P}_m = 1 + 0.07 \ V^{0.82} \tag{39}$$

due to Wimmer (1975) for square box columns with 50 mm x 50 mm cross-sections and 1.5 mm thick and V in m/s. Wierzbicki, Molnar and Matolscy (1978) developed a linear viscoplastic analysis and predicted that

$$\overline{P}_m^{\,d}/ \ \overline{P}_m = 1 + 0.1 \ V^{0.714} \tag{40}$$

for steel box columns impacted axially.

Equations (19) and (29) for the symmetric collapse mode predict

$$\overline{P}_m^{\,d}/\overline{P}_m = 1 + 0.183 \ V^{0.256} \tag{41}$$

and

$$\overline{P}_m^{\,d}/\overline{P}_m = 1 + 0.170 \ V^{0.256} \tag{42}$$

for square tubes with c = 37.07 mm and c = 49.31 mm, respectively. Similarly, equations (21) and (31) for the asymmetric collapse mode B predict that

$$\overline{P}_m^{\ d}/\overline{P}_m = 1 + 0.193 \ V^{0.256} \tag{43}$$

and

$$\overline{P}_m^{\ d}/\overline{P}_m = 1 + 0.180 \ V^{0.256} \tag{44}$$

for square tubes with c = 37.07 mm and c = 49.31 mm, respectively. The impact velocity V in equations (41) to (44) is measured in m/s.

It is interesting to compare the static progressive crushing behaviour of circular and square tubes having the same values of cross-sectional area A_1 and wall cross-sectional area A. The solidity ratio ϕ for circular and square tubes is therefore equal and the corresponding values of η are given in Fig. 7(b) for circular tubes

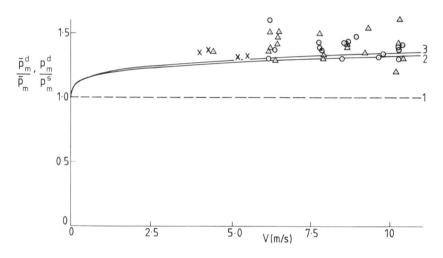

Fig. 11.   Variation of the dynamic to static mean
            crushing load ratio with impact velocity
            for 37.07 mm square tubes.

                 o: experimental results with sym-
                      metric collapse mode (Abramowicz
                      and Jones (1984a)).
                 Δ: experimental results with asym-
                      metric collapse mode (Abramowicz
                      and Jones (1984a)).
                 X: experimental results of Wierzbicki,
                      Molnar and Matolscy (1978).
        - - - 1:   equations (29) and (31) for a
                      strain rate insensitive material
        ——— 2:   equation (41)
        ——— 3:   equation (43).

and Fig. 10 for square tubes. It is evident from these two figures that a square tube is weaker and that for a given structural effectiveness ϕ has an associated value of η which is about two-thirds of the corresponding circular tube result. This observation agrees with the conclusion of Van Kuren and Scott (1977) for spot-welded high strength steel tubes.

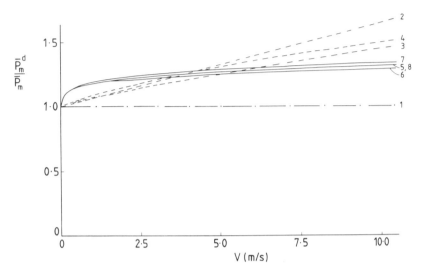

Fig. 12.   Comparison of theoretical predictions with previous empirical relations

—·—·— 1:   strain rate insensitive behaviour
- - - 2:   Equation (38) due to Ohkubo, Akamatsu and Shirasawa (1974)
- - - 3:   Equation (39) due to Wimmer (1975)
- - - 4:   Equation (40) due to Wierzbicki, Molnar and Matolscy (1978)
———5-8:   Equations (41)-(44), respectively.

CONCLUSIONS

A modified version of Alexander's (1960) theoretical solution for axisymmetric, or concertina, progressive buckling of circular tubes, which includes a correction for the effective crushing distance, (equation (10)), gives good agreement with the mean of the experimental static axial crushing loads. If the influence of material strain rate sensitivity is also introduced into this theoretical method, then the predictions (equation (27)) agree reasonably with the mean dynamic axial crushing loads observed in experimental tests.

The approximate theoretical predictions developed by Abramowicz and Jones (1984a) for the static progressive buckling of square box columns give reasonable agreement with the corresponding experimental results. It was also shown that this approximate theoretical work predicted satisfactory agreement with dynamic axial crushing tests on square box columns, provided the influence of material strain rate sensitivity was retained as well as the effective crushing distance.

ACKNOWLEDGEMENTS

The authors wish to acknowledge the Science and Engineering Research
Council for their support of this study through grant number GR/B/
89737.

One author (W.A.) obtained an SERC Visiting Fellowship (grant number
GR/C/40930) and visited the Department of Mechanical Engineering at
Liverpool University during Autumn 1983. The authors are indebted
to Mr. R. S. Birch and Mr. W. S. Jouri for their assistance with the
experimental equipment, Mrs. M. White for her typing and Mr. F. C.
Cummins and Mrs. A. Green for their preparation of the drawings.

REFERENCES

Abramowicz, W. (1983). The effective crushing distance in axially
    compressed thin-walled metal columns. *International Journal of
    Impact Engineering*, 1, No. 3, 309-317.
Abramowicz, W. and Jones, N. (1984a). Dynamic axial crushing of
    square tubes. *International Journal of Impact Engineering*
    2, No. 2, 179-208
Abramowicz, W. and Jones, N. (1984b). Dynamic axial crushing of
    circular tubes. *International Journal of Impact Engineering*
    2, No. 3, 263-281
Abramowicz, W. and Wierzbicki, T. (1979). A kinematic approach to
    crushing of shell structures. Proc. 3 Int. Conf. on Vehicle Struc-
    tural Mechanics, SAE, 211-223.
Alexander, J. M. (1960). An approximate analysis of the collapse of
    thin cylindrical shells under axial loading. *Quart. J. Mechs. and
    Applied Math.*, 13, 10-15.
Andrews, K. R. F., England, G. L. and Ghani, E. (1983). Classifi-
    cation of the axial collapse of cylindrical tubes under quasi-
    static loading. *Int. J. Mech. Sci.*, 25, 687-696.
Campbell, J. D. and Cooper, R. H. (1966). Yield and flow of low-
    carbon steel at medium strain rates. Proc. Conf. on the Physical
    Basis of Yield and Fracture, Inst. of Physics and Physical Soc.,
    77-87.
Davies, G. A. O. (Ed.) (1984). *Structural Impact and Crashworthiness.*
    Elsevier Applied Science Publishers, England
Ezra, A. A. and Fay, R. J. (1972). An assessment of Energy Absorbing
    Devices for prospective use in aircraft impact situations. In G.
    Herrmann and N. Perrone (Eds), *Dynamic Response of Structures*,
    Pergamon Press, pp. 225-246.
Hayduk, R. J. and Wierzbicki, T. (1982). Extensional collapse modes
    of structural members. Proc. Symp. Advances and Trends in Struc-
    tural and Solid Mechanics, Washington, 405-434.
Johnson, W. (1972). *Impact Strength of Materials*. E. Arnold, London
    and Crane Russak, New York.
Johnson, W. and Mamalis, A. G. (1978). *Crashworthiness of Vehicles*.
    Mechanical Engineering Publications Ltd., London.
Johnson, W. and Reid, S. R. (1978). Metallic energy dissipating
    systems. *Applied Mechanics Reviews*, 31, No. 3, 277-288.
Jones, N. (1983). Structural aspects of ship collisions. In N.
    Jones and T. Wierzbicki (Eds), *Structural Crashworthiness*, Butter-
    worths Publishers. Chap. 11, pp. 308-337.
Jones, N. (1984). Dynamic elastic and inelastic buckling of shells.
    In J. Rhodes and A. C. Walker (Eds), *Developments in Thin Walled
    Structures*, Vol. 2. Elsevier Applied Science Publishers, Chap. 2,
    pp. 49-91.
Jones, N. and dos Reis, H. L. M. (1980). On the dynamic buckling of
    a simple elastic-plastic model. *Int. J. Solids and Structures*, 16,
    969-989.

Jones, N. and Wierzbicki, T. (1983). (Eds), *Structural Crashworth-iness*, Butterworths Publishers

Lowe, W. T., Al-Hassani, S. T. S. and Johnson, W. (1972). Impact behaviour of small scale model motor coaches. Proc. I. Mech. E., Auto. Div., 186, 409-419.

Macaulay, M. A. and Redwood, R. G. (1964). Small scale model railway coaches under impact. *The Engineer*, 1041-1046, 25 Dec.

Mahmood, H. F. and Paluszny, A. (1981). Design of thin walled columns for crash energy management - their strength and mode of collapse. Proc. 4th Int. Conf. on Vehicle Structural Mechs., 7-18.

Mamalis, A. G. and Johnson, W. (1983). The quasi-static crumpling of thin-walled circular cylinders and frusta under axial compression *Int. J. Mech. Sci.*, 25, 713-732.

Meng, Q., Al-Hassani, S. T. S. and Soden, P. D. (1983). Axial crushing of square tubes. *Int. J. Mech. Sci.*, 25, No. 9-10, 747-773.

Ohkubo, Y., Akamatsu, T. and Shirasawa, K. (1974). Mean crushing strength of closed-hat section members. Soc. of Automotive Engineers paper no. 740040.

Postlethwaite, H. E. and Mills, B. (1970). Use of collapsible structural elements as impact isolators with special reference to automotive applications. *Journal of Strain Analysis*, 5, No. 1, 58-73.

Pugsley, A. (1960). The crupling of tubular structures under impact conditions. Proc. Symp. The Use of Aluminium in Railway Rolling Stock, Inst. Loco. Engrs., The Aluminium Development Association, London, 33-41.

Pugsley, A. G. (1979). On the crumpling of thin tubular struts. *Quart. J. Mechs. App. Maths.*, 32, 1-7.

Pugsley, A. and Macaulay, M. (1960). The large scale crumpling of thin cylindrical columns. *Quart. J. Mechs. and App. Maths.*, 13, 1-9.

Rawlings, B. (1974). Response of structures to dynamic loads. In J. Harding (Ed.), *Mechanical Properties at High Rates of Strain*, Inst. of Phys. Conf. Series No. 21, London, pp. 279-298.

Thornton, P. H., Mahmood, H. F. and Magee, C. L. (1983). Energy absorption by structural collapse. In N. Jones and T. Wierzbicki (Eds.), *Structural Crashworthiness*, Butterworths Publishers, Chap. 4, pp. 96-117.

Van Kuren, R. C. and Scott, J. E. (1977). Energy Absorption of High-Strength Steel Tubes Under Impact Crush Conditions, Int. Auto. Eng. Cong., Detroit, SAE paper 770213.

Vaughan, H. (1969). The Response of a Plastic Cylindrical Shell to Axial Impact, *ZAMP*, 20, pp. 321-328.

Wierzbicki, T. (1983a). Crushing behaviour of plate intersections. In N. Jones and T. Wierzbicki (Eds.), *Structural Crashworthiness*, Butterworths Publishers, Chap. 3, pp. 66-95.

Wierzbicki, T. (1983b). Crushing analysis of metal honeycombs. *International Journal of Impact Engineering*, 1, No. 2, 157-174.

Wierzbicki, T. (1983c). Optimum design of integrated front panel against crash. Report for Ford Motor Company, Vehicle Component Dept., 15 July.

Wierzbicki, T. and Abramowicz, W. (1981). Crushing of thin-walled strain rate sensitive structures. *Rozprawy Inzynierskie*, Polska Akademia Nauk, 29, No. 1, 153-163.

Wierzbicki, T. and Abramowicz, W. (1984). On the crushing mechanics of thin-walled structures. *Journal of Applied Mechanics*, paper number 83-WA/APM-12.

Wierzbicki, T. and Akerstrom, T. (1977). Dynamic crushing of strain rate sensitive box columns. Proc. 2nd Int. Conf. on Vehicle Structural Mechanics, SAE paper no. 770592, 19-31.

Wierzbicki, T., Molnar, C. and Matolscy, M. (1978).   Experimental-
   theoretical correlation of dynamically crushed components of bus
   frame structure.   Proc. XVII Int. FISITA Congress, Budapest, June.
Wimmer, A. (1975).   The effect of rate of load application on the
   mechanical resistance and deformation characteristics of sheet
   metal structures for motor vehicles (in German).   *Automobiltech-
   nische Zeitschrift*, <u>77</u>, No. 10, 281-286.

METAL TUBES AS IMPACT ENERGY ABSORBERS

S. R. REID

Department of Mechanical Engineering, University of Manchester Institute of Science and Technology, Manchester, UK

ABSTRACT

Recent work concerned with the plastic crushing of metal tubes is reviewed. The three basic modes of collapse, lateral compression, axial compression and transverse loading of tubular beams are examined. Significant contributions have been made since these were reviewed by Johnson and Reid (1978). Attention is then focused on to the behaviour of filled tubes under axial compression and transverse bending as examples of interactive modes. As the tube crushes, the filling is compressed and reacts against the tube wall tending to produce a greater stretching contribution which in turn leads to higher mean crushing loads. This technique for improving the energy absorbing capacity of tubes can lead to tensile failure or column buckling which reduce the effective stroke of the device. Tube splitting and the tubular ring energy absorber are then considered as examples in which the interaction is between different types of deformation fields (bending and fracture or bending and membrane deformation) within the tube wall.

INTRODUCTION

During the past ten years considerable attention has been paid to the large plastic deformation of metal structures. This has contributed to a better understanding of the capacity of such structures to absorb energy during impact. Research into the behaviour of structural elements per se as well as into the design of components to meet specific energy absorbing and impact requirements has produced a corpus of knowledge which is now available for use in a wide variety of engineering situations. Applications include aspects of the crashworthiness of vehicles (cars, lifts, helicopters, ships etc), crash barrier design, aspects of nuclear reactor safety and ship collision damage to offshore structures. Each area of application has its own requirements and special features but underpinning these is the need to understand how metal structural components respond to loads which are applied impulsively and have sufficient intensity and duration that gross plastic deformation results.

Johnson and Reid (1978) (referred to as JR below) reviewed the work

249

published in this area placing emphasis on the behaviour of simple
structural elements. A range of dominant modes of deformation (tension, bending, buckling, torsion, extrusion and cutting) was identified and load-deflection characteristics of a number of elements
were described. Since 1978 work in this area has continued and indeed expanded. Many of the more recent developments in the field
were presented at the First International Symposium on Structural
Crashworthiness held in Liverpool in September 1983, the proceedings
of which can be found in the book edited by Jones and Wierzbicki
(1983) (referred to as JW below) and the special issues of the International Journal of Mechanical Sciences ($\underline{25}$, Nos. 9-10, 1983) and
the International Journal of Impact Engineering ($\underline{1}$, No. 3, 1983).
In addition there have been further contributions to the subject in
the proceedings of several other recent conferences which will be
referred to below as well as a wide range of individual papers appearing in the literature. This level of activity reflects the
growing interest in this area of research.

The aim of this chapter is to up-date the review by Johnson and Reid
(1978). In view of the number of conferences to which reference has
been made above, a comprehensive review is unwarranted since many of
the papers appearing in the proceedings of these conferences are
themselves either review papers in specific areas or contain references to previous work on particular problems. Rather the aim will
be to draw attention to developments which have taken place since
1978 in much the same style as in the original review, emphasising
the dominant mechanisms of deformation.

One limitation is that the components considered herein are metal
tubes of various cross-sections constrained in different ways. They
are subjected to different types of loading usually applied locally
(in contrast to blast loading) which leads to plastic collapse followed by progressive crushing of the structure or large global deformations.

The reason for examining this class of structures is that almost all
of the applications with which we shall be concerned involve structures made of tubular components. Furthermore the ability of the
structure to crush extensively is often a pre-requisite in this
field and one naturally turns to shell-type components to meet this
requirement since solid section beams or plates have a limited deflection range which terminates abruptly when Euler buckling or tensile failure occurs. In designing an energy absorbing device we
need to ensure that the transmitted load levels lie below certain
critical values in order to limit the deceleration of the occupant
of a vehicle. Coupled with this a sufficient level of deformation
(stroke) must be provided so that the kinetic energy involved is
dissipated during the crushing of the structure. These two factors
usually imply that the response of the structure well beyond the
point of initial plastic collapse is of primary interest.

                    LATERALLY COMPRESSED CIRCULAR TUBES

An extensive review of lateral tube compression was provided in
chapter 1 of JW by Reid (1983). The tube is loaded along its whole
length so that it undergoes two-dimensional deformation which is
dominated by bending about discrete plastic hinges which run parallel
to the axis of the tube. The behaviour is essentially the same as
that of a ring under radial loads and many of the characteristics
have been examined in the context of ring compression. Figures 1
and 2 show one way of increasing the energy absorbing capacity of

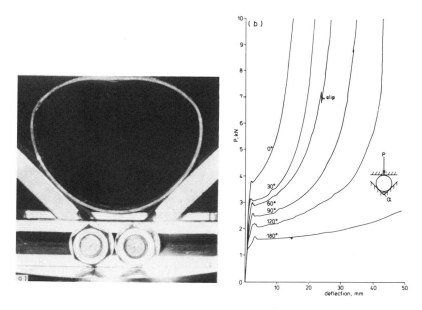

Fig. 1. (a) Ring on inclined supports under flat
plate compression.
(b) Load-deflection curves.

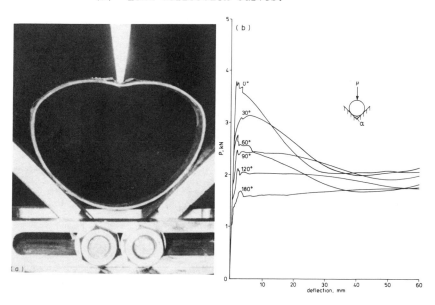

Fig. 2. (a) Ring on inclined supports under radial
point loading.
(b) Load-deflection curves.

a tube under diametral compression. This case (corresponding to $\alpha = 180°$) was examined by Reid and Reddy (1978) for compression between flat plates and by Reid and Bell (1982) for opposed point loading. The cases shown in Figs. 1 and 2 span the range between the diametral compression of a free tube and that of a laterally constrained tube ($\alpha = 0°$) which was investigated by Reddy and Reid (1979).

As can be seen, the initial collapse load, P_α, for a tube of radius R increases as the angle α between the inclined supports reduces from 180° in a manner which agrees closely with the following formula derived from limit analysis:

$$P_\alpha = \frac{4M_0}{R} \cot [(\pi+\alpha)/8].\tag{1}$$

$M_0 = Yh^2/4$ is the fully plastic bending moment per unit length of tube or ring where h is the wall thickness and $Y = \sigma_0$ for a ring (length not greater than a few thicknesses) or $2\sigma_0/\sqrt{3}$ for a tube (length not less than the diameter of the tube), σ_0 being the tensile yield stress. The post collapse behaviour depends critically upon the shape of the indenter which governs the way in which the loads are applied as the geometry of the tube alters.

Similar enhancement of the collapse load and energy absorbing capacity is achieved by constraining the mode of deformation of the tube by using tensile bracing across various diameters of the tube as described by Reid, Drew and Carney (1983). These two examples provide illustrations of how the energy absorbing capacity can be increased by generating what Jones and Wierzbicki (1976) described as higher modal plastic response. They used the term in the context of the dynamic loading of simple beams but the principle is the same in that a mode of plastic collapse is generated which contains more plastic hinges than occur in the fundamental ($\alpha = 180°$) mode. The higher modes are generated in the examples quoted by an interaction between the tube and passive constraints, i.e. constraints which, whilst they improve the performance of the deforming element, absorb no energy themselves. A few examples of more complex cases in which all the interacting components deform are given below.

An interesting application of the use of laterally compressed tubes has been in new designs for modular crash cushions. These are clusters of tubes used to prevent errant vehicles from striking rigid structures on freeways in the United States. Figure 3 shows a typical layout for one of these devices and Fig. 4 shows the performance of two versions struck obliquely as indicated in Fig. 3(b). The system in Fig. 4(a) used plain tubes and was incapable of preventing serious damage to the vehicle. That shown in Fig. 4(b) utilised braced tubes and the improved performance led to acceptable levels of deceleration and damage in the vehicle (Reid, Drew and Carney (1983)).

It has been observed that assemblies of tubes or rings respond to dynamic loads applied at one end in a wave-like manner. The final deformation results from the transmission and reflection of structural waves through the system. This has been explored experimentally in the context of simple one-dimensional ring systems by Reid and Reddy (1983). A simple shock theory was described by Reid, Bell and Barr (1983) which was based upon an analogy with plastic waves

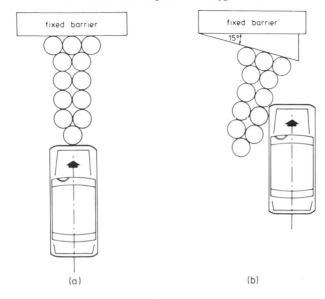

Fig. 3.   Modular crash cushion subjected to (a) axial
and (b) oblique impact.

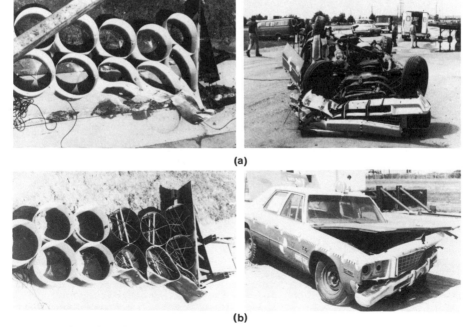

(a)

(b)

Fig. 4.   (a) Response of modular crash cushion comprised of plain tubes to oblique impact.
(b) Response of system containing reinforced
tubes.

propagating in a rigid-plastic bar. This model has been extended
recently by Reid and Bell (1984) to include the effects of elastic
deformation, strain-rate effects and non-uniformity in the mode of
deformation of individual rings. Figure 5 shows the type of system
examined by Reid and Bell (1984) together with a comparison between
the observed and predicted transient response. Whilst this analysis
is somewhat crude is does highlight the significant role played by
elastic deformation during dynamic structural behaviour and relates
to work of a more fundamental nature by Symonds and Fleming (1984).

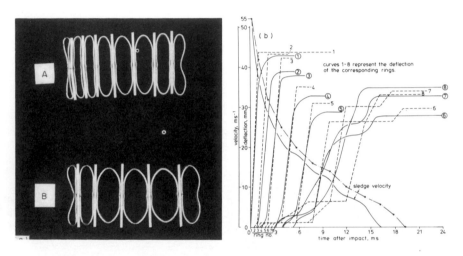

Fig. 5.   (a) Systems of steel rings separated by
              plates subjected to end impact. Ring
              + plate mass = 48 g;   projectile mass
              = 133 g;   impact speed A: 55 m/s B:
              40 m/s.
          (b) Comparison between observed transient
              response of system and the predictions
              of the shock theory due to Reid and
              Bell (1984).

AXIAL COMPRESSION OF METAL TUBES

In JR emphasis was placed upon the fact that axially compressed tubes
provide a very efficient means of absorbing energy and since 1978
the mechanics of axial crushing of circular, square and rectangular
tubes has continued to attract attention. Only passing reference
will be made to it here since much of the recent work is referred
to in the article by Jones and Abramowicz. Perhaps the most
interesting development in this area has been the kinematically
consistent method of analysing the crushing of plate structures
(including square and rectangular section thin-walled tubes)
devised by Wierzbicki and his co-workers. This work has
begun to clarify the interaction between bending and membrane

stretching in thin-walled structures as they crush. It is this
which controls the geometry of the buckling region in the component
and the mean loads sustained during collapse.

The axisymmetric bellows mode of collapse of relatively thick-walled
circular tubes (R/h<15) involves a clear interaction between circumferential stretching and bending at discrete hinges. For thinner
tubes the role of extensional deformation is less clear. Analyses
based upon overall modes of collapse which are completely inextensional have been produced (Johnson, Soden and Al-Hassani (1977)) and
these provide reasonable estimates for the mean collapse load. The
use of mechanisms which do not contain stretching has however been
questioned by Reid (1978) on the grounds that during the crumpling
process the Gaussian curvature of the shell changes and compatibility
therefore requires a non-zero membrane strain field. This problem
has yet to be resolved for a circular cross-section tube but progress
has been made for structures comprised of plane elements.

Wierzbicki and Abramowicz (1983) analysed the crushing of thin walled
structures made from plate elements. By insisting on kinematic admissibility they constructed fold mechanisms which consist of patterns of stationary plastic hinges and narrow toroidal regions of
circumferential stretching and bending which travel through the
material. It is the presence of the latter that ensures compatibility. The mean crushing load P_m takes the general form

$$\frac{P_m}{M_0} = A_1 \frac{b}{h} + A_2 \frac{C}{H} + A_3 \frac{H}{b}. \tag{2}$$

A_1, A_2 and A_3 are constants dependent on the geometry of the structures, b is the smaller radius of curvature of the toroidal surface,
h is the wall thickness, C is a quarter of the circumference of the
shell and 2H is the axial extent of the fold mechanism. Minimising
P_m with respect to b and H in the spirit of the upper bound theorem
of limit analysis leads to the optimum value

$$\frac{P_m}{M_0} = 3 \sqrt[3]{A_1 A_2 A_3} \; \sqrt[3]{C/h} \tag{3}$$

The three terms of the right hand side of equation (2) represent,
respectively, contributions from stretching in the toroidal regions,
bending in the stationary and travelling horizontal hinges and bending in the hinges bounding the toroidal regions. The optimum corresponds to equal contributions from these three sources. Thus one
third of the energy is absorbed in membrane stretching despite the
fact that extension is confined to a small area of the shell.

If in a particular mode of deformation the ratio of bending to extension is $\alpha:\beta$ then the mean load depends on the wall thickness, h,
as follows

$$P_m = A \, h^{\frac{2\alpha+\beta}{\alpha+\beta}} \tag{4}$$

This result encompasses the above result ($\alpha=2$, $\beta=1$) as well as the
classical axisymmetric mode ($\alpha=\beta=1$), inextensional deformation ($\alpha=1$,
$\beta=0$) and pure extension ($\alpha=0$, $\beta=1$).

More complex fold mechanisms than the basic ones discussed by Wierz-

bicki and Abramowicz (1983) are summarised in the article by Jones
and Abramowicz.

Square and rectangular tubes are important in view of their extensive
use in automobile structures. An important practical consideration
in their use is whether they behave in a stable manner since there
is a tendency for the mode of collapse to change from progressive
crumpling to an Euler type of failure which is very inefficient.
This has been discussed in detail by Thornton (1983) in JW. This
paper also contains numerous references to the behaviour of sheet
metal structures again of particular significance in the context of
automobile crashworthiness.

            TUBULAR BEAMS

Early work by Thomas, Reid and Johnson (1976) and Watson, Reid and
Johnson (1976) concerned with the lateral loading of simply supported
tubular beams was reported in JR. The tubes were loaded by wedge-
shaped indenters and the sequence of events, shown in Fig. 6, is
that local denting under the applied load is followed by a denting/
bending phase in which rotation occurred about a central hinge leading to ultimate collapse and loss of load carrying capacity. These
tests could be thought of as crude models for automobile bumpers and
indeed many of the features resemble the response of commercial
bumpers measured by Johnson and Walton (1983) although presumably
most of these were of open rather than closed section.

The conclusion of this work on bumpers was that the majority were
singularly ineffective as energy absorbers. The problem has however
taken on a wider significance in the context of the ability of steel
braced offshore structures to withstand collisions from supply boats.
Several detailed studies of tubular beams with a variety of end
support conditions have been published.

Soreide and Amdahl (1982) considered the behaviour of tubes which
had fully-fixed (no axial or rotational end movements) or axially
free but rotationally fixed ends. Figure 7 summarises the behaviour
of tubes supported in this way. P_0 is the collapse load for the

tubular beam based on the assumption that the fully plastic collapse
moment M_p is attained at the two end sections and under the centrally

applied load. Thus

$$P_0 = \frac{8M_p}{L} = \frac{8D^2 t \sigma_0}{L} \tag{5}$$

Several features are apparent in Fig. 7. Because of local denting
the ideal initial collapse load is not attained by either beam. The
beam which is free to move axially never attains this load level but
actually goes unstable rather like a simply supported tube (see Fig.
10 below). The axially fixed beam on the other hand develops significant membrane tension which leads to a dramatic increase in the
load carried and, consequently, the energy absorbed. Early work by
Hodge (1974) has been used to estimate the large deformation behaviour of fully-fixed and partially fixed tubular beams by Soreide
and Amdahl (1982) and De Oliveira (1981). The ovalisation of the
cross-section under the load and local buckling on the compression
side of the tube at the fixed supports both need to be taken into
account when assessing the performance of a particular tube and
computational techniques have been developed which do this in an
appropriate way, see for example Soreide (1981). As noted by De

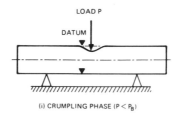

(i) CRUMPLING PHASE (P < P$_B$)

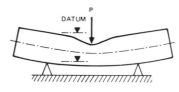

(ii) CRUMPLING & BENDING PHASE (P ⩾ P$_B$)

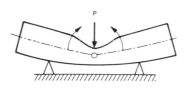

(iii) STRUCTURAL COLLAPSE (P = P$_{MAX}$)

Fig. 6.   Three stages in the mode of deformation of
a simply supported tube subjected to a
central transverse load.

Oliviera (1981) the response of tubular beams is limited by tensile
failure at the ends of the beam.   A degree of axial constraint is
necessary in order to generate the substantially improved performance evident in Fig. 7.   However if there is no axial flexibility
the energy absorbing capacity is somewhat limited.

The impact speeds for offshore collisions are such that the problem
is almost quasi-static although dynamic effects stemming principally
from the strain rate sensitivity of the material yield stress have
been detected and are most apparent during the initial denting phase
of the deformation.   (Soreide and Amdahl (1982)).

S. R. Reid

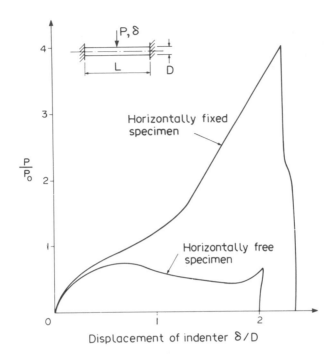

Fig. 7.   Effect of axial restraint on the behaviour
          of tubes loaded transversely at the centre.

INTERACTIVE MODES OF DEFORMATION

It was noted in JR that, on the basis of simple calculations, an axially compressed metal tube will absorb almost ten times the energy
of a similar tube compressed laterally. This results from the fact
that in the former case the mode of deformation involves an interaction between bending and stretching within the shell wall whereas
the latter involves only bending. Stretching alone is not a particularly effective mechanism for impact energy absorbing applications
since the stroke is limited by tensile failure. Whilst axial buckling and similar modes of collapse such as tube inversion (see JR)
are efficient, they do require close control over the direction in
which the loads are applied otherwise Euler-type instability intervenes and the potential energy absorbing capacity is not realised.
Generally speaking this is not a problem with laterally compressed
elements as demonstrated for instance in the modular crash cushions
shown in Fig. 4.

Earlier we examined ways in which passive constraints led to a more
efficient performance from a laterally compressed tube. It should
be pointed out that such a modification would only be utilised if
the extra weight of the supports did not carry a design penalty.
In this final section a number of examples are described in which
different types of interaction modify the response of tubular elements. They are presented as an indication of the widening range
of energy absorbing components based upon metal tubes.

(a) Polyurethane Foam-Filled Tubes Under Axial Compression

The chemicals needed to produce polyurethane foam are readily available and can be used to conduct tests on foam-filled structures. By
mixing the chemicals in different proportions and confining them, a
range of initial foam densities can be produced. An indication of
the variations in the strength of foams of different densities is
shown in Fig. 8. These data were acquired from tests in which cylindrical specimens of foam (ISO foam RM 118) were confined in a metal
tube and compressed axially so that any lateral expansion was prevented. The shape of the force-displacement (reduction in volume)
curve is reminiscent of that for a metal ring compressed between
rigid plates (see $\alpha=180°$ curve in Fig. 2(b)). The similarity presumably reflects the collapse of the spheroidal bubbles in the foam
and the subsequent rapid stiffening as consolidation occurs analogous
to corresponding stages in the deformation of a ring. The foam is
usually characterised by its crush strength σ_f which fulfills a role
similar to that of the collapse load of a laterally compressed ring.
It is interesting to note that the higher the density of the foam,
the sooner the stiffness begins to increase at a noticeably more
rapid rate. This behaviour was noted by Wilsea, Johnson and Ashby
(1975) who examined the measurement of foam properties from sphere
indentation tests.

Thornton (1980), Thornton, Mahmood and Magee (1983) and Lampinen and
Jeryan (1982) have discussed the effects of filling sheet metal fab-
ricated tubes with polyurethane foam. Lampinen and Jeryan conducted
a wide range of axial compression tests on circular and square seam
welded tubes and on square spot welded tubes (termed Z-columns) and
produced design formulae using regression analysis to encompass the
data they produced. Comparisons between different components can be
made on the basis of specific energy absorbing capacity, E_{sa}, the

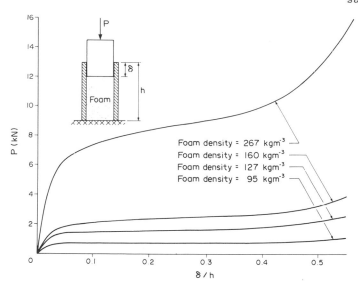

Fig. 8. Compression curves for confined cylinders
of foam (specimen diameter = 50 mm) having
different initial densities.

energy absorbed per unit weight of the collapsed portion of the tube.
Using their notation

$$E_{sa} = \frac{E_a}{W_1 \delta_a} \qquad (6)$$

where E_a = area under the force-displacement curve

W_1 = mass per unit length of the component

δ_a = total tube deformation before instability

If P_m = mean collapse load, equation (6) becomes

$$E_{sa} = \frac{P_m}{A_s \rho_s} \qquad (7)$$

where A_s is the cross-sectional area of the tube-wall and ρ_s the
mean density of the material used. However E_{sa} does not provide a
complete picture of how an axially compressed component behaves since
the highly efficient progressive crumpling mode can be interrupted
by the onset of global instability if the tube buckles as an Euler
column. This can arise if too dense a foam is used and the consolidation causes the buckling process to lock, transforming the tube
into a column of solid section. Lampinen and Jeryan introduced a
modified specific energy absorbing capacity E_{sa}^* defined by

$$E_{sa}^* = \eta_c E_{sa} \text{ where } \eta_c = \delta_a/l_0. \qquad (8)$$

l_0 is the original length of the tube. This parameter ascribes
equal importance to increases in P_m and δ_a which may or may not be
the relevant weighting for all structural components. However for
foam filling this approach does highlight the need to be aware of
some of the detrimental effects of changes which simply increase P_m.

Lampinen and Jeryan performed tests on tubes which typically had a
wall thickness of 0.75 mm, widths (diameter or side length) of 100
mm made from sheet steel with a yield stress of 190 MPa and an ultimate stress of 310 MPa. The foams they used conformed approximately
to the following relationship between the crushing strength at 5%
reduction in volume σ_f (MPa) and initial density ρ_f (kg/m^3) proposed
by Thornton (1980)

$$\sigma_f = 7.4 \times 10^{-4} \rho_f^{1.6}. \qquad (9)$$

For low density foam (of the order of 50 kg/m^3) the increase in mean
load required to compress the composite structure is simply that required to crush the foam. As such it could be argued that there is
no genuine interaction in this case. For high foam densities (in
excess of 300 kg/m^3) the axial force generated after a relatively
small deformation reaches the Euler collapse load and column failure
intervenes at an early stage. This is presumably as a result of the
rapid stiffening in the foam response evident in Fig. 8. For intermediate densities several interesting points emerge. For tubes of
all dimensions P_m increases and δ_a generally reduces with increasing
foam strength. The performance of filled circular tubes as reflected
by E_{sa}^* is always inferior to that of an empty tube of the same dimensions, whereas square section and Z-columns both show some improvement

and optimum performance corresponds to a foam density of around 200 kg/m³.

Thornton et al (1983) and Lampinen and Jeryan (1982) question the benefits of foam reinforcement for energy absorbing purposes despite the evidence (Table 12 in Lampinen and Jeryan (1982)) that higher mean loads, P_m, can be achieved in a more weight-effective manner than using thicker sheet metal. Doubts about the efficiency of foam filling rest largely on the reduction in δ_d (which may not always be important) and concern about producing foam of consistent strength.

In terms of the mechanics of the deformation process there are several interesting points to note. Tables 8 and 10 in Lampinen and Jeryan (1982) show clearly that the enhancement in P_m is not merely due to the crushing of the foam. There is a significant extra load stemming from a change in the mode of deformation of the tubes. There is a shortening in the wavelength of the buckles which corresponds to higher loads. This same type of behaviour is apparent in some recent tests on rectangular tubes performed by Gray (1984). The shorter wavelength is shown in Fig. 9(a) and the behaviour of similar tubes filled with foams of different density is shown in Fig. 9(b). The shorter wavelengths are evident in the higher frequencies of the oscillations in the load-displacement diagram. In this case the empty tube actually went unstable after only a few buckles whereas the filled tubes all responded in a stable manner.

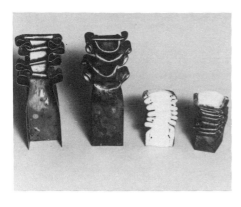

(a)

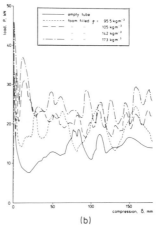

(b)

Fig. 9.   (a) Sheet metal square tube and foam-filled
              rectangular tube after compression.
          (b) Load-compression curves for rectangular
              tubes (51 mm x 102 mm x 0.8 mm) filled
              with foams of different densities.

It appears from the evidence in Lampinen and Jeryan (1982) that tubes of non-circular cross-section show a greater capacity for improvement by foam filling than those of circular section (especially those deforming in the axisymmetric mode). As noted in the section on axial compression of tubes, the analysis of Wierzbicki and Abramowicz (1983) shows that one of the differences between these two categories of tube is the proportion of energy absorbed by stretching and bending. In the classical axisymmetric bellows mode for a circular tube the contribution from each of these two deformation components is the same whereas for rectangular and square tubes only one third of the energy is absorbed in stretching. If the axisymmetric circular tube represents the optimum performance it seems a reasonable hypothesis that the internal pressure produced by the foam as it crushes generates a greater stretching contribution and leads to a more significant improvement in these tubes.

Most of the discussion of the effect of foam filling above relates to their static behaviour. Under impact loading conditions two effects are evident. First there is a tendency towards greater overall stability arising from lateral inertia as discussed by Reid and Austin (1984). The second is a strain rate effect which increases the yield stress of the materials. The incorporation of this effect into the analysis of empty tubes has been described in chapter by Jones and Abramowicz. In the tests reported by Gray (1984) a 25% increase in mean load was measured for empty tubes crushed at about 10 m/s. However the dynamic enhancement factor for foam-filled tubes was noticeably higher at about 45%. The tests reported by Lampinen and Jeryan (1982) were conducted at impact speeds of 13.5 m/s and 20 m/s and their results indicate dynamic enhancement factors generally between 45-70% for empty and filled tubes alike.

(b)   Sand-Filled Tubular Beams Under Central Load

Figure 7 shows that a tubular beam without axial constraint has a limited energy absorbing capacity. As an extension to the work by Watson, Reid and Johnson (1976) a few tests were conducted on simply supported tubes filled with different quantities of fine sand. As can be seen in Fig. 10 the unstable behaviour, which results from the loss of bending strength due to severe denting of the tube under the indenter is arrested to a degree which depends on the amount of sand contained in the tube. The early behaviour in which local denting without bending occurs is unaffected although the reduction in the volume of the tube causes the sand to become pressurised. This pressure limits the ovalisation at the centre and generates membrane tension in the tubes. Failure eventually occurred by fracture at the central cross-section on the opposite side to the indenter. It should be noted that it was found to be more effective to fill the tubes with sand which has a significant degree of bulk compressibility than other materials having higher bulk moduli (e.g. water). Under dynamic loading conditions it was found that the loads generated in the sand-filled tubes were significantly higher than those predicted by applying a strain rate factor to the observed static behaviour of a similar tube. Under dynamic loading the effect of the presence of the sand was also evident at lower deflection levels than those shown in Fig. 10. This suggests that there is a tendency for the sand to lock under dynamic loads. The deformation is much more localised around the indenter, the crushing of the cross-section is arrested earlier and the bending strength of the tube is consequently enhanced.

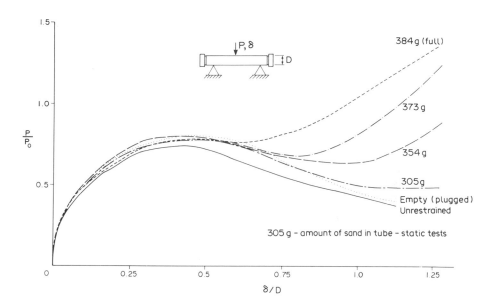

Fig. 10.   Simply supported tubes containing different
           quantities of sand subjected to central
           transverse load.

(c)   Tube Splitting

The invertube is described in JR.   It consists of a circular tube
supported on a radiused die and compressed axially.   The die causes
the tube to flare generating circumferential stretching and bending.
Provided the die radius is chosen suitably, the tube inverts.   Circumferential bending and unbending is produced close to the die and
a tube of increased radius is produced.   If the die radius is bigger
than that which leads to stable inversion the tube either fragments
(the frangible tube, see McGehee (1966)) if it is made of brittle
material or it splits under the influence of a series of cracks which
propagate axially.   Stronge, Yu and Johnson (1983) examined the
splitting of square section tubes.   Two modes of splitting were
identified.   In the first the four edges split at the die, the walls
of the tube then curl upwards and move parallel to the sides of the
undeformed part of the tube in much the same way as in tube inversion.
This is not a very efficient mode since the stroke is then limited
to approximately half the length of the tube.   If the radius of the
die is increased this mode changes to curling in which the sides
curl up in a scroll-like manner.   This type of deformation is also

apparent in Fig. 11(a) although here the tube is a circular one and
relates to recent work by Reddy (1984). The load-deflection curve
for the latter is shown in Fig. 11(b) and, once the stable set of
fractures have become established, the load remains virtually constant for the remainder of the stroke.

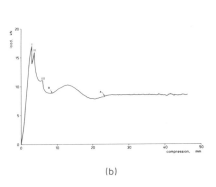

(b)

(a)

Fig. 11.   (a) Splitting of a circular aluminium tube
                by axial compression on to a radiused
                die.
            (b) Load-deflection curve for a brass tube.
                I, II and III represent points at which
                fractures are initiated. At B the frac-
                tures begin to bifurcate and by A all
                the cracks were propagating axially.

The deformation mode for both the circular and the square tubes is
an interaction between axial plastic bending in the strips, axial
fracture and friction at the die tube interface. A simple formula
which fits the data for the steady state load P_s for a circular

tube of radius R, wall thickness h and with n axial cracks (Reddy
(1984)) is

$$P_s = (2\pi RM_0/a + n\, G_c h)/(1-\sqrt{2}\,\mu),$$ (10)

where a is the radius of the die, G_c is the fracture toughness of
the material and μ is the coefficient of friction between the tube
and the die.

(d)   Tubular Ring

A final example of a tubular device in which an interaction is pro-
moted in a deliberate way is the tubular ring shown in Fig. 12. The
mode of deformation and behaviour of this element under static and
dynamic loads have been described by Reid, Austin and Smith (1984).

Each ring is made up of four sections of tube cut at 45° to the axis and welded together at four mitred joints. The ring is crushed by lateral compression as shown in Figs. 12(a) and (b) and, as can be seen from Fig. 12(c), the load-deflection characteristics show marked increases in energy absorbing capacity over an equivalent free-tube (curve 1). The improved performance stems from the fact that the edges of the individual tubes would naturally warp as shown in Fig.

(a)   (b)

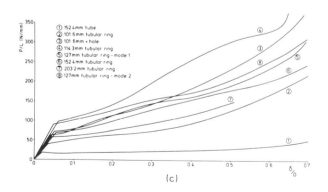

(c)

Fig. 12.   Tubular ring (steel tube, D = 50.8 mm,
            t = 1.6 mm) of side 152.4 mm (a) before
            and (b) after compression. (c) load per
            unit length versus deflection for range
            of side lengths.

13 but are prevented from doing so at the welds. This constraint generates a type of in-plane bending on the tube. Thus the side walls of the tubular ring are in a state of tension or compression which prevents the individual sides from collapsing in the simple lateral compression mode. As Fig. 14(a) shows, the energy absorbing capacity is 4.5 to 9 times that of an equivalent free tube. It also shows that there is an optimum configuration where the performance is maximised. Dynamic tests have also been performed on layered systems of these components. The results show that the transmitted force pulse tends to be fairly flat. This is a reflection of material strain rate effects evident in the dynamic behaviour of a single tubular ring shown in Fig. 14(b) as well as the non-uniform mode of collapse which resembles that produced in the ring systems studied by Reid and Reddy (1983) which can be seen in Fig. 15.

CONCLUSION

A wide range of modes of plastic crushing are available using metal tubes many of which are relevant to impact energy absorption. Efficient use of the material would seem to depend upon the ability to generate both bending and stretching deformation in the tube wall. A number of the interactive modes of collapse have been described which aim to do this and they would seem to provide promising ways of improving the performance of basic tubular elements.

Fig. 13.   Edge of a tube cut at 45° to the axis warping as a result of lateral compression.

ACKNOWLEDGEMENTS

The author would like to express his gratitude to Dr. T. Y. Reddy
for useful discussions during the preparation of this chapter and
to Mrs. L. Coull, Mr. D. Bain and Mr. I. McKinnon for help in preparing the manuscript.   Some of the work described was supported by
grants from the Departments of Energy and Transport (administered
by the Transport and Road Research Laboratory), NATO (Grant No.
04981) and the Marine Technology Directorate of SERC all of which
are gratefully acknowledged.

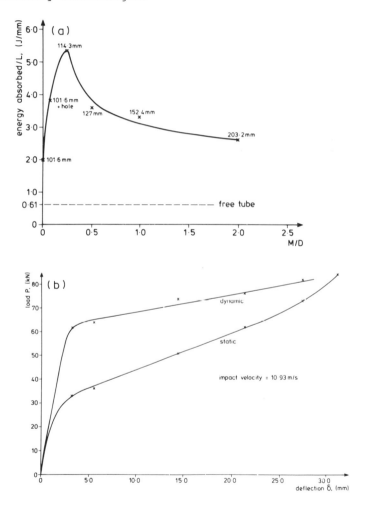

Fig. 14.   (a) Variation of energy absorbed per unit
                length at 53% compression versus M/D
                where M = length of side of central
                hole.
            (b) Comparison between static and dynamic
                load-deflection for ring shown in Fig.
                12(a).

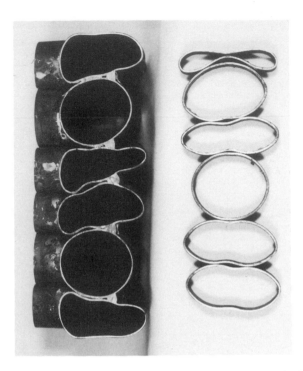

Fig. 15.   Cross-section of a dynamically crushed
six-layer system compared with a one-
dimensional ring system.

REFERENCES

De Oliviera, J. G. (1981). The behaviour of steel offshore struc-
    tures under accidental collisions. *Proc. 13th Offshore Technology
    Conference*, Houston, Paper OTC 4136, 187-198.
Gray, M. D. (1984). The performance of sheet metal tubes as impact
    energy absorbers. Honours Thesis, Department of Engineering,
    University of Aberdeen, Scotland.
Hodge, P. G. (1974). Post-yield behaviour of a beam with partial-
    end fixity. *Int. J. Mech. Sci.*, 16, 385-388.
Johnson, W., Soden, P. D. and Al-Hassani, S. T. S. (1977). Inexten-
    sional collapse of thin-walled tubes under axial compression. *J.
    Strain Analysis*, 12, 317-330.
Johnson, W. and Reid, S. R. (1978). Metallic energy dissipating
    systems. *Applied Mechanics Review*, 31, No. 3, 277-288.
Johnson, W. and Walton, A. C. (1983). An experimental investigation
    of the energy dissipation of a number of car bumpers under quasi-
    static lateral loads. *Int. J. Impact Eng.*, 1, 301-308.
Jones, N. and Wierzbicki, T. (1976). A study of the higher model
    dynamic plastic response of beams. *Int. J. Mech. Sci.*, 18, 533-
    542.
Jones, N. and Wierzbicki, T. (Eds.) (1983). *Structural Crashworth-
    ness*. Butterworths.
Lampinen, B. E. and Jeryan, R. A. (1982). Effectiveness of polyure-
    thane foam in energy absorbing structures. Annual Meeting, Detroit,

Michigan, *SAE*, Paper 820494.

McGehee, J. R. (1966). Experimental investigation of parameters and materials for fragmenting-tube energy absorption process. *NASA Report* TN D-3268, Langley, Virginia.

Reddy, T. Y. and Reid, S. R. (1979). Lateral compression of tubes and tube-systems with side constraints. *Int. J. Mech. Sci.*, 21, 187-199.

Reddy, T. Y. (1984). Private communication.

Reid, S. R. (1978). Communication: Inextensional collapse of thin-walled tubes under axial compression. *J. Strain Analysis*, 13, 240-241.

Reid, S. R. and Reddy, T. Y. (1978). Effects of strain hardening on the lateral compression of tubes between rigid plates. *Int. J. Solids and Structures*, 14, 213-225.

Reid, S. R. and Bell, W. W. (1982). Influence of strain hardening on the deformation of thin rings subjected to opposed concentrated loads. *Int. J. Solids and Structures*, 18, 643-658.

Reid, S. R. (1983). Laterally compressed metal tubes as impact energy absorbers. Chapter 1 of *Jones and Wierzbicki* (1983), 1-43

Reid, S. R., Bell, W. W. and Barr, R. (1983). Structural shock model for one-dimensional ring systems. *Int. J. Impact Eng.*, 1, 175-191.

Reid, S. R., Drew, S. L. K. and Carney, J. F. (1983). Energy absorbing capacity of braced metal tubes. *Int. J. Mech. Sci.*, 25, 649-667.

Reid, S. R. and Reddy, T. Y. (1983). Experimental investigation of inertia effects in one-dimensional metal ring systems subjected to end impact. I: Fixed-ended systems. *Int. J. Impact Eng.*, 1, 85-106.

Reid, S. R. and Austin, C. D. (1984). Influence of inertia in structural crashworthiness. *Proc. Int. Conf. on Vehicle Structures.* I. Mech. E., C173/84, 63-70

Reid, S. R., Austin, C. D. and Smith, R. (1984). Tubular rings as impact energy absorbers. *Structural Impact and Crashworthiness.* Vol 2, ed J. Morton, Elsevier Applied Science Publishers, 411-426

Reid, S. R. and Bell, W. W. (1984). Response of one-dimensional ring systems to end impact. *Proc. Third Int. Conf. on Mechanical Properties of Materials at High Rates of Strain.* Institute of Physics. Conference Series No. 70, 471-478

Soreide, T. H. and Amdahl, J. (1982). Deformation characteristics of tubular members with reference to impact loads from collisions and dropped objects. *Norwegian Maritime Research*, 10, 3-12.

Stronge, W. J., Yu, T. X. and Johnson, W. (1983). Long stroke energy dissipation in splitting tubes. *Int. J. Mech. Sci.*, 25, 637-647.

Symonds, P. S. and Fleming, W. T. (1984). Parkes revisited: on rigid-plastic and elastic-plastic dynamic structural analysis. *Int. J. Impact Eng.*, 2, 1-36.

Thomas, S. G., Reid, S. R. and Johnson, W. (1976). Large deformations of thin-walled circular tubes under transverse loading, Part I. *Int. J. Mech. Sci.*, 18, 325-333.

Thornton, P. H. (1980). Energy absorption by foam filled structures. *Transactions SAE*, 89, 529-539.

Thornton, P. H., Mahmood, H. F. and Magee, C. L. (1983). Energy absorption by structural collapse, Chapter 4 of *Jones and Wierzbicki* (1983), 96-117

Watson, R. Reid, S. R. and Johnson, W. (1976). Large deformations of thin-walled circular tubes under transverse loading, Part III, *Int. J. Mech. Sci.*, 18, 501-509.

Wierzbicki, T. and Abramowicz, W. (1983). On the crushing mechanics of thin-walled structures. *Journal of Applied Mechanics*, 50, 727-734.

Wilsea, M., Johnson, K. L. and Ashby, M. F. (1975). Indentation of foamed plastics. *Int. J. Mech. Sci.*, 17, 457-460.

INITIATION OF PERFORATION
IN THIN PLATES BY
PROJECTILES

W. GOLDSMITH

*Department of Mechanical Engineering, University of California,
Berkeley, CA 94720, USA*

ABSTRACT

This paper is primarily concerned with the phenomena that are observed
when thin, metallic plates are struck normally by projectiles just
below and just above the ballistic limit. The tips of the strikers,
which are composed of hard steel that does not deform are either
blunt, hemispherical or conical, prefacing a cylindrical shank. A
few tests involving thick targets were also conducted.

INTRODUCTION

During the past three decades, a substantial amount of effort has
been devoted to both theoretical and experimental investigations of
the penetration of plates by projectiles, primarily at normal incidence. These plates have been generally considered either as thin,
where all field variables exhibit uniform values across the thickness
at any particular time, or else as being of moderate thickness where
the presence of the distal surface produces some modification of the
phenomena during the impact process. In addition, a considerable
number of studies were concerned with penetration into semi-infinite
media; these will not be further considered here. A recent comprehensive survey of the penetration of projectiles into targets has
been undertaken by Backman and Goldsmith (1978); in addition, a
series of selective topics involving this subject has been presented
by Eringen (1978) and Zukas and colleagues (1982).

Analytical approaches to the plate perforation problem can be subdivided into groups. First there are numerical procedures utilizing
finite-difference, finite-element or boundary element methods where
both projectile and target, if deemed deformable, are divided into a
a series of discrete cells with equations of motion applied in turn
to each of these small units. Initial velocity, projectile geometry
and angle of attack must be specified as well as constitutive relations and strength characteristics for each material involved. The
analysis leads to large-scale computations capable of predicting the
history of the deformation process, but at substantial expense for
very short time intervals and incorporates significant inaccuracies
when inadequate or even incorrect material information is included

in the code. Serious errors may also accrue when some existing behavioral characteristics are not incorporated in the program.

Alternatively such developments can take the form of phenomenological
descriptions leading to one or a series of models each of which is
considered to be appropriate to a certain régime of penetration.
Such models are usually very simple and lead to relatively rapid
methods of numerical evaluation, but frequently require incorporation
of certain simplifying assumptions that restrict their general applicability. Nevertheless, this approach has found considerable favor
in recent years because of the ease of computation (Awerbuch and
Bodner, 1974a; Woodward and DeMorton, 1976; Woodward, 1982; Recht,
1978; Ravid and Bodner, 1983; Dienes and Miles, 1977; Kelly and
Wilshaw, 1968; Liss, Goldsmith and Kelly, 1983; Landkof and Goldsmith, 1985). Furthermore, comparison of such predictions with
corresponding experimental data has been generally satisfactory,
particularly in domains far removed from limiting system behavior,
such as the vicinity of the ballistic limit or the onset of fragmentation and/or vaporization of target and/or projectile.

While enormous amounts of empirical information from all conceivable
types of tests concerning plate penetration have undoubtedly been
assembled in government installations throughout the world, data of
this type published in the open literature are extremely sparse
(Backman and Goldsmith, 1978; Goldsmith and Finnegan, 1971; Awerbuch
and Bodner, 1974b; Awerbuch and Bodner, 1977; Liss, Goldsmith and
Kelly, 1983). Furthermore, the results reported in these references
are primarily concerned with the velocity drop of projectiles during
penetration and the establishment of ballistic limits. The author
is not aware of any previous work specifically designed to illuminate
the mechanisms that govern the initial phases of the perforation process. In view of the extreme scientific and practical aspects of
such incipient fracture phenomena, the present contribution is primarily devoted to an experimental investigation of this domain in
relatively thin metallic targets due to the normal impact of hard-
steel cylinders with either blunt, hemispherical or conical nose
shapes. A few test results in targets of the same material and of
sufficient thickness so that no permanent effect from the impact is
manifested at the distal surface will also be included.

ANALYTICAL DEVELOPMENTS

Models have recently been developed to describe the normal penetration
and perforation of thin metallic plates and those of intermediate
thickness by projectiles with these three different nose configurations when striking in the neighborhood of the ballistic limit (Liss,
Goldsmith and Kelly, 1983; Landkof and Goldsmith, 1985; Levy and
Goldsmith, 1984a). For a blunt projectile, this involves a five-
stage penetration process consisting in order of occurrence of indentation, plug formation, separation, slipping and post-perforation
deformation. The striker can be considered either as rigid or deformable, and the material behavior is described by a strain-rate independent rigid-plastic constitutive relation involving both compression
and shear, the latter representing the response of the target outside
the region of contact in terms of a travelling plastic hinge. The
target and projectile are subdivided into a number of rigid bodies
separated by propagating shock waves. The equations of motion obtained for the relatively small number of system components defined
by these shock waves and hinges are solved on a computer, and these
predictions compare well with the results of corresponding experiments
(Liss and Goldsmith, 1984). Two representative examples of this

concurrence are portrayed in Figs. 1 and 2 that exhibit the variation
of the terminal projectile momentum or the velocity drop as a function
of the initial momentum or velocity for the normal impact of a 40 g,
12.7 mm diameter steel cylinder on 2024-0 aluminium plates of 3.2 and
6.4 mm thickness, respectively. The results are in excellent agreement except in the immediate vicinity of the ballistic limit where
the effect of target bending, neglected in the analysis will play a
significant role in the energy distribution of the phenomenon.

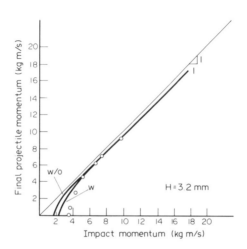

Fig. 1.   Final projectile momentum from model (solid
          lines) with (w) and without (w/o) plastic
          target shear deformation compared to test
          results (circles) as a function of initial
          momentum for a 3.2 mm thick 2024-0 aluminium
          plate struck by a 40 g projectile.

The analysis for the case of the normal impact of an undeformable
conically-tipped cylindrical projectile on a softer metallic plate
has been accomplished by means of an energy balance accounting for
crack propagation followed by a plastic hinge arrested at the end of
this fracture; in addition, the plastic bending of the petals so
formed due to projectile passage and the rotational energy of these
petals upon complete perforation are included. Predicted terminal
projectile velocities are in excellent agreement with measurements
at initial speeds substantially higher than the ballistic limit
(Landkof and Goldsmith, 1985). However, in the neighborhood of this
limit, discrepancies exist due to the absence of a complete accounting
of the target response outside the area of the crater. The effect of the
presence of a predrilled hole struck at the exact center was also
investigated; both theory and experiment predict the existence of
an "optimal" initial hole radius (smaller than that of the striker)
where a maximum amount of energy is absorbed, greater than that for

an intact plate. The reason for this unusual circumstance is a meta-
morphosis of the process from petalling to extrusion for certain
ranges of the impact parameters that appears to require more energy
than the combination of crack initiation and propagation as well as
the bending of the resulting petals.

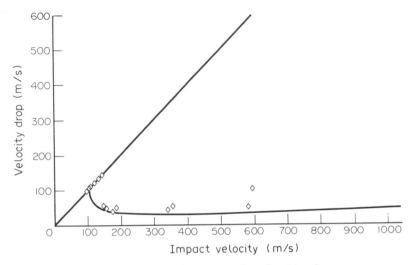

Fig. 2.   Predicted and measured velocity drop as a
           function of impact velocity for the normal
           impact of a hard-steel blunt cylindrical
           projectile of 40 g mass on a 6.4 mm thick
           2024-0 aluminium target.

An extremely simple phenomenological model has been proposed for the
prediction of the impact parameters when a hemispherically-tipped
cylindrical projectile is fired normally against a metallic plate,
expressed in terms of lumped masses. For the régime below the ballistic limit, the plate response is represented by an equivalent
mass subject to viscous damping due to striker contact, with elastic
action neglected. The parameters of the system can be completely
calculated from an energy balance for plate and projectile and an
assumed plate deformation shape, chosen as an exponentially decaying
deflection in accordance with the test data of Calder and Goldsmith
(1971). For plate perforation conditions, the system is portrayed
by two masses, one representing the plug and the rest an effective
mass for the remainder of the deformed target. Two dashpots are
employed as resistive elements. A closed-form solution was obtained
by use of some of the techniques for obtaining the system parameters
in the non-perforation cases and a suitable choice of initial conditions when the plug starts moving relative to the remainder of the
target (Levy and Goldsmith, 1984a and b). The projectile is presumed
be undeformed throughout. In spite of the simplicity of this analysis
which can only be considered as an interim description conceivably
representing a special case of a more complicated model based on
continuum mechanical considerations, excellent correlation with test
results were noted below the ballistic limit; less satisfactory
agreement was found in the perforation region. The sudden drop in
the peak force generated at the ballistic limit was found to be
present both in the results obtained from the model and in the test
data.

TEST ARRANGEMENT

Plate samples consisting of 2024-0 aluminium and mild steel (about SAE 1015) were cut from sheets of various thicknesses to an outside diameter of 138 mm (5.45 in) and clamped in a steel frame with an interior diameter of 113 mm (4.5 in). Some rectangular blocks with a thickness of 50 mm (2 in) were also employed on occasion when thick or semi-infinite target response was desired. All targets were subjected to the normal impact of 12.7 mm ($\frac{1}{2}$ in) diameter cylindrical projectiles featuring various nose shapes, composed of hard steel, with masses ranging from 29 to 40 g; a few blunt-nosed cylindrical strikers of soft aluminium with a mass of 16 g and a diameter of 12.7 mm ($\frac{1}{2}$ in) were also fired. Initial velocities of the strikers were ascertained by the interruption due to projectile passage of two light beams directed horizontally through two slots in the barrel onto two photodiodes located a known distance (15 cm) apart. When desired, camera coverage was obtained by means of a Photek IV camera (Photonic Systems, Inc., Sunnyvale, CA) with framing rates from 100 to 10,000 per second or a Beckman-Whitley WB-2 high-speed device capable of recording up to one million pps. Lighting was provided either by an eight-bulb circular unit with a power of 2400 W or a xenon flash tube activated from a capacitor bank (Levy and Goldsmith, 1984b). When a plug was separated from the target traversing the space beyond ahead of the projectile, a combination of a single laser just behind the target, a laser light grid placed a short distance (20 cm) down range orthogonal to the trajectory of the bullet, a counter and a digital oscilloscope were required to ascertain the terminal striker velocity (Liss and Goldsmith, 1984). The general test arrangement without the counter and oscilloscope is depicted in Fig. 3.

Fig. 3.   Experimental arrangement.

Each sequence of plates with the same thickness was successively subjected to increasing projectile speed for a particular tip configuration with special reference to the initial occurrence of a minute
crack on the distal side and the minimum velocity required to obtain
complete perforation as manifested by striker passage through the
complete target. The target response was carefully noted including
the use of a profilometer to delineate its terminal configuration,
and an electron microscope was utilized to examine the detailed
characteristics of the cracks generated. Several target thicknesses
were investigated including the replication of the semi-infinite
medium where only entry phenomena could be detected. No deformation
of the hard-steel projectiles was observed in any of the tests conducted for this particular investigation.

A parallel investigation conducted on thicker plates at higher velocities and at obliquity, using a powder gun and slightly different
methods of speed measurement will be described in another publication.

RESULTS AND DISCUSSION

Table 1 lists the type of projectiles and targets employed, the target
thickness, the initial projectile velocity and the target response
that is also photographically portrayed in the sequel. The plate
deformation characteristics are sketched in Figs. 4 and 5 for the
cylindro-conical and blunt projectiles, respectively; targets struck
by the sphere and hemispherically-tipped cylinders were not available
for sectioning. Values of these parameters for some of the runs are
presented in Tables 2 and 3 respectively.

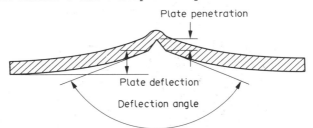

Fig. 4.   Plate deformation profile due to the normal
          impact of an undeformable 60° cylindro-
          conical projectile.

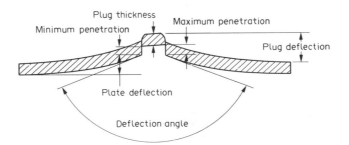

Fig. 5.   Plate deformation profile due to the normal
          impact of an undeformable blunt cylindrical
          projectile.

TABLE 1 Projectile and Target Parameters

Run No.	Projectile Type	Projectile Mass, g	Target Material	Target Thickness, mm	Initial Speed, m/s	Remarks
1	H	37.7	Al	1.27	59.4	No perforation*
2	H	37.7	Al	1.27	64.0	Incipient cracking*
3	H	37.7	Al	1.27	68.2	Perforation
4	H	37.7	Al	1.27	82.0	Projectile passage*
5	S	8.4	Al	1.27	127.0	Projectile passage*
6	CC	28.9	Al	1.59	19.4	No perforation*
7	CC	28.9	Al	1.59	21.4	Incipient cracking*
8	CC	28.9	Al	1.59	26.5	Perforation*
9	CC	28.9	Al	1.59	27.2	Perforation
10	CC	28.9	Al	3.18	45.7	No perforation
11	CC	28.9	Al	3.18	57.0	Incipient cracking, EM*
12	CC	28.9	Al	3.18	60.7	Perforation
13	B	37.4	Al	3.18	87.2	Plug shearing
14	B	28.9	Al	3.18	101.0	No shearing
15	B	28.9	Al	3.18	104.0	Plug shearing*
16	B	28.9	Al	3.18	105.0	Plug separated except for a 60° arc*
17	CC	28.9	Al	6.35	128.0	Incipient cracking*
18	CC	28.9	Al	6.35	128.0	Incipient cracking*
19	B	39.7	Al	12.7	184	Plug displaced, projectile lodged*
20	CC	28.9	Mild Stl.	1.18	23.1	Incipient cracking*
21	CC	28.9	Mild Stl.	1.18	31.4	Perforation
22	CC	28.9	Mild Stl.	3.18	90.5	Incipient cracking*
23	CC	28.9	Mild Stl.	3.18	119.0	Perforation*
24	CC	29.0	Mild Stl.	6.35	316	Petalling, projectile lodged*#

H = hemispherically-tipped projectile; S = 12.7 mm diameter hard-steel sphere
CC = 60° cylindro-conical projectile; B = blunt projectile
EM = electron microscopic photography

* Refer to photographs for this run; # Projectile fired from a powder gun

TABLE 2   Target Deformation due to the Conical-nosed
          Projectile

Run No.	Plate Penetration, mm	Plate Deflection, mm	Deflection Angle, deg
6	0.91	4.56	156
7	1.47	4.84	153
8	--	5.32	147
9	--	5.48	133
10	2.95	--	---
12	--	8.04	126
17	--	5.30	161
18	--	5.20	153
19	--	3.20	---
20	--	1.58	154
21	--	2.46	137
22	--	5.02	148
23	--	7.00	129
24	--	5.48	---

TABLE 3   Target Deformation due to the Blunt Projectile

Run No.	Plug Thickness, mm	Plate Deflection, mm	Plug Deflection, mm	Penetration Max. mm	Min.	Deflect. Angle deg
13	2.90	13.04	15.02	2.74	0.98	137
14	3.02	12.70	13.68	1.25	1.25	133
15	3.04	13.42	15.86	2.96	0.92	125
16	2.92	13.58	---	--	1.00	122

The most effective representation of the macroscopic phenomena is
revealed by the photographs of the impact and distal target sides
and of the target cross-sections. These pictures are shown in Figs.
6-9 for five thicknesses of the 2024-0 aluminium targets and in Fig.
10 for two thicknesses of the mild steel plates utilized.   The
terminal target/projectile configuration for Run No. 24 is shown in
Fig. 11; this test was conducted at an angle of obliquity of 10°
by means of a powder gun.   The cylindro-conical projectile generated
four major petals and lodged in the target.   The slight angularity
is not expected to produce significantly different results relative
to normal impact.

Various definitions of the ballistic limit exist; according to one
concept, this parameter is defined as the speed when the first
evidence of fracture is manifested on the distal side of the plate,
whereas another stipulation involves the velocity where 50% of the
strikers for that particular combination of projectile and target
exit.   According to the first definition, the ballistic limit may
be regarded as the state of incipient cracking in Table 1.   The ballistic limit is further described by Backman and Goldsmith (1978)
and has been experimentally determined by a number of investigators
(Goldsmith and Finnegan, 1971, 1985).

Figure 12 portrays the craters produced in a thick (50 mm) block of
2024-0 aluminium by a 60° cylindro-conical and blunt hard-steel projectile at an initial velocity of 202 and 179 m/s, respectively, as

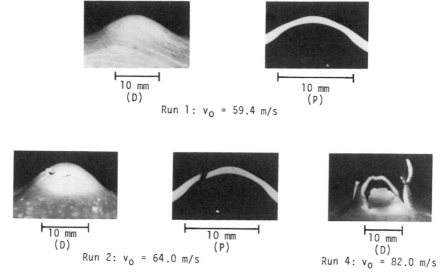

Run 1: v_0 = 59.4 m/s

Run 2: v_0 = 64.0 m/s   Run 4: v_0 = 82.0 m/s

Hemispherically-tipped projectile

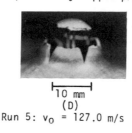

(D)
Run 5: v_0 = 127.0 m/s

Sphere

Fig. 6.   Photographs of the deformation of 1.27 mm
           thick 2024-0 aluminium plates struck nor-
           mally by hemispherically-tipped and spheri-
           cal hard-steel projectiles
           (D): distal side   (P): profile.

well as the dent formed in this material by the normal impact of a
soft aluminium blunt cylinder at a velocity of 186 m/s. All projectiles had a diameter of 12.3 mm. Figure 13 shows corresponding im-
prints of the conical-nosed hard-steel projectile in a similar thick
block of mild steel, fired at similar velocities. The test involving
a soft aluminium projectile on this steel generated a virtually in-
visible crater.

The results for the tests involving the thick blocks are presented
in Table 4. The material displaced by the strikers in the aluminium
block consisted of a series of small petals, of the order of twenty-
five, emanating from a thin rim at the edge of the crater for the
60° cylindro-conical steel bullets, similar to those manifested in
the thin plates at low striking velocities (Figs. 8-10). In contrast,

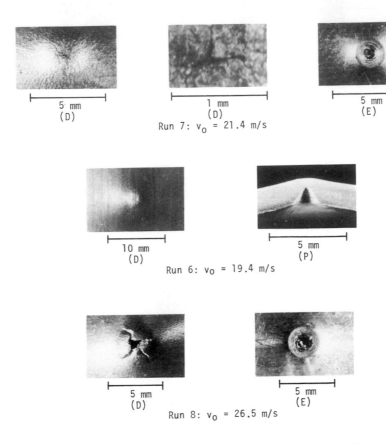

Fig. 7. Photographs of the deformation of 1.59 mm
         thick 2024-0 aluminium plates struck nor-
         mally by 60° cylindro-conical hard-steel
         projectiles.
         (D): distal side (E): entrance side
         (P): profile.

the blunt steel projectiles created a mound around the region of indentation without manifestation of any petalling. A distinct shortening and concomitant bulging near the impact face was observed in
the initially cylindrical soft aluminium bullets, similar to that
noted by Liss and Goldsmith (1984), even at these relatively low
impact speeds. At velocities of the order of 900 m/s and similar
targets, such aluminium bullets will mushroom involving reductions
to about one-third their original length (Goldsmith and Finnegan,
1984).

The phenomena representing the target response just below, at and just
above incipient perforation are significantly influenced by nose
shape, as is evident from an examination of Figs. 6-13. A hemispheri-

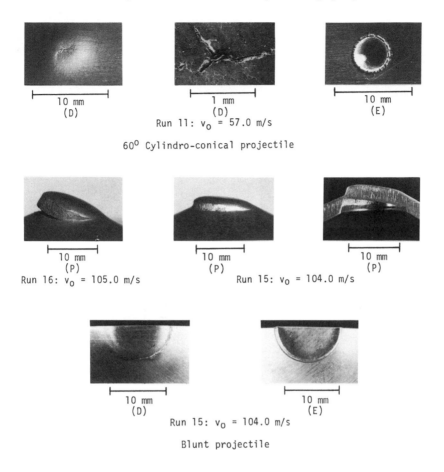

Run 11: v_0 = 57.0 m/s

60° Cylindro-conical projectile

Run 16: v_0 = 105.0 m/s   Run 15: v_0 = 104.0 m/s

Run 15: v_0 = 104.0 m/s

Blunt projectile

Fig. 8.   Photographs of the deformation of 3.18 mm
           thick 2024-0 aluminium plates struck nor-
           mally by hard-steel bullets.
           (D): distal side   (E): entrance side
           (P): profile.

cally-tipped projectile first develops a spherical cap in the target
primarily as the result of tensile stresses and involves a substantial
reduction in target thickness. At slightly higher impact speeds, a
circumferential crack is initiated as the result of this tension,
leading to shear failure at a 45° angle to the direction of principal
stress, just as is the case in a uniaxial tension test. At still
higher velocities, radial cracks are developed with their origin at
the initial circumferential discontinuity; these produce roughly
trapezoidal-shaped petals with the original cap joined to one of
these protrusions. A complete separation of the cap from the remainder of the crater occurs invariably only when the projectile is
a sphere rather a hemispherically-tipped cylinder; the formation
of such separate caps has also been noted in other investigations
(Calder and Goldsmith, 1971; Levy and Goldsmith, 1984b).

W. Goldsmith

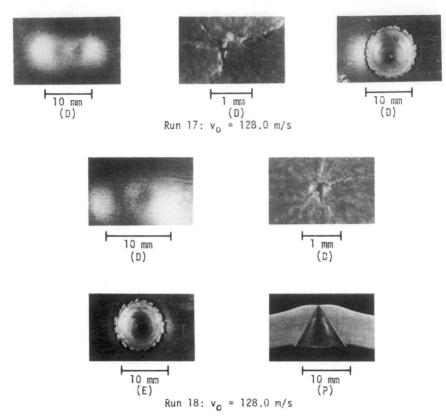

Run 17: v_0 = 128.0 m/s

Run 18: v_0 = 128.0 m/s

6.35 mm thick targets
60° cylindro-conical projectile

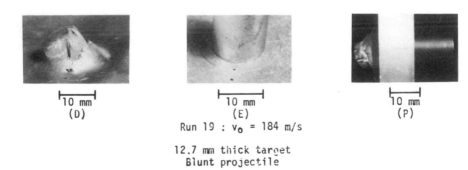

Run 19 : v_0 = 184 m/s

12.7 mm thick target
Blunt projectile

Fig. 9. Photographs of the deformation of 6.35 and
12.7 mm thick 2024-0 aluminium plates struck
normally by hard-steel bullets.
(D): distal side   (E): entrance side
(P): profile.

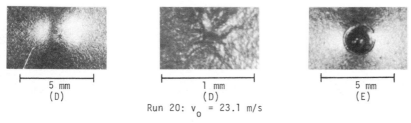

5 mm
(D)

1 mm
(D)

5 mm
(E)

Run 20: v_0 = 23.1 m/s

1.18 mm thick mild steel targets

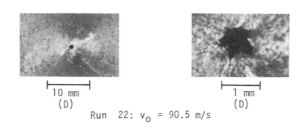

10 mm
(D)

1 mm
(D)

Run 22: v_0 = 90.5 m/s

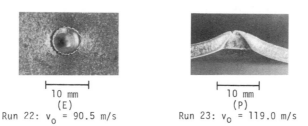

10 mm
(E)

10 mm
(P)

Run 22: v_0 = 90.5 m/s   Run 23: v_0 = 119.0 m/s

3.18 mm thick mild steel targets

Fig. 10.   Photographs of the deformation of mild
            steel targets struck normally by 60°
            cylindro-conical hard-steel projectiles.
            (D): distal side   (E): entrance side
            (P): profile.

TABLE 4   Results of Penetration Tests in Thick
           Metallic Targets

| Run | | Projectile | | | | | Target and Crater | | | |
No.	Type	Material	Mass g	Bulge mm	ΔLength mm	Material	Depth, mm Max.	Min.	Lip mm	Petal Hgt. mm
25	CC	Steel	28.9	--	--	2024-0 Al	11.35	---		3.3
26	B	Steel	28.9	--	--	2024-0 Al	2.55	3.12	0.2	---
27	B	Al	12.6	1.61	2.40	2024-0 Al	0.16	---		0.2
28	CC	Steel	28.9	--	--	Mild Stl.	8.80	---		2.6
29	B	Steel	28.9	--	--	Mild Stl.	2.26	1.98	0.2	---
30	B	Al	12.3	2.14	2.80	Mild Stl.	0.10	0	-	---

CC = 60° cylindro-conical projectile;   B = blunt projectile

20 mm   20 mm
(D)   (E)

Fig. 11.   Photograph of the terminal state of a 6.35
           mm thick mild steel plate struck by a hard-
           steel 60° cylindro-conical projectile fired
           at an initial velocity of 316 m/s at an
           angle of obliquity of 10°. The bullet
           lodged in the target. (E): entrance;
           (D): distal side.

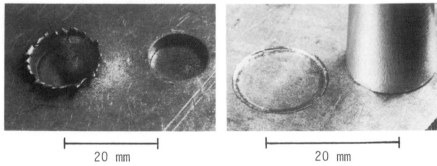

20 mm   20 mm
Fig. 12.   Photograph of the craters generated by
           cylindro-conical and blunt hard-steel
           projectiles (left) and by an aluminium
           projectile (right) on a 2024-0 aluminium
           block (Runs 25-27).

An analytical model of this type of perforation phenomenon must be
developed from the principles of continuum mechanics. For a sufficiently ductile substance such as represented by the two target
materials employed in the present investigation, and for a sufficientl
thin target relative to the projectile diameter, the process can be
described by an axisymmetric deformation of a rigid/work-hardening
membrane. For example, for the aluminium utilized in the present
tests, an appropriate linear relation for the uniaxial nominal stress-
strain curve is given by

$$\sigma = 165.5 + 836.4\ \varepsilon \text{ MPa} \tag{1}$$

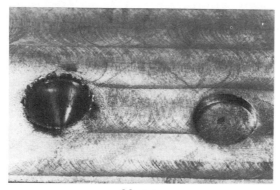

20 mm

Fig. 13.   Photograph of the craters generated by a
60° cylindro-conical (Run 28) and a blunt
(Run 29) hard-steel projectile striking a
thick block of mild steel normally at an
initial speed of 202 m/s.

with σ as stress, ε as strain, and assumed identical behavior in tension and compression (Landkof and Goldsmith, 1985).   Circumferential
fracture delineating the size of the cap occurs upon attainment of
the ultimate tensile strength;   this failure will initially cause a
reduction in the tensile stresses of the crater walls which, however,
will again increase upon further incursion of the projectile.   This
feature will quickly re-establish the failure stress level, resulting
in radial cracking to provide the required stress relief.   The author
is not aware of an analytical development describing this phenomenon
at the present time.   As in the case of conical-nosed strikers under
conditions of normal impact, the number of petals produced with a
rounded tip is either four or five.

A normally-impinging sharp-pointed projectile will carve out a conical
region of the impact side of the target and concomitantly produce a
circular dimple on the distal face prior to the first evidence of
cracking.   This is accompanied by the formation of a myriad of very
small, thin, irregularly-shaped petals bent outward from the thin,
short hollow cylindrical backward extension of the crater.   The configuration is created by the material displaced from the crater proper
requiring an energy W that can be approximated by the relation

$$W = (1/3)\pi Y d^3 \tan^2 \alpha \tag{2}$$

Here, Y is an average yield stress, d is the depth of penetration and
α is the half-cone angle of the nose.   As an example, a value of
W = 1.5 J is obtained for a penetration depth of 3.18 mm (1/8 in) and
an average yield stress Y = 133.7 MPa.   This penetration phenomenon
is also accompanied by significant flexure of the target, generating
a deflection angle approximately twice that of the total cone angle
2α of the striker.   These energy components are not included in an
energy balance describing this type of impact configuration (Landkof
and Goldsmith, 1985) which otherwise takes account of crack propagation and petal bending.

Most of the cracks visible at minimum impact velocities, including
those examined under an electron microscope (second photograph, Fig.
8) exhibit four quite irregular discontinuities. However, in at leas-
one experiment it was found that a single, nearly linear crack of
substantial extent relative to those produced at similar velocities
was the only manifestation of the failure condition on the distal
side. Since the direction of these cracks are almost certainly controlled by the microstructure and residual stresses, created in part
by the technique of manufacture such as the direction of rolling of
sheets, the rather unique situation of but one tensile failure exhibi-
should not be taken as characteristic in any modelling process. In
spite of many inconsistencies, it seems reasonable to base any analysis of the crack formation process on a symmetric pattern involving
four sets of linear fracture surfaces, at least as a first approximation.

Finally, the blunt projectile induces first compression and subsequently shear, as detailed in the analysis by Liss, Goldsmith and
Kelly (1983). The circular rims on the distal side of the thin plate
and the upward-sloping contour around the impact region of the thick
blocks represent the material displaced by target compression. It
is considered that this mechanism also strongly influences the deformation of the target outside the crater zone and should be added to
the current analysis, particularly in the vicinity of the ballistic
limit for thin plates. It appears to be the only major mechanism
that is not incorporated in the model cited and, as already stated,
should substantially decrease the discrepancy between the predicted
and measured kinematic values that are exhibited in Fig. 1.

        CONCLUSIONS

Two relatively soft metals, 2024-0 aluminium with a Brinell hardness
of 100 and mild steel with a Brinell hardness of 128 have been examined to determine the predominant phenomena that occur when cylindrical hard-steel projectiles with three different nose shapes im-
pinge at normal incidence both on thin plates and on thick targets
composed of these substances. Ballistic limits, defined here as the
first indication of fracture on the distal side, were established for
the aluminium target as follows: 64 m/s for the hemispherically-
nosed striker impinging on a 1.27 mm thick plate, 21.4, 57 and 128
m/s for a 60° conical tip contacting 1.59. 3.18 and 6.35 mm thick
targets, respectively, and 103 m/s for the flat cylinder impacting
a 3.18 mm thick specimen. A limit of 23.1 m/s for a 1.18 mm thick
plate and a value of 90.5 m/s for a 3.18 mm thick sample, both composed of mild steel, were ascertained for a 60° conical-nosed hard-
steel cylinder. Blunt aluminium cylinders deformed substantially
under similar normal impact conditions, but produced much smaller
craters in thick blocks of the two metals.

A continuum model of the perforation of a hemispherically-nosed projectile awaits development that incorporates an initial contoured
deformation with attendant reduction in the plate thickness, an initial tensile failure resulting in a circumferential crack, and a subsequent radial cracking process generating petals and permitting the
passage of the striker. The current analysis for the conical-tipped
cylinder should also include the energy required to displace the target material during entry into a thin hollow ring extending backwards
from the distal side of the target and terminating in a multitude of
small petals. Both this configuration and that for a blunt cylinder
must also be modified to account for the bending of the plate outside
the crater region; this deformation characteristic absorbs a signifi

cant amount of the total energy near the ballistic limit, but is relatively unimportant at much higher impact speeds.

ACKNOWLEDGEMENT

The author would like to express his sincere appreciation to Messrs. S. Vorestek and S. Unnasch for significant assistance in the conduct of the tests. This work was sponsored by the U. S. Army Research Office under Contract DAAG 29-80-K-0052.

REFERENCES

Awerbuch, J., and S. R. Bodner (1974a). *Int. J. Solids Struct.*, 10, 671-684.
Awerbuch, J., and S. R. Bodner (1974b). *Int. J. Solids Struct.*, 10, 685-699.
Awerbuch, J., and S. R. Bodner (1977). *Exp. Mech.*, 17, 147-153.
Backman, M. E., and W. Goldsmith (1978). *Int. J. Eng. Sci.*, 16, 1-99.
Calder, C. A., and W. Goldsmith (1971). *Int. J. Solids Struct.*, 7, 863-881.
Dienes, J. K., and J. W. Miles (1977). *J. Mech. Phys. Solids*, 25, 237-256.
Eringen, A. C., Editor (1978). *Int. J. Eng. Sci.*, 16, 793-920.
Goldsmith, W., and S. A. Finnegan (1971). *Int. J. Mech. Sci.*, 13, 843-866.
Goldsmith, W., and S. A. Finnegan (1985). Normal and Oblique Impact of Cylindro-conical and Cylindrical Projectiles on Metallic Plates. U. S. Naval Weapons Center, China Lake, Calif. NWC TP 6479.
Kelly, J. M., and T. R. Wilshaw (1968). *Proc. R. Soc. A*, 306, 435-477.
Landkof, B., and W. Goldsmith (1985). *Int. J. Solids Struct.* In press.
Levy, N., and W. Goldsmith (1984a). Normal impact and perforation of thin plates by hemispherically-tipped projectiles. I: Analytical considerations. *Int. J. Impact Eng.* 2, 209-239, 1984.
Levy, N., and W. Goldsmith (1984b). Normal impact and perforation of thin plates by hemispherically-tipped projectiles. II: Experiments. *Int. J. Impact Eng.* 2, 299-324, 1984.
Liss, J., and W. Goldsmith (1984). *Int. J. Impact Eng.*, 2, 37-64.
Liss, J., W. Goldsmith and J. M. Kelly (1983). *Int. J. Impact Eng.*, 1, 321-341.
Ravid, M., and S. R. Bodner (1983). *Int. J. Eng. Sci.*, 21, 577-591.
Recht, R. F. (1978). *Int. J. Eng. Sci.*, 16, 809-827.
Woodward, R. L. (1982). *Int. J. Mech. Sci.*, 24, 73-87.
Woodward, R. L., and M. E. DeMorton (1976). *Int. J. Mech. Sci.*, 18, 119-127.
Zukas, J. A., T. Nicholas, H. F. Swift, L. B. Greszczuk, and D. R. Curran (1982). *Impact Dynamics*. J. Wiley & Sons, New York.

IMPACT AND PERFORATION OF CYLINDRICAL SHELLS BY BLUNT MISSILES

W. J. STRONGE

University of Cambridge, Department of Engineering Trumpington Street, Cambridge CB2 1PZ, UK

ABSTRACT

Experiments on perforation of thin-walled, mild steel tubes by non-deforming spherical missiles that impact at normal obliquity are compared with previous studies of thin-plate perforation. In these experiments, the missile radius R_m is larger than the tube wall thickness h but substantially smaller than the tube radius R_t. For both plates and shells, impact by blunt-nosed missiles results in two simple modes of deformation near the impact point - plugging and dishing. Spherical missiles penetrating thin shells primarily dish the surface; the initial fracture leading to perforation is caused by tensile radial stress around the contact region. This brittle fracture propagates from the distal surface through part of the thickness before linking with a ductile fracture that completes the plug formation process. The brittle fracture is first apparent on the distal surface at a substantially smaller impact velocity than is required for perforation of the tube wall. Following plug formation, the hole enlargement proceeds by petalling or plastic expansion until the missile passes through the tube wall. Both fracture and the ballistic limit of steel tubes are related to the relative size of the spherical missile and its kinetic energy at impact.

NOTATION

c_t	material parameter ($c_t^2 = Y/\rho$)
h	plate or shell thickness
v_f	velocity for first fracture
v_0	impact velocity
v_x	ballistic limit
D_f	diameter of first observable fractures
D_m	missile diameter
D_t	cylindrical shell diameter
K_f	missile kinetic energy at first fracture

K_x missile kinetic energy at ballistic limit

M_m missile mass

M_t projected plug mass ($M_t = \rho_t \pi h R_m^2$)

R_m missile radius

R_n missile nose radius

R_t cylindrical shell radius

Y yield stress in uniaxial tension (350-500 N/mm² for mild steel)

γ shear strain

ρ density of plate or tube material

τ shear stress

θ temperature

INTRODUCTION

One of the most challenging problems in impact mechanics is relating
deformation and fracture to the minimum speed required for a missile
to perforate a structure. When a plate or shell is hit by a high-
speed missile the area around the impact point is deformed; the
radius of this deformed region may be several missile diameters when
the plate is thin. In general, deformation increases with speed up
to a critical speed where perforation is complete; the deformation
decreases at higher speeds. For a particular structure and missile,
the critical impact speed where 50% of the missiles impacting at a
normal angle of obliquity just pass through the structure is known
as the *ballistic limit*[1].

Blunt missiles penetrate and finally perforate plates and shells by
driving material ahead of the missile in the direction of the impact
velocity. Above the ballistic limit, a plug of material with a
diameter somewhat smaller than the projectile diameter is sheared
from the structure by the penetrating missile. In contrast, sharp
missiles pierce the structure and radially expand the surrounding
material.

If a blunt, non-deformable missile impacts at a speed less then the
ballistic limit, most of the kinetic energy of the missile is dissipated in plastic shear, flexure and stretching within the surrounding
structure. Near the ballistic limit, strains become sufficiently
large to initiate fractures; these also dissipate energy. Above
the ballistic limit, fractures form soon after impact at the periphery of the developing plug. These limit the impulse that may be
transmitted to the adjacent structure through the narrow band of
highly sheared material surrounding the plug. For impact speeds in
this range, the acceleration and plastic deformation of the plug and
shear in the narrow band around the plug absorb most of the energy
lost by the missile during perforation.

Analyses of ballistic perforation have focussed on metal plates -
primarily plates hit at a normal angle of obliquity - and they have
almost all considered only a single mode of structural deformation

---

[1]The review of plate perforation investigations by Backman and Goldsmith (1978) identifies this definition as the Navy ballistic limit.
The Army ballistic limit is a slightly slower speed where 50% of the
missiles will have a terminal nose penetration of one plate thickness

(Yuan and co-workers, 1983;   Liss, Goldsmith and Kelly, 1983;
Lethaby and Skidmore, 1974).   Although some of these models do consider structural deformation outside the plug, they are only appropriate for limited ranges of plate thickness and missile nose shape.
They have missed the rich interplay between the various possible
modes of deformation which complicates plate response to perforation
near the ballistic limit.   Only a few recent analyses have examined
transitions between modes.   These separate into thin and thick plate
analyses;   in a thin plate the transverse displacements are assumed
to be uniform through the thickness at any time whereas, in a thick
plate the analysis considers propagation of a disturbance through
the thickness.   Woodward (1984) has identified at least five modes
of failure for thick plates and made a rough estimate of the range
of geometric parameters wherein each may be active.   A dynamic thin
plate analysis that incorporates shear, flexure and stretching deformations with a critical shear strain fracture criterion was
developed by Shadbolt, Corran and Ruiz (1983).   This analysis shows
a transition in the principal deformation mode for energy dissipation
as the ratio of plate thickness to projectile diameter is increased.
This transition is a feature of experimental perforations by flat-
nosed cylindrical missiles at the ballistic limit of mild steel,
aluminium alloy and stainless steel plates (Corran, Shadbolt and
Ruiz, 1983;   Zaid and Travis, 1974).

Cylindrical shell structures are initially stiffer under concentrated
transverse forces than equal thickness plates.   A few experiments
on high-speed perforations of thin shells by conical-nosed missiles
have been analysed by Johnson, Reid and Ghosh (1981).   In comparison
with static tests on similar shells, the dynamically perforated
structure had deformation more localised around the hole.   The edge
of the hole was expanded by plastic hole enlargement whereas in
static perforations this usually occurred by petalling (Johnson and
co-workers, 1980).   Measurements of perforation speed were not made
in these tests.

The present investigation aims to relate the ballistic limit for thin
shells to the modes of deformation and subsequent fracture in these
structures.   Damage to metal shells from spherical missiles that
impact at speeds near the ballistic limit has been examined.   The
critical speeds for perforation have been measured within small
ranges of missile diameter and tube thickness;   these speeds are
compared with both data and theory for the ballistic limit of plates.
As the shell thickness is increased, these shell and plate perforations show similar transitions in the specific ballistic limit
energy.   These transitions indicate a change in the principal mode
of deformation for energy dissipation preceding perforation.

EXPERIMENTAL OBSERVATIONS

Materials

Hardened steel spheres were fired from a smooth bore cartridge gun
and impinged against the convex surface of a mild steel tube at a
normal angle of obliquity.   The impact velocity, measured by two
pairs of photocells at the muzzle, was varied between $v_0 = 50$ m/s

and 200 m/s.   At the lower speed, the sphere merely indented the
tube, whereas at $v_0 = 200$ m/s, the sphere perforated and passed

through the near tube wall - subsequently, it hit the far wall.

Three different sphere diameters were used in these tests, D_m = 6.35, 9.53 and 12.7 mm; the masses were 1.04, 3.50 and 8.32 g respectively. There was no observable deformation of the missile during any of these impacts.

The impact occurred midway between the ends of the 400 mm lengths of 51 mm diameter mild steel, seamless tubing. The ends of the tubes were clamped in blocks that constrained transverse and angular displacements rather than axial displacements. Standard gauge tubes with wall thicknesses h = 1.23, 2.1 and 3.0 mm were tested in the as-received condition. This was drawn tubing; hence, some properties such as fracture toughness are anisotropic.

Measured Deformations

Both a cross-section and an axial-section of 2.1 mm thick tubes that have been hit by 12.7 mm diameter spheres are shown in Fig. 1 for a range of impact speeds that span the ballistic limit. Photographs of tube sections are shown for impact speeds where (a) there is only indentation, (b) the first fracture is visible, (c) the sphere first passes through the wall, and (d) the impact velocity is larger than the ballistic limit. The deformation generally increases with impact speed up to the ballistic limit v_x, where the sphere first passes through the cylinder wall. The deformation increases most in magnitude and to a small degree in extent. Above the ballistic limit the deformation diminishes as the missile transit time through the wall decreases. Calder and Goldsmith (1971) have reported similar effects for thin mild steel and aluminium alloy plates. The portion of the impact energy that is expended in plastic deformation of the surrounding plate or cylinder is a maximum at the ballistic limit. For tubes impacted by a sphere, the curvature of the deformed surface is everywhere positive towards the cylinder axis except for the small contact region under the sphere.

Transverse final displacement of the surface surrounding the impact point of plates has been termed "dishing". Dishing of the surface has been measured in mild steel (Calder and Goldsmith, 1971; Lethaby and Skidmore, 1979; Zaid and Travis, 1974) and aluminium alloy plates (Goldsmith and Finnegan, 1971; Goldsmith, Liu and Chulay, 1965). In all of these investigations, fractures only occur after normal displacements of the plate under the missile are of the order of several plate thicknesses when the thickness h < 0.5 D_m. Impact of spherical missiles on cylindrical shells also results in dishing but it is neither as large nor as extensive as that in plates because the shell wall is stiffer than a plate of equal thickness within the relevant range of deflection. Figure 2 compares the meridional and axial transverse deformations of a cylinder with the deformations of a mild steel plate of equal thickness. Both the cylinder and the plate were hit by a 12.7 mm diameter steel sphere at the impact velocity for first fracture of the cylinder, v_f = 73 m/s.

Figure 3 shows deformations of cylindrical shell sections at the impact velocity of first fracture for three values of wall thickness. The fracture velocity, v_f increases with wall thickness, h whereas the largest deflection under the sphere decreases. The first fracture locations for these three cylinders are shown on the deformed cross-sections (Fig. 3). The fractures always appear first in the meridional (rather than the axial) direction from the point of impact.

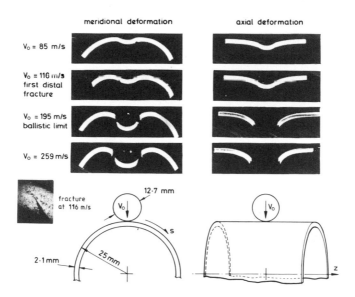

Fig. 1.   Perforation of 51 mm dia. steel tube by
          normal impact of 12.7 mm dia. sphere.

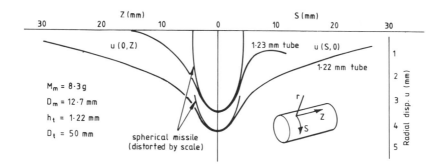

Fig. 2.   Comparison of steel plate and tube defor-
          mation from impact of non-perforating
          sphere, v_0 = 73 m/s.

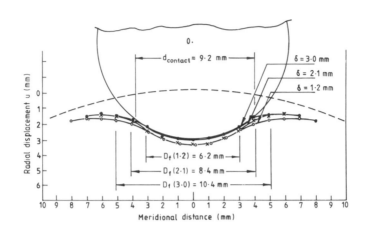

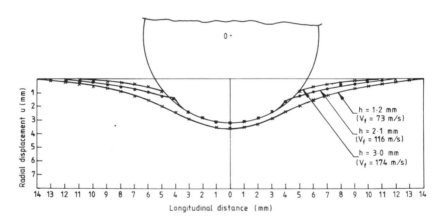

Fig. 3.   Deformation of 51 mm dia. steel tube from
          sphere impact at fracture velocity.   (a)
          Axial section   (b)   Cross section.

Plug Formation and Discing

Blunt missiles that perforate thin plates or shells push a 'cap' of
plate material ahead of the missile that is called a plug.   When
flat-nosed, cylindrical missiles impact flat plates, plug formation
(i.e. separation by fracture from the surrounding plate) is primarily
the result of shear deformation at the periphery of the impressed
region.   The process of shearing a cap from the plate is called
plugging.   However, a process called discing can also contribute to
plug formation in thin plates (Woodward, 1979).   Discing is a circumferential fracture near the periphery of the impressed region that results from radial tensile stress associated with dishing of the plate.

During impact of flat-nosed cylinders on plates at normal incidence, the stress field favours plugging. The flat-nosed projectile profile results in a large shear stress gradient at the edge of the contact region. Consequently, the plastic shear strain is highly localised in a band around the projectile when the strain hardening coefficient is small (Bai and Johnson, 1982).[2]

During impact of hemispherical-nosed missiles on plates, shear strain localisation within the early stage of penetration is prevented by the continuing deformation of thin plates. This spreads the contact region and limits the shear strains. If the plate survives the initial stage of acceleration where shear is the dominant force on the surrounding plate, the following stage will involve dishing and the development of large, in-plane tensile forces.

These experiments on cylindrical shells all had $h < 0.25\ D_m$. Visible fractures did not appear until after the displacement of the impact point was larger than the wall thickness. The first apparent fracture in these tubes was always on the distal surface in the meridional direction from the point of impact. This circumferential fracture had a diameter that was slightly less than the smallest diameter of the contact region; however, it was not always continuous around the circumference before plug projection proceeded. The brittle fracture on the distal surface only propagated through part of the thickness before it joined with a ductile fracture that continued to the impact surface. In these tubes, it appears that plug formation initiated as discing and was completed by plugging.

Impact Energy for Fracture and Missile Perforation

The kinetic energies of a spherical missile at the impact speeds for first fracture and perforation of a steel tube with 1.23 - 3.0 mm wall thickness are shown in Fig. 4a. The missile energies at the fracture and ballistic limits are functions of the size of the impacting sphere (see Fig. 4b). A power law which best fits this data for fracture energy, $K_f(J)$ is

$$K_f = M_m v_f^2/2 = 1.7\ h^{2.0}\ D_m^{0.8} \qquad (1)$$

where the tube thickness h (mm) and the projectile diameter D_m (mm) have ranges [1.23 - 3.0 mm] and [6.35 - 12.7 mm], respectively. For a spherical missile, the ballistic limit is significantly larger than the impact speed for fracture. The power law for missile kinetic energy at the ballistic limit $K_x(J)$ is

$$K_x = M_m v_x^2/2 = 1.10\ h^{1.63}\ D_m^{1.48} \qquad (2)$$

---

[2] The strain-hardening coefficient is a function of the plastic strain as a result of thermo-mechanical coupling. This can cause material instabilities at high strain rates (adiabatic shear) which are commonly observed after impact of certain titanium and high-strength steel alloys. Nevertheless, plugging can occur without adiabatic shear.

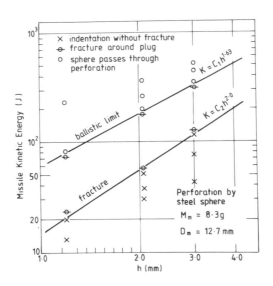

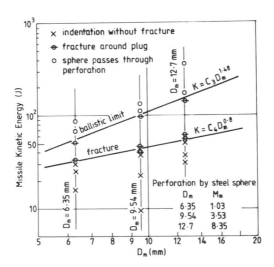

Fig. 4.   Missile energy at fracture and the ballistic
          limit of tubes
          (a)   12.7 mm dia. sphere impact on 1.23 –
                3.0 mm thick tubes
          (b)   6.35 – 12.7 mm dia. sphere impact on
                2.1 mm thick tubes.

In Table 1 the empirical constants for these alternative perforation
criteria are compared with similar equations determined for the ballistic limit of a blunt missile perforating a thin mild steel plate
at a normal angle of obliquity.

TABLE 1 Steel Tube and Plate Perforation by Blunt Missiles

Missile Energy for Failure: $K_x = Ch^\alpha D_m^\beta$ where C is a constant with

units $J/(mm)^{\alpha+\beta}$

Failure Criterion	Missile	C	α	β	Source
tube fracture	sphere	1.7	2.0	0.8	
tube ballistic limit	sphere	1.1	1.63	1.48	
plate ballistic limit	sphere	–	1.43	1.57	de Marre
plate ballistic limit	flat cyl.	1.48	1.5	1.5	BRL
plate ballistic limit	flat cyl.	3.1	1.5	1.5	Ohte
plate ballistic limit	flat cyl.*	42.5	2.0	–	Corran

*Flat cylinders with D_m = 12.7 mm and M_m = 34 g impacting at normal
   obliquity.

Standard empirical equations for the ballistic limit of steel plate
have been listed in Table 1. Early work by Jacob de Marre fit an
equation for the kinetic energy at the ballistic limit to data obtained by nineteenth century French ballisticians. de Marre recognised plate thickness, missile diameter and kinetic energy as the
only significant parameters for determining the ballistic limit of
plates perforated by non-deforming spherical missiles. A more recent
equation by the US Army Ballistic Research Laboratory (BRL) has a
similar form (Russell, 1962). Experiments by Ohte and co-workers
(1981) have confirmed applicability of this form of equation for
thin plates (h = 0.08 D_m) at a large scale. These experiments showed
no effects of shape for a conical nose until the half-angle between
the axis and the surface was less than 62°. Corran, Shadbolt and
Ruiz (1983) measured a stronger dependence on plate thickness in the
range h < 0.35 D_m than the previous authors.

The comparison of tube and plate perforation energies in Table 1
involve two nose shapes that have slightly different failure modes.
Whereas flat cylinders perforate by plugging, hemispherical-nosed
missiles partially fail by discing before plugging. Corran and co-
workers (1983) investigated the transition from plugging to discing
failure in 1.3 mm thick mild steel plates. Their results for a
range of nose shapes are shown in Fig. 5. For a thickness-to-
diameter ratio h/D_m = 0.10, discing dominated the energy dissipation
for R_m/R_n > 0.75 (i.e. an almost hemispherical nose).

A missile with a hemispherical or shallow conical nose, perforating
a thin plate from a speed slightly above the ballistic limit, will

form a plug by first discing and then plugging. The plug is smaller than the missile diameter. The deformed tube profiles in Fig. 3 show the diameter of the initial distal fracture which forms a disc; this decreases with plate thickness.

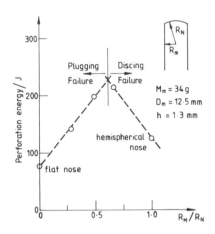

Fig. 5.   Effect of nose radius on missile energy at the ballistic limit (after Corran, Shadbolt and Ruiz).

DISCUSSION

Plugging and discing are the principal modes of fracture for thin plates perforated at normal obliquity by blunt missiles. Early analyses of perforation focussed on plugging and ignored all deformations that took place outside the narrow band of highly sheared plate material at the edge of the contact region. When the elastic and plastic deformations of the surrounding plate and the plug are neglected, the missile kinetic energy at the ballistic limit can be equated to the work done in shearing the plug from the plate. If the uniaxial yield stress is a constant Y and if the fracture is assumed to occur at a displacement h of the impact surface, the energy required for plugging is[3]

---

[3]A sophisticated model of plugging by Bai and Johnson (1982) for flat-nosed missiles analyses the shear mode of deformation during penetration of plates. Using a constitutive equation for shear stress $\tau = \tau_*(1 + b\theta)\gamma^n$ where γ is the shear strain and θ is temperature, they have determined a critical shear strain γ_* for thermoplastic shear instability. This model results in

$$K_f \approx h^2 D_m, \qquad h < 0.3 D_m$$

$$K_f \approx h D_m^2, \qquad h > 0.5 D_m$$

for material properties characteristic of mild steel. In thicker plate, the rate of increase in energy dissipation with plate thickness is limited by a thermoplastic shear instability. For Titanium and Aluminium alloys, the transition is predicted at substantially smaller plate thicknesses.

$$K_f = M_m v_f^2/2 = (\pi/2\sqrt{3})\ Y\ h^2\ D_m. \tag{3}$$

In contrast, the energy dissipation associated with failure by disc-ing depends on the magnitude and extent of dishing. This energy has been calculated only for particular cases; nevertheless, it is a function of the same terms that affect hole enlargement in plates by tapered missiles;

$$K_f = B\ Y\ h\ D_m^2 \tag{4}$$

where B is a constant (Brown, 1964).

Consequently, plugging or discing failure modes for thin plates can be distinguished by the rate of increase in energy dissipation with plate thickness during perforation. A *specific ballistic limit energy* is defined as the ratio of the missile kinetic energy at the ballistic limit to the energy dissipation per unit strain of the projected plug.[4] In Fig. 6, specific ballistic limit energies calculated from data of several investigations are compared with two empirical and two theoretical equations for plate perforation. Most of the data shown on this Figure are for flat-nosed cylindrical missiles perforating flat plates. The specific ballistic limit energy for other nose shapes is noted. In comparison with flat-nosed missile perforation, a sharp conical nose will pierce and enlarge a hole through a plate and this requires less energy for hole formation; conversely, a hemispherical nose will dish the plate and require somewhat more energy for perforation (as was shown in Fig. 5). The data for spheres perforating cylindrical shells are shown as open circles for a range of the missile diameter-to-cylinder diameter ratio $0.12 < D_m/D_t < 0.25$. The specific ballistic limit energy for perforation of cylindrical shells is not appreciably different from that for plates although the plastic deformation and the initial modes of fracture are different.

The data on Fig. 6 have been compared with the energy requirements for plugging, discing and two empirical equations for steel plate perforation:

(i)   Equation (3) for plugging

$$M_m v_x^2/M_t c_t^2 = 2.30\ (h/D_m) \tag{5}$$

(ii)   BRL equation (defined as first breach of the distal surface)

$$M_m v_x^2/M_t c_t^2 = 9.42\ (h/D_m)^{0.5} \tag{6}$$

(iii)   Ballistic limit for flat-nosed cylinders $(h/D_m < 0.12)$ by Ohte and co-workers.

$$M_m v_x^2/M_t c_t^2 = 15.8\ (h/D_m)^{0.5} \tag{7}$$

---

[4]The projected plug mass M_t, is equal to $\pi \rho_t h R_m^2$ where ρ_t is the density of the shell or plate material. The ratio of yield stress to density is a material property defined as $c_t^2 = Y/\rho_t$.

(iv)   Equation (4) for discing or a thermoplastic shear insta-
        bility

$$M_m v_x^2 / M_t c_t^2 = \text{fcn. indep. of thickness} \qquad (8)$$

Equation (5) considerably underestimates the ballistic limit energy,
even for flat-nosed missiles. The theory of Bai and Johnson (1982)
also underestimates the energy requirement since it includes only
shear deformation; furthermore, the transition from fully developed
to thermoplastic shear predicted by this theory occurs at a larger
thickness than is indicated by the data in Fig. 6. In this figure,
each set of data shows an increase in the specific ballistic limit
energy for h < 0.25 D_m and no further increase above this thickness.

This indicates a change in the mode of failure which may be the onset of either thermoplastic shear or discing. Plugs collected from
tubes with h > 0.23 D_m show a decrease in the per cent of shear

fracture surface with increasing plate thickness. However adiabatic
shear bands were not observed in these mild steel plugs formed by
impact near the ballistic limit speed.

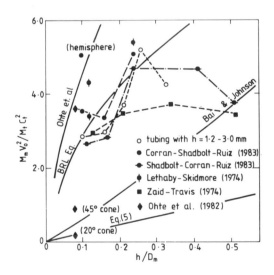

Fig. 6.   Specific ballistic limit energy for plates
           and shells compared with theories and
           empirical equations.

CONCLUSIONS

The perforation of thin, mild steel plates and shells by large
hemispherical-nosed missiles (h < 0.2 D_m) is initiated by discing
rather than plugging. The initial fracture is on the distal side of
the plate and has a diameter that is smaller than the missile diameter; consequently, plug formation is succeeded by hole enlargement prior to perforation by the missile. In comparison with a

flat-nosed missile of equal mass and diameter, a hemispherical nose
requires slightly more energy for perforation.

Flat-nosed missiles shear a plug from the plate or shell. When the
plate is thin and the missile is large, there appears to be significant energy dissipated in both shearing and dishing during plug
formation. Dishing decreases in importance with increasing plate
thickness but it continues to be a measurable source of energy
dissipation throughout the range h < 0.45 D_m.

Both hemisperical and flat-nosed missiles which are not so large
(h > 0.23 D_m) perforate thin mild steel plates and shells by plugging.
There is a change in the energy requirement for perforation at this
thickness that appears similar to the onset of adiabatic shearing;
yet, no adiabatic shear bands have been observed from impact near
the ballistic limit. This change is caused by the rapidly diminishing
importance of surrounding plate deformations after a critical shear
strain is reached through most of the thickness around the developing
plug.

ACKNOWLEDGEMENT

The experiments on tube perforation were performed by Ma Xiaoqing,
a visiting scholar from Beijing Institute of Technology. The author
is grateful for his patient assistance.

REFERENCES

Backman, M. E. and W. Goldsmith (1978). The mechanics of penetration
    of projectiles into targets. *Int. J. Eng. Sci.*, 7, 1-99.
Bai, Y. L. and W. Johnson (1982). Plugging: physical understanding
    and energy absorption. *Metals Tech.*, 9, 182-190.
Brown, A. (1964). A quasi-dynamic theory of containment. *Int. J.
    Mech. Sci.*, 6, 257-262.
Calder, C. A. and W. Goldsmith (1971). Plastic deformation and per-
    foration of thin plates resulting from projectile impact. *Int. J.
    Solids Structures*, 7, 863-881.
Corran, R. S. J., P. J. Shadbolt and C. Ruiz (1983). Impact loading
    of plates - an experimental investigation. *Int. J. Impact Engr.*,
    1, 3-22.
Goldsmith, W. and S. Finnegan (1971). Penetration and perforation
    processes in metal targets at and above ballistic velocities.
    Int. J. Mech. Sci., 13, 843-866.
Goldsmith, W., T. W. Lui and S. Chulay (1965). Plate impact and
    perforation by projectiles. *Exp. Mech.*, 5, 385-404.
Johnson, W., S. K. Ghosh, A. G. Mamalis, T. Y. Reddy and S. R. Reid
    (1980). The quasi-static piercing of cylindrical tubes or shells.
    Int. J. Mech. Sci., 22, 9-20.
Johnson, W., S. R. Reid and S. K. Ghosh (1981). Piercing of cylin-
    drical tubes. *J. Press. Vessel Tech.*, 103, 255-260.
Lethaby, J. W. and I. C. Skidmore (1974). The deformation and
    plugging of thin plates by projectile impact. In *Mechanical
    Properties at High Rates of Strain*, Inst. of Physics, London,
    pp. 429-441.
Liss, J., W. Goldsmith and J. M. Kelly (1983). A phenomenological
    penetration model of plates. *Int. J. Impact Engr.*, 1, 321-342.
Ohte, S., H. Yoshizawa, N. Chiba and S. Shida (1982). Impact strength
    of steel plates struck by projectiles. *Jap. Soc. Mech. Eng.*, 25,
    1226-1231.

302 W. J. Stronge

Shadbolt, P. J., R. S. J. Corran and C. Ruiz (1983). A comparison
of plate perforation models in the subordnance impact velocity
range. *Int. J. Impact Engr.*, 1, 23-49.
Russell, C. R. (1962). *Reactor Safeguards.* Pergamon, Oxford.
Woodward, R. L. (1979). Penetration behaviour of a high-strength
aluminium alloy. *Metals Tech.*, 6, 106-110.
Woodward, R. L. (1984). The interrelation of failure modes observed
in the perforation models in the subordnance impact velocity range.
Int. J. Impact Engr., 2.
Yuan, W., L. Zhou, X. Ma and W. J. Stronge (1983). Plate perforation
by deformable projectiles - a plastic wave model. *Int. J. Impact
Engr.*, 1, 393-412.
Zaid, A. I. O. and F. W. Travis (1974). A comparison of single and
multi-plate shields subjected to impact by a high speed projectile.
In *Mechanical Properties at High Rates of Strain*, Inst. of Physics,
London, pp. 417-428.

THE SHEAR-CONTROL FRAGMENTATION OF EXPLOSIVELY-LOADED STEEL CYLINDERS

J. PEARSON

Michelson Laboratories, Naval Weapons Center, China Lake, California, USA

ABSTRACT

The shear-control method is a fragmentation control technique which utilizes grid systems formed into the inner surface of the case. This paper briefly reviews the mechanics of the process when it is used with explosively loaded steel cylinders, and describes many of the theoretical and parametric relationships which have been established for this method. The results of high-speed photographic studies are used to illustrate the behavioral features associated with shear-control and random fragmentation processes both at normal and low temperatures.

INTRODUCTION

Since the advent of high explosives and their use in fragmenting munitions there has been a need on the part of ordnance researchers to understand the mechanics of the fragmentation process which occurs in the surrounding metal case when an explosive charge is detonated. The modern era of study for this problem probably dates from the World War II period and the early development of high-speed photography as an experimental tool in the study of the fragmentation process.

The works of Taylor (1963a, 1963b), Gurney (1943), Mott (1947), Pearson and Rinehart (1952), Rinehart and Pearson (1954), Hoggatt and Recht (1968), Banks (1968, 1969), and Al-Hassani and Johnson (1969, 1971), are representative of the many theoretical and experimental studies conducted in this field; studies which started in the early days of World War II and have continued to the present day. From studies such as these has evolved a fairly comprehensive understanding of the normal fragmentation behavior of cased munitions.

The present paper deals primarily with the fragmentation behavior of explosively-loaded steel cylinders when the normal fragmentation process has been overridden by a fragmentation control technique based on the use of families of grids formed or machined into the inner surface of the steel cylinder. This technique, known as the shear-control method of fragmentation, was originally developed for

use in fragmentation munitions (Pearson, 1968, 1978). However, it
has also found use in industrial applications where it is used to
provide controlled cutting and separation actions for explosively-
loaded devices.

This paper reviews the mechanics of the shear-control method of
fragmentation, and compares the controlled failure processes which
occur when this method is used in explosively-loaded steel cylinders
to the random failure processes which occur in similar cylinders
without the control method. Details of the failure processes were
obtained by relating high-speed photographic studies of the cylinder
expansion and fragmentation process to postmortem studies of the
fragments. The parametric aspects of such factors as grid design,
geometry of grid profiles, material properties of the cylinder, and
low temperatures are discussed in terms of the failure processes
produced. Unless otherwise stated, experimental data in this paper
are given for cylinders of SAE 1015 steel with dimensions of 5-inch
O.D. x 4½-inch I.D. x 10 inches long. Each cylinder contained about
9 pounds of Composition C-3 explosive which was detonated using
single-point, end initiation with electric detonator and tetryl
booster. The data for this paper come from a number of separate
studies conducted by the writer at the Naval Weapons Center over a
period of many years.

                    THE NATURE OF SHEAR-CONTROL FRAGMENTATION

The shear-control method of fragmentation derives its name from the
ability of the technique to control both the initiation locations
of shear fractures in the metal case and the orientation of the
planes along which the fractures propagate. This method uses families of mechanical stress raisers in the form of a grid system pro-
cessed into the inner surface of the metal case. The elements of
the grid system control the initiation of shear fractures at the
root of each individual grid element, and the shear fractures then
propagate outward through the cylinder wall along fracture trajectories established by the stress field existing in the metal cylinder
after detonation of the explosive and during the initial phase of
cylinder expansion.

A typical diamond pattern grid, as used in ordnance applications,
is shown in Fig. 1. The grid consists of intersecting families of
left-hand and right-hand spiral grooves. In the use of such a grid
system, the shear trajectories emanate in mutually orthogonal pairs
from the root of each grid element, extending outward through the
wall of the cylinder in the form of logarithmic spirals. The geometrical pattern of these shear trajectories remains the same whether
the metal is in the elastic or fully yielded condition (Van Iterson,
1947). It is along these trajectories that the shear fractures
propagate. Whether the controlled fractures tend to propagate along
both trajectories, or are restricted to only certain trajectories
of a specific orientation is determined by the cross-sectional profile of the grid element. This effect is shown in Fig. 2 for two
basic grid profiles. These profiles are shown as they would appear
in the cross section of the cylinder if the grids were parallel
grooves running the length of the cylinder. For symmetrical V-
grooves (Fig. 2A) fractures tend to propagate equally along both
trajectories, while for nonsymmetrical profiles of the type shown
in Fig. 2B, the fractures tend to propagate along one specific trajectory only. This trajectory, shown as the primary trajectory in
Fig. 2B, always maintains the orientation shown relative to the shape
of the grid profile.

Fig. 1.   Steel cylinder showing diamond pattern grid.

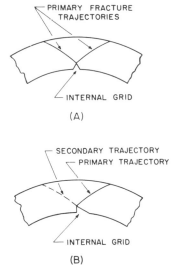

Fig. 2.   Fracture trajectories for (A) symmetrical,
and (B) nonsymmetrical grid profiles.

Figure 3 shows typical fully formed fragment shapes produced by diamond pattern grids of the type shown in Fig. 1, with grid profiles
of the two types shown in Fig. 2. The top two horizontal rows show
the inner surfaces of the fragments, while the bottom two rows show
the outer surfaces. Figure 4 gives the design dimensions for a nonsymmetrical profile, diamond-pattern grid of the type used to produce
fragments of the type shown in the left-hand portion of Fig. 3. A
nonsymmetrical profile grid such as this tends to produce fragments
of one uniform size and shape. As indicated by the small arrows,
each fragment is formed by four shear fractures; two wing fractures
(diverging arrows) and two undercut fractures (converging arrows),
producing a fragment shape as shown. The fragments shown in the
right-hand portion of Fig. 3 were produced with a grid having the
same 3/8-inch grid spacing dimensions, but with a 60-degree symmetrical profile of about the same depth. Each of the diamond pyramidal
shaped fragments was produced by four undercut shear fractures. The
remainder of the cylinder wall appeared as long, thin fragments, as
shown on the extreme right.

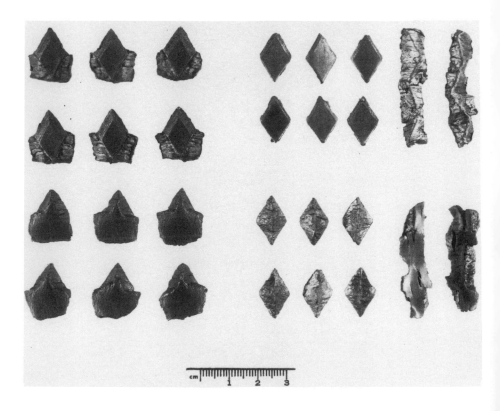

Fig. 3. Types of fragments produced by (left) non-
         symmetrical profile grid, and (right) sym-
         metrical profile grid.

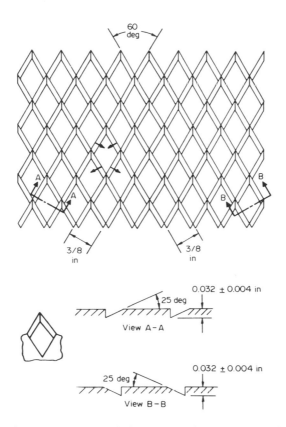

Fig. 4.   Diamond grid design with nonsymmetrical
          profile.

For industrial applications where the control of straight axial line
cuts is frequently required, the grid systems usually consist of one
or more axially oriented, parallel grooves in the inner surface of
the cylindrical part. Figure 5 shows the geometry of fracture activation for families of such grooves using the grid profile geometries
of Fig. 2. Figure 6 shows a cylinder with the grid design of Fig.
5A, and Fig. 7 shows the long, rodlike fragments produced by such
a system. In Fig. 7, the upper row of fragments came from the outer
portion of the cylinder, while the fragments in the bottom row came
from the inner portion of the cylinder, that is, fragment areas shown
as B and A, respectively, in Fig. 5A.

Based on a large number of studies conducted with diamond pattern
grids in cylinders of plain, low-carbon steel, the following fracture
activation results have been noted. Here, a grid element represents
one side of a diamond in the grid pattern. For symmetrical profile
grids approximately 75 percent of all the theoretical trajectories
emanating from the roots of all elements in the grid (two trajectories

per grid element) are activated. That is, the trajectories support
complete shear fracture. For nonsymmetrical profile grids somewhat
over 50 percent of all theoretical trajectories are activated, with
close to 100 percent activation on the primary trajectories and only
minimal activation on the secondary trajectories. For other types
of relatively ductile, but higher strength steels, the same general
pattern holds, but with a tendency for slighly less activation on
the primary trajectories, and somewhat greater activation on the
secondary trajectories.

For a system of axially-oriented, parallel grooves with symmetrical
profiles, essentially 100 percent of all theoretical trajectories
are normally activated (see Fig. 7). For the same type of system
with nonsymmetrical profiles it is normal to have 100 percent activation on the primary trajectories and zero percent activation on
the secondary trajectories.

Another interesting feature of this method is the activation sensitivity of the metal based on the orientation of the nonsymmetrical
grid profile relative to the direction of detonation. Best control
for diamond pattern grids of the type shown in Fig. 4 is exercised
if the explosive is detonated such that the detonation front impinges
against the sloping or shallow side of the grid elements. This orientation maximizes primary activation and minimizes secondary activation

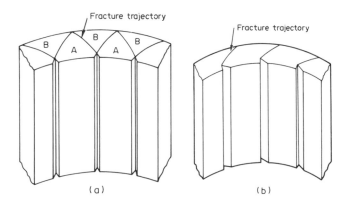

Fig. 5.   Fracture trajectories for parallel grid
          system: (A) with symmetrical profile, and
          (B) with nonsymmetrical profiles.

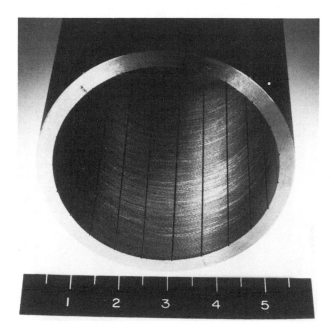

Fig. 6. Cylinder showing parallel grid system.

However, if the detonation front impinges against the steep side of
the grid elements, while primary activation remains about the same,
secondary activation may be increased from essentially zero to as
much as 8 or 9 percent. Table 1 gives fracture activation data for
diamond pattern grids in SAE 1015 cylinders for both geometries of
detonation front orientation.

TABLE 1   Fragmentation Data for Detonation Front Orientation

Explosive	Grid spacing (inch)	Grid orientation	Fracture activation, %	
			Primary trajectory	Secondary trajectory
Comp C-3	3/8	Shallow	100.0	0.7
Comp C-3	3/8	Steep	100.0	8.5
Comp B-3	3/8	Shallow	99.5	4.0
Comp B-3	3/8	Steep	100.0	7.0
Comp C-3	1/4	Shallow	98.0	3.0
Comp C-3	1/4	Steep	95.0	9.0

Two primary requirements in planning the geometry of an effective
grid system are (1) the grid should be designed so that it utilizes
the maximum strain conditions which exist in the expanding case, and

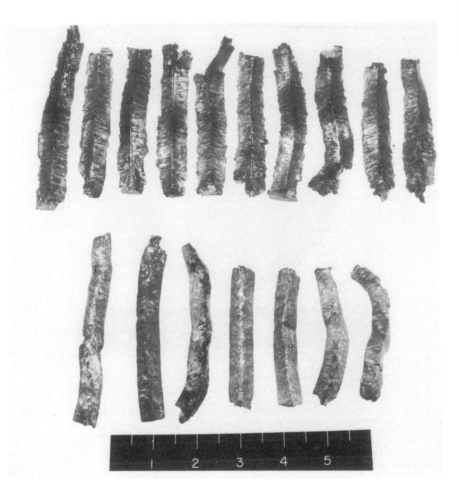

Fig. 7.   Rodlike fragments from parallel grid system
          with symmetrical profiles.

(2) that the grid geometry maintains symmetry with respect to the
strain field. The diamond pattern grid in Fig. 1 meets both of these
requirements. In the expanding cylinder, circumferential strain is
considerably greater than the axial strain. Therefore, the grid of
Fig. 1 has the diamonds elongated in the axial direction to more
effectively use the circumferential strain, while it remains symmetrical with respect to the strain field.

Grids of the type discussed here are normally produced with a depth
of about 0.030 to 0.045 inches. While such depths represent a minimum requirement for nonsymmetrical profile grids, symmetrical profile
grids work well with considerably smaller depths. In fact, the controlled cutting of explosively-loaded cylindrical devices can be
activated by the presence of blemishes such as axially-oriented

scratches or tool marks which have penetrated the inner surface by
only a few thousandths of an inch, or even less in some instances.

The most important factor in the use of the shear-control method is
the behavioral property of the case material. Since this method is
based on the control of shear fracturing, it works best when used
with ductile steels, and is least effective with brittle steels.
Therefore, it is necessary to relate the fracture process to the
ductility/brittleness behavior of the cylinder material under the
extreme loading conditions imposed by explosive detonation.

A good guide to the effectiveness of the method with any specific
steel is the normal fragmentation characteristics of the case material
when no control feature is used. If the cylinder fractures completely
through the wall in shear, or essentially so, then that steel should
work well with this method for a cylinder of that size. If, on the
other hand, the cylinder fractures predominantly in tension, that
steel is generally not suited to this type of control.

Many steels will behave somewhere between the two extremes of completely ductile and completely brittle behavior, with the case fracturing through in some ratio, say 80/20, where the inner 80 percent
of the wall thickness fractures in shear, and the outer 20 percent
fractures in tension. While such a material does not behave as well
as the 100 percent shear type for precision control, it may be
acceptable for many applications. At what ratio the steel would be
deemed "not acceptable" to use with shear-control is a figure which
will vary with the requirements of the application.

The shear-control method has been used effectively with a variety
of plain carbon and alloy steels representing a hardness range of
from about Rockwell "B" 75 to "C" 40, with the ultimate strength
ranging from about 60,000 psi to 200,000 psi. At the higher strength
values some judicious selection of properties and heat treatment may
be required to maintain sufficient dynamic ductility for effective
utilization of the method. Table 2 gives a list of some of the
steels which have been studied for use with this method, and their
relative acceptability.

TABLE 2. Steels Used in Shear-Control Studies

Type	Hardness	Ultimate strength, psi	Suitability
SAE 1015	R_B 75	65,000	Acceptable
SAE 1026	R_B 95	100,000	Acceptable
SAE 1040	R_C 22	110,000	Acceptable
SAE 4142	R_C 22	118,000	Acceptable
AISI 52100	R_C 28	120,000	Acceptable
SAE 4340	R_C 31	155,000	Acceptable
Hy Tuf	R_C 40	190,000	Acceptable
AISI 52100	R_C 46	237,000	Marginal
AISI 52100	R_C 60	310,000	Not acceptable

Table 3 gives typical fracture activation data for several different
steels using the standard 5-inch O.D. cylinder dimensions, diamond
grids of the type shown in Fig. 4 with 3/8-inch grid spacings, and
shallow profile orientation. A general comparison of such results
shows that for the higher strength steels there is a noticeable
tendency to increase activation on the secondary trajectories. It
should be noted that for some of the tests a portion of that increase
is due to the use of Composition B explosive which produces slightly
greater secondary activation than does Composition C-3. This effect
for explosive type can be seen by comparing the two tests for SAE
1015 steel in the table.

TABLE 3   Fragmentation Data for Different Cylinder
                      Materials

| Material | Hardness | Explosive | Fracture activation, % | |
			Primary trajectory	Secondary trajectory
SAE 1015	R_B 75	Comp C-3	100.0	0.7
SAE 1015	R_B 75	Comp B	99.5	4.0
AISI 52100	R_C 28	Comp C-3	97.3	16.0
SAE 4340	R_C 31	Comp B	99.0	10.6
Hy Tuf	R_C 40	Comp B	97.3	9.1

LOW TEMPERATURE EFFECTS

The effect of decreasing temperature on the shear-control process
has been studied for a plain, low-carbon steel (SAE 1015) at temperatures ranging from +80 to -110°F. Comparison studies were conducted with plain wall cylinders, and with cylinders having diamond
grids of the type shown in Fig. 4, with both 3/8-inch and 1/4-inch
grid spacings. The cylinders were loaded with Composition C-3 explosive, and the tests were conducted with loaded test units at
normal temperature (about 80°F), -60 and -110°F. The low test temperatures were obtained using refrigeration and dry ice techniques
(Pearson, 1970).

It is well known that under conventional loads the ability of
ferritic steels to absorb energy is substantially lowered with decreasing temperature (Parker, 1957; Tipper, 1962). This characteristic loss of toughness with decreasing temperature is accompanied
by a change in the deformation patterns from a ductile to a brittle
mode, and the primary fracture type changes from one of the shear
to one of the tensile. This change in behavior is well illustrated
by the temperature-toughness curves published by Rinebolt and Harris
(1951). Interpolation of these curves showed that under conventional
loading SAE 1015 steel would behave in a substantially brittle manner
at both of the low temperatures. The cylinder tests, however, demonstrated that the effect of the explosive loading in this geometry
of system was to override the conventional low-temperature behavior
pattern so that the dominant mode of fracture for all test cylinders
remained one of shear, even at the lowest test temperature. At
normal temperatures the fragments were formed by shear fractures
extending completely through the cylinder wall. At the lower temperatures a tensile skin was active in the outer region of the cylinder wall in which a number of tensile fractures extended inward

from the outer surface. At -60°F the general behavior was about 75/
25;   that is, the inner 75 percent of the wall fractured in shear,
while the outer 25 percent fractured in tension. At -110°F the ratio
of the fracture zone measurement was about 65/35. These fracture
zone patterns were essentially the same for cylinders both with and
without grids for the respective temperature. For the cylinders
with grids, shear-control was the primary mechanism of failure, with
the tensile skin providing only a secondary mode of failure.

Fracture patterns for the plain-wall test cylinders are shown in the
cross-sectional views of Fig. 8. Fracture zone measurements for
these cylinders are given in Table 4. Figure 9 shows the cross section of a recovered fragment from a low temperature (-110°F), plain-
wall cylinder. The ductile and brittle regions, and the associated
shear and tensile fractures, are clearly evident. Fracture activation
data for the shear-control cylinders are presented in Table 5. The
overall consistent behavior of shear-control at all three temperatures
is indicated by the remarkable consistency of the primary and secondary activation data.

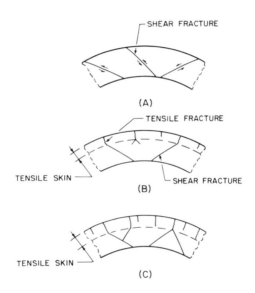

Fig. 8.   Fracture patterns in cylinder walls for
(A) +80°F, (B) -60°F, and (C) -110°F.

TABLE 4   Fracture Zone Measurements at Different Temperatures

Temperature, °F	Depth of tensile skin, inch	Fragment thickness, inch
Normal	None	0.14-0.18
-60	0.02-0.06	0.16-0.18
-110	0.04-0.10	0.19-0.21

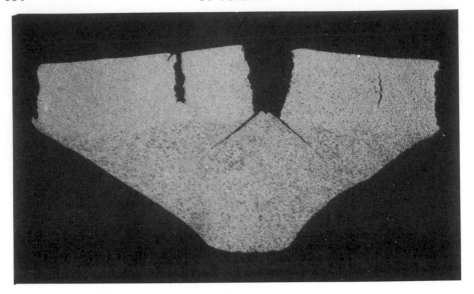

Fig. 9. Cross section of fragment from low temper-
         ature (-110°F) plain-wall cylinder.

TABLE 5   Fracture Activation with Diamond Pattern
          Grids at Different Temperatures

Temperature, °F	Grid spacing, inch	Activation of shear fracture, %	
		Primary trajectory	Secondary trajectory
Normal	3/8	100.0	0.7
-60	3/8	97.7	6.8
-110	3/8	99.3	5.0
Normal	1/4	98.0	3.0
-60	1/4	98.0	2.9
-110	1/4	98.1	2.1

Figure 10 shows grouping of representative fragments for test cylinders with the 3/8-inch spacings in the diamond grid. The fragments
are grouped in three vertical rows each for each different temperature,
with the top two horizontal rows showing the inner surfaces of the
fragments, and the bottom two rows the outer surfaces. The effect
of the low-temperature tensile skin is evident by comparing the outer
surfaces of the fragment groups.

HIGH-SPEED PHOTOGRAPHIC STUDIES: EXPERIMENTAL DETAILS

General

The manner in which a shear-control grid influences the deformation
and fragmentation process of explosively loaded steel cylinders both
at normal (+80°F) and low (-110°F) temperatures has been studied by

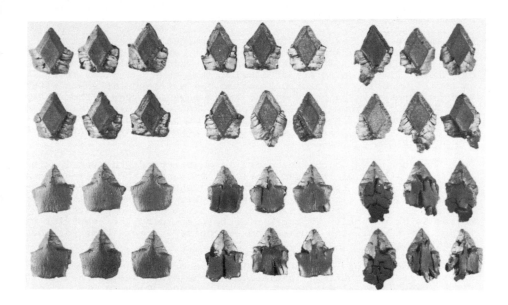

Fig. 10. Control fragments from steel cylinders
          tested at (left) normal, (middle) -60°F,
          and (right) -110°F temperatures.

means of high-speed photography. The studies which will be described
here were all conducted with cylinders of SAE 1015 steel with dimensions of 5-inch O.D. x 4½-inch I.D. x 10-inch long. Each cylinder
was loaded with 9 pounds of Composition C-3 explosive, with detonation
initiated at one end (upper end in the pictures) using a detonator
and tetryl booster. These studies covered the behavior of plain
wall cylinders, and cylinders with both symmetrical and nonsymmetrical
profile, diamond grids. Representative examples of several of these
study areas will be treated later in this paper.

Photographic Techniques

Details of cylinder expansion, initial fracture growth, and the start
of the fragmentation process have been studied by means of a Cordin
high-speed framing camera. For studies of this type the camera was
mounted in a barricade enclosure with the outer walls covered with
armor plate to provide protection against fragments. The camera
viewed the event through a port in the enclosure wall with the use
of a front-surface mirror external to the enclosure. Lighting of
the event was provided by means of two tubular argon flash bombs
located on either side, and slightly forward of the cylinders.
Colored backboards set behind the test round were used to accent
the expansion behavior of the cylinder.

For recording these tests the Cordin camera was operated at 333,000
frames per second. At this framing rate the interframe time between

pictures was 3 microseconds, with the interframe time being measured
from peak to peak illumination between successive pictures. This
camera gives a sequence of 26 frames on 35-mm film.

The exposure time for each frame is determined by the rotational
sweep rate of a high-speed rotating mirror, and the interdiction of
the light rays by a fixed optical shutter which involves two apertures
or gates. If the gates are of equal width, then the light intensity
to the film during the exposure period has a triangular shape with
maximum intensity reached when the light rays from the mirror are
so oriented that they see both gates in the completely overlapped or
open position. The total exposure time for any single frame consists
of the total time during which the light rays see any amount of over-
lap between the gates.

From the manner in which the camera is constructed, the total exposure
time is one-third of the framing time. Thus, for the present studies
the total exposure time for each frame in the sequence was 1 microsecond. However, since the film has a lower limiting level of response
to the light intensity, only about 3/4 of the total exposure time
serves as an effective exposure time for recording the event for each
frame. In the present studies the effective exposure time was about
3/4 microsecond.

Proper recording of the event requires synchronization between the
camera, external shutter, initiation of the explosive charges in the
flash bombs, and initiation of the test round. Only 10 percent of
the rotational time of the camera is actual recording time, the other
90 percent is blank since the mirror sweep does not engage the film.
The blank period is used for starting and stopping the event. The
detonators in the argon flash bombs and the test round are all appro-
priately "pulsed" during a blank period so that full light intensity
is present, and initiation of the main charge just taking place, at
the start of the recording period. After the recording period is
completed, a primacord shutter located outside the camera port is
detonated to cap the event, and thus prevent any double exposure of
the film. The original pictures were all taken in color using Kodak
High-Speed Echtachrome film for greater detail in reading. In
transferring the pictures from color to black and white for publication some of the original quality has been lost.

Viewing Times

At a framing rate of 333,000 frames per second the recording time of
each event was about 75 microseconds. This was an excellent time
period in which to view the overall process from the start of detonation until the fragmentation process was obscured by detonation
products. However, to study the behavior of any specific location
along the cylinder wall there is a shorter "viewing window" which
starts when the passage of the detonation front within the cylinder
starts to radially displace that section of the cylinder, and stops
when the detonation products emanating from fractures in that location
obscure any further detailed viewing of the event. In the present
studies, for a location about one-half way down the cylinder this
time was about 36 microseconds.

HIGH-SPEED PHOTOGRAPHIC STUDIES: CYLINDER BEHAVIOR PATTERNS

General

In the following sections the behavioral patterns of plain wall

cylinders, and cylinders with diamond pattern grids, at both normal
(about +80°F) and low (about -110°F) temperatures are discussed based
on the study of high-speed photographic records. The normal temperature rounds were stored for several days at an air temperature between
76 and 82°F. These rounds were fired within 10 minutes after removal.
The low temperature rounds were chilled using a dry ice and chiller
method while-sample rounds were monitored with multiple thermocouple
probes and recording temperature units. A minimum time of about 8
hours was required to stabilize the temperature of the complete
cylinder-explosive system. All low temperature rounds were fired
within 1½ minutes after removal from the chiller. Even with that
short exposure, and the low humidity of the Mojave desert, light
coverings of frost and, in some instances, patches of glaze ice would
form on the outer surface of the cylinder. However, this did not
constitute much of a viewing problem since much of this covering was
removed as the detonation front swept down the cylinder.

Plain Wall Cylinders at Normal Temperature

Typical behavior for a plain wall cylinder tested at normal temperature
was as follows. As the detonation front passed a given section of
the cylinder, the wall began to move outward. When that section
had increased about 18 to 20 percent on the outside diameter shear
fractures first appeared on the outside surface and then propagated
inward through the wall of the cylinder. Starting as axially-oriented
hairline fractures, with successive frames these fractures increased
in number, and each fracture grew in length and width until it became
obscured by the products emerging from the mouth of the fracture.
Detonation products first appeared in the fracture opening when the
case section had increased about 60 percent on the diameter. From
the first sign of case expansion to the first sign of fracture took
about 9 microseconds, with the detonation products appearing 15 to
18 microseconds after the start of the fracture process. The pictures
of Fig. 11 show a plain wall cylinder at about 33 and 45 microseconds
after initiation of detonation.

Fig. 11.   Behavior of plain wall cylinder (+80°F) at
about 33 and 45 microseconds after detonation.

Table 6 lists fracture dimensions for the early growth of three
prominent shear fractures visible on the outer surface of a plain
wall cylinder. These fractures were located at distances of about
2¼, 5 and 8 inches from the detonator end. Starting as an axially-
oriented hairline crack, each fracture grew with a narrow elliptical
shape which had fairly sharp ends. The table shows the initial lengt
and width of each fracture as it first appeared, and the fracture
dimensions at the time that detonation products began emerging from
the fracture. Additional growth of the fractures, and the subsequent
joining of fractures to form discrete fragments, were largely obscure
by the detonation products.

TABLE 6   Fracture Growth Dimensions for Plain Wall Cylinder

Distance from detonator end (inch)	Initial dimensions		Dimensions when detonation products appeared	
	Length (inch)	Width (inch)	Length (inch)	Max. width (inch)
2 1/4	5/8	Hairline	1	3/16
5	3/8	Hairline	1 1/8	3/16
8	9/16	Hairline	1 3/16	5/32

Plain Wall Cylinders at Low Temperatures

For a plain wall cylinder tested at about -110°F the fracture behavior was influenced by the presence of the outer tensile skin.
The axially-oriented hairline fractures which first appeared on the
outer surface of the cylinder at about 20 percent increase on the
diameter were all tensile breaks. Detonation products began to
emerge from the fracture openings at about 65 percent increase on
the diameter. Corresponding times were again approximately 9 microseconds from the start of metal displacement to first fracture, and
about 18 microseconds additional to the first sign of detonation
products in the fracture.

The surface fractures which initially formed in any given section
of the low temperature cylinder were far greater in number, and
individually less distinct than for the normal temperature case.
While the shear fractures which formed at normal temperature grew
in length as individual, rather easy-to-follow fractures during the
early growth stage, the low-temperature tensile fractures grew as
an overall interacting field through the extensive and hard-to-follow
linking together of many short sections. Meaningful measurements
of individual fracture growth were not obtainable.

Cylinders with Diamond Grids: Normal Temperature

For normal temperature tests, cylinders with diamond grids followed
an unusual and very distinctive behavior pattern, one which was
quite different than that of the plain wall cylinder. This distinctive behavior pattern was most pronounced for nonsymmetrical profile
grids, and to a much less extent for symmetrical profile grids.
This distinctive behavior can be seen in the pictures of Fig. 12
which show a cylinder with a nonsymmetrical profile diamond grid
with 3/8 inch grid spacings at about 33 and 45 microseconds after
the start of detonation. The overall behavior was as follows.

Early in the expansion phase of the cylinder a highly visible, transient diamond pattern appeared on the outside surface which had the
same geometry as the control grid. A large number of symmetrically
arranged, small indentations subsequently appeared in this pattern
from each of which there emanated small, dark, spherical puffs of
detonation products. It is thought that the small indentations in
the transient pattern are localized regions of heavy surface strain
which are associated with the outer terminus of localized deformation
zones (i.e., bands of concentrated shear) which the shear fractures
follow from the roots of the grid element to the outer surface of
the cylinder (Pearson and Finnegan, 1981). The opening of such
fractures at the outer surface of the cylinder was accompanied by
the immediate discharge of detonation products which obscured the
formation of small shear fractures on the surface. Detonation products first appeared when the case section had increased about 45
percent on the diameter, and about 21 microseconds after the start
of case expansion.

Fig. 12.   Behavior of cylinder with nonsymmetrical
           profile, diamond grid (+80°F) at about
           33 and 45 microseconds after detonation.

For a symmetrical profile grid the cylinder behavior combined features
of both the plain wall and nonsymmetrical profile behavior. Early
in the expansion process a transient diamond grid was visible on the
outer surface, but was not as distinct, nor did it last as long as
for the nonsymmetrical profile grid. Also, axially-oriented shear
fractures appeared on the outside surface prior to the appearance
of the detonation products. These short shear fractures tended to
form in a somewhat symmetrical fashion indicating that their locations
were strongly influenced by the inner surface grid. The transient
grid first appeared at about 3 percent increase on the diameter, the
first fractures at about 18 to 20 percent, and the detonation products

appeared at about 45 percent. Some of this behavior is evident in
the pictures of Fig. 13, which show a cylinder of this type with
3/8-inch grid spacings at about 33 and 39 microseconds after the
start of detonation. In this test the detonator wire was accidently
positioned in front of the cylinder, and the pictures also show the
"impact splatter" when the expanding case engages the wire.

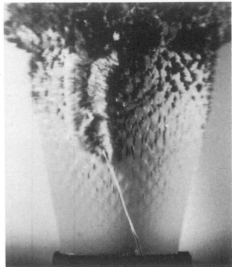

Fig. 13. Behavior of cylinder with symmetrical pro-
          file, diamond grid (+80°F) about 33 and 39
          microseconds after detonation.

Cylinders with Diamond Grids: Low Temperature

The behavioral pattern for test cylinders with a nonsymmetrical profile, diamond pattern grid tested at about -110°F is shown in the
pictures of Fig. 14. These pictures show a cylinder with a 1/4-inch
grid at about 36 and 42 microseconds after the start of detonation.
The transient grid on the outer surface was only briefly evident,
and to a much lesser degree than at normal temperature. For this
cylinder a large number of short, relatively independent fractures
appeared during the early case expansion. The arrangement of these
fractures was largely symmetrical, rather than random; with rows
of somewhat uniformly spaced fractures following the lines of the
transient grid. Even at low temperatures such as this, the presence
of an inner surface grid was influencing the geometry of the fracture
field formed on the outer surface. These fractures appeared at about
17 to 20 percent increase on the diameter, and about 12 microseconds
after the start of cylinder expansion. Detonation products started
emanating from these fractures when the case had increased between
about 43 and 50 percent, and about 9-12 microseconds after first
sign of fracture. This fracture data, as for all the cylinders
discussed, was obtained by following the initiation and growth of
several specific and prominent fractures located at different points

along the cylinder. Also, it should be recognized that all data
presented for these high-speed photographic studies are based on
the 3 microsecond interframe times. No attempt was made to inter-
polate behavior between frames.

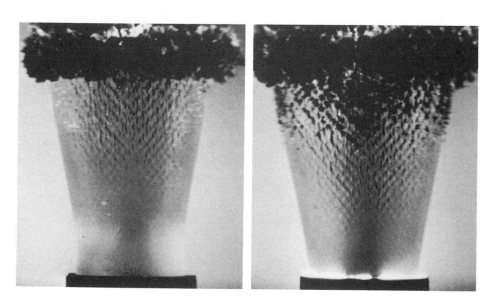

Fig. 14. Behavior of cylinder with nonsymmetrical
          profile, diamond grid (-110°F) at about
          36 and 42 microseconds after detonation.

The pictures of Fig. 15 show a test cylinder with a symmetrical
profile, diamond grid with 3/8-inch spacings, which was fired at
about -110°F. The pictures show the cylinder at about 39 and 45
microseconds after the start of detonation. The transient grid
effect was almost nonexistent, although it could be seen faintly
at several locations in the light frost which had formed on the
cylinder.

A large number of short, axially-oriented fractures appeared with
the expansion of each case section. The fracture field has a slight
symmetry which was easier to see when the detonation products began
to emerge. The fractures appeared at about 15 to 18 percent increase
in the diameter, with the products appearing at about a 40 to 45
percent increase.

ACKNOWLEDGEMENT

The author would like to express his appreciation to the following
personnel of the Naval Weapons Center who have contributed to these
studies: Mr. S. A. Finnegan for metallographic evaluation; Mr.
L. N. Cosner for the ordnance test coordination; Mr. D. J. Fischer
and Mr. R. Gallup for high-speed photography; and Mr. K. L. Auster-
man for photographic processing.

Fig. 15.   Behavior of cylinder with symmetrical
           profile, diamond grid (-110°F) at about
           39 and 45 microseconds after detonation.

REFERENCES

Taylor, G. I. (1963a). Analysis of the explosion of a long cylin-
    drical bomb detonated at one end. In G. K. Batchelor (Ed.), *The
    Scientific Papers of G. I. Taylor*, Vol. 3, Cambridge Univ. Press,
    Cambridge, pp. 277-286.
Taylor, G. I. (1963b). The fragmentation of tubular bombs. In G.
    K. Batchelor (Ed.), *The Scientific Papers of G. I. Taylor*, Vol.
    3, Cambridge Univ. Press, Cambridge, pp. 387-390.
Gurney, R. W. (1943). The initial velocities of fragments from
    bombs, shells and grenades. *BRL Report No. 405.* BRL, Aberdeen,
    MD.
Mott, N. F. (1947). Fragmentation of shell cases. *Proc. R. Soc. A*,
    189, 300-308.
Pearson, J., and Rinehart, J. S. (1952). Deformation and fracturing
    of thick-walled steel cylinders under explosive attack. *J. Appl.
    Phys.*, **23**, 434-441.
Rinehart, J. S., and Pearson, J. (1954). *The Behavior of Metals
    Under Impulsive Loads.* Am. Soc. Metals, Metals Park, Ohio.
Hoggatt, C. R., and Recht, R. F. (1968). Fracture behavior of tubula
    bombs. *J. Appl. Phys.*, **39**, 1856-1862.
Banks, E. E. (1968). The ductility of metals under explosive loading
    conditions. *J. Inst. Metals*, **96**, 375-378.
Banks, E. E. (1969). Fragmentation behavior of thin-walled metal
    cylinders. *J. Appl. Phys.*, **40**, 437-438.
Al-Hassani, S. T. S., and Johnson, W. (1969). The dynamics of the
    fragmentation process for spherical shells containing explosives.
    Int. J. Mech. Sci., **11**, 811-823

Al-Hassani, S. T. S., and Johnson, W. (1971). Dynamics deformation and fragmentation of strain-hardening, strain-rate sensitive shells containing high explosives. In S. A. Tobias and F. Koenigsberger (Ed.), *Advances in Machines Tool Design and Research*, Pergamon, London, 957-979.

Pearson, J. (1968). Parametric studies for fragmentation warheads. *NWC Tech. Pub.* <u>4507</u>, Naval Weapons Center, China Lake, CA (Unclassified).

Pearson, J. (1978). The shear-control method of warhead fragmentation. *Proceedings Fourth International Symposium on Ballistics*, ADPA, Monterey, California.

Van Iterson, F. K. Th. (1947). *Plasticity in Engineering*, Hafner, New York.

Pearson, J. (1970). Low temperature fragmentation of mild steel cylinders. *NWC Tech. Pub.* <u>4877</u>, Naval Weapons Center, China Lake, CA (Unclassified).

Parker, E. R. (1957). *Brittle Behavior of Engineering Structures*. John Wiley and Sons, New York.

Tipper, C. F. (1962). *The Brittle Fracture Story*. Cambridge Univ. Press, Cambridge.

Rinebolt, J. A., and Harris, W. J., Jr. (1951). Effects of alloying elements on notch toughness of pearlitic steels. *Trans. Am. Soc. Metals*, <u>43</u>, 1175-1214.

Pearson, J, and Finnegan, S. A. (1981). A study of the material failure mechanisms in the shear-control process. In M. A. Meyers and L. E. Murr (Ed.), *Shock Waves and High-Strain-Rate Phenomena in Metals*, Plenum, New York, 205-218.

FRACTURING OF EXPLOSIVELY LOADED SOLIDS

S. T. S. AL-HASSANI

Department of Mechanical Engineering, U.M.I.S.T.,
Manchester M60 1QD, UK

ABSTRACT

The formation of cracks and fracture damage in solids of axisymmetric
shapes subjected to surface explosive loading is described and the
controlling mechanisms are considered. High speed photography of
model perspex specimens was employed to observe the initiation and
growth of the internal fissures. The results are used to verify the
predictions of ray theory. The location and the rate of growth of
the cracks are found to depend mainly upon the geometry of the surface of the specimens. Solids of spherical, cylindrical, parab-
oloidal, hyperboloidal and elliptical shapes are considered.

INTRODUCTION

Spalling, scabbing, cracking and fracture damage is often encountered
when geological and engineering structures are subjected to intense
impulsive loading. An understanding of the development of such damage
can aid in the design of structures and missiles to resist failures
and in devising novel methods for demolition purposes.

The effect of subjecting model, brittle engineering structures to
point explosive loads has recently received considerable attention.
The geometries so far investigated include long, thin, straight and
curved rods [1 and 2], spheres [3-8], thick spherical shells [9],
bars with transverse discontinuities [10], tapered bars and plates
[11], cones [12], paraboloids and hyperboloids of revolution [13],
ellipsoids of revolution [14] and hemispherically-ended rods loaded
at their plane end [15] and at their curved end [16]. The latter
paper includes a discussion of the formation of cardioid shaped fractures stemming from the growth of "Hertzian" cone cracks emanating
from the impact zone.

Particularly interesting features in axisymmetric solids are fractures
resulting from the focussing of tensile waves reflected from the
curved boundaries. These features can be understood by reference to
"ray theory", commonly used in optics and acoustics, which enables
the stress wave fronts propagating through the solid to be traced.
The theory for the transmission and reflection of stress discontinuities in a solid of revolution is outlined in refs. [8] and [13].

In this chapter, the relevant concepts of ray theory are used to show
what happens when a spherical pulse is reflected from the curved
boundaries of an axisymmetric solid. In the light of the general
case, and with the aid of high-speed photography, the behaviour of
a stress pulse inside "Perspex" (PMMA) models of various shapes is
examined. The stress pulse is produced by a small explosive charge
on the surface of each model. In view of its practical importance,
the case of a thick-walled cylinder subjected to an axially moving
internal impulsive pressure is also considered.

REFLECTION OF SPHERICAL PULSES WITHIN AN AXISYMMETRIC SOLID

Full details of the theoretical analysis of the behaviour of wavefronts in bounded solids is given in ref. [8]. However, an appreciation of the general concepts of the propagation and reflection of
stress wavefronts in solids may be obtained by considering figure 1.

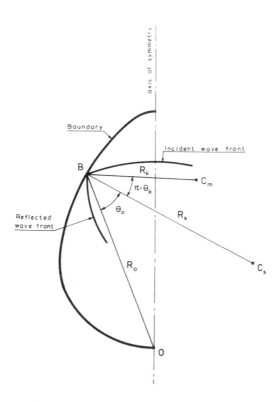

Fig. 1.   (a)   Wavefront geometry. C_s is the centre
          of curvature of the surface at B and C_m is
          the centre of curvature of the reflected
          wavefront.

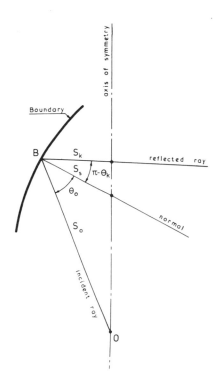

Fig. 1.   (b)   Tangential radii of curvature.

When a homogeneous, isotropic, elastic axisymmetric solid is subjected
to a localised pressure at point 0 on the surface, two spherical wavefronts will emanate from the point of loading;   that is, a dilatational wave (P-wave) and a slower moving shear wave (S-wave).
Only one of the incident waves and only one of its reflected waves
are shown in Fig. 1.   On reflection a P wave will produce a P wave,
P_p, and an S wave, S_p.   An incident S wave will also produce a P wave,
P_s and an S wave, S_s.

Rayleigh surface waves also propagate over the surface.   A detailed
study of their behaviour as they travel over the surface of a sphere
subjected to impact is given by Silva Gomes [17].   If the surface is
flat another type of shear wave will be generated due to the grazing
incidence of the P wave.   It is usually referred to as a "head wave"
or H-wave and is discussed by Pao and Mow [13].

Before meeting the free boundary, the surfaces of stress discontinuity $[\sigma_{ij}]$ travel with constant speeds.   The jump in the stress is
related to the jump in the particle velocity, $[\partial u_i/\partial t]$, through the
momentum equation

S. T. S. Al-Hassani

$$n_j \ [\sigma_{ij}] = - \rho c \ [\frac{\partial u_i}{\partial t}], \qquad (1)$$

where n_j is the jth component of the unit vector normal to the surface of discontinuity, ρ is the density and c the speed of propagation. The jump in particle velocity, V, across the wavefront at any time is related to its initial magnitude, V_0, by

$$V = V_0 \ (\frac{R_0 S_0}{RS})^{\frac{1}{2}} \qquad (2)$$

where R and S are the instantaneous principal radii of curvature of the wavefront and R_0 and S_0 are values corresponding to $V=V_0$.

In a two-dimensional situation, $S = S_0 \to \infty$, and equation (2) gives $V = V_0 \ \sqrt{R_0/R}$. For a spherically symmetric wavefront, as in our case, $R = S$, and we have $V = V_0 R_0/R$. On encountering the free curved boundary of the body, the wavefronts are reflected back into the body, resulting in a change in direction of propagation as well as in the radii of curvature of the wavefront. The angles of the incident and reflected rays are related through Snell's law:

$$\frac{\sin \theta_0}{c_0} = \frac{\sin \theta_k}{c_k} \qquad (3)$$

for k = 1,2. The value k = 1 refers to the reflected dilatational wave, and k = 2 refers to the reflected shear wave. Denoting by $V^{(0)}$, $V^{(1)}$, and $V^{(2)}$ the amplitudes of the velocities associated with the incident wave, reflected dilatational waves and reflected shear waves, respectively, it may be shown that

(i)   for an incident P wave:

$$V^{(1)} = V^{(0)} \ [\frac{c_1^2 \cos^2 2\theta_2 + c_2^2 \sin 2\theta_2 \sin 2\theta_0}{c_1^2 \cos^2 2\theta_2 - c_2^2 \sin 2\theta_2 \sin 2\theta_0}]$$

and $\qquad (4)$

$$V^{(2)} = V^{(0)} \ [\frac{2c_1 c_2 \cos 2\theta_2 \sin 2\theta_0}{c_1^2 \cos^2 2\theta_2 - c_2^2 \sin 2\theta_2 \sin 2\theta_0}];$$

(ii)   for an incident S wave:

$$V^{(1)} = V^{(0)} \ [\frac{2c_1 c_2 \cos 2\theta_0 \sin 2\theta_0}{c_1^2 \cos^2 2\theta_0 - c_2^2 \sin 2\theta_0 \sin 2\theta_1}]$$

and $\qquad (5)$

$$V^{(2)} = V^{(0)} \ [\frac{c_1^2 \cos^2 2\theta_0 + c_2^2 \sin 2\theta_0 \sin 2\theta_1}{c_1^2 \cos^2 2\theta_0 - c_2^2 \sin 2\theta_0 \sin 2\theta_1}].$$

In an axisymmetric situation, the reflected wavefronts are all surfaces of revolution about the axis of symmetry. At the boundary, the radius of meridional curvature, R_k, of the reflected wavefront has been shown to be given by [8]

$$R_k = - \ [\frac{\tan \theta_k}{\tan \theta_0} \ (\frac{\cos \theta_0}{R_0} - \frac{1}{R_s}) + \frac{1}{R_s}]^{-1} \cos \theta_k, \quad k=1,2 \qquad (6)$$

where R_s is the meridional radius of curvature of the boundary at the point of incidence, and R_0 is the radius of curvature of the incident wavefront at the same point. Equation (6) gives the position of the centre of curvature, C_m, of the reflected wavefront at point B on the boundary surface (figure 1(a)). This centre of curvature lies on the reflected ray at an algebraic distance, R_k, measured from B towards the direction of propagation of the reflected wavefront. The centre of tangential curvature of the reflected wavefront lies on the axis of symmetry (figure 1(b)), at a distance S_k from point B given by the equation

$$S_k = - S_2 \left[\cos \theta_k - \sin \theta_k \frac{S_0 \cos(\theta_k - \theta_0) - S_s \cos \theta_k}{S_0 \sin(\theta_k - \theta_0) - S_s \sin \theta_k}\right] \quad (7)$$

(k = 1,2), where S_s and S_0 are the tangential radii of curvature of the boundary surface and incident wavefront at point B, respectively.

In wavefront analysis nomenclature, a caustic is an envelope of a set of converging rays. In general, in a three-dimensional situation, the caustic consists of two distinct sheets and the reflected wavefront has an edge of regression where it touches either sheet of the caustic. When there is axial symmetry, one of the sheets of the caustic degenerates into the axis of symmetry and, instead of an edge of regression, there is a singular point on the axis of the solid. For this case, the only sheet of the caustic is given by equation (6), which defines a surface of revolution about the axis of symmetry. The amplitude of the discontinuity at the wavefront becomes larger and larger as the caustic is approached. This would imply, of course, that the linear theory breaks down before the caustic is reached. It is, however, reasonable to infer that as the reflected rays converge on to the caustic, the amplitude of the stress (whether tensile or shear) increases dramatically such that a fracture may result close to the caustic.

DESCRIPTION OF EXPERIMENTS

The investigations to be reported are mainly conducted on Perspex models machined from commercial grade Perspex acrylic stock. The material properties of Perspex are: E = 3.0 x 10^9 N/m^2, ν = 0.35 and ρ = 1.19x10^3 Kg/m^3, giving dilatational wavespeed c_L = 2011 m/sec and shear wavespeed c_T = 966 m/sec. The impulsive loading is obtained by energising ICI No.6 electrical detonators bonded to the surface of the specimen. The location of internal fractures are identified and the times of their occurrence measured by using ultra-high-speed photography. Typically framing rates of 5 x 10^5 - 10^6 frames per second were used to obtain records of dynamic fractures in spheres, paraboloids, hyperboloids and ellipsoids of revolution.

SPHERES

An experiment on a perspex sphere subjected to a surface explosion at one pole showed that three distinct regions of fracture (F-1,

F-2 and F-3) occur inside the sphere and one (F-4) concentrated on the surface at the antipole, see Fig. 2. Each one appeared at a different time and was associated with a different type of stress wave. Figure 3 shows three frames from a sequence taken at 670,000 frames/sec. (i.e. ≅ 1.5 microseconds per frame). The perspex sphere was 7.37 cm diameter and was subjected to an explosion at the lower pole. The top picture corresponds to 49 microseconds, the middle to 62.5 microseconds and the lower to 85 microseconds after detonation. The times at which the three internal fracture regions F-1, F-2 and F-3, their position in relation to the point of loading and their instantaneous shape, have been shown [8] to agree remarkably well with the predictions of simple ray theory. F-1 is thought to be initiated by the reflected P initiated by the incident P wave, but later reinforced by the reflected S_s wave. Its position matches well with the upper regions of the caustics of the P_p and S_s waves. F-2 was associated with the reflected shear wave generated by the incident P-wave, i.e. S_p, and fracture F-3 with the caustic of the reflected P-wave originating from the incident S-wave, i.e. P_s wave.

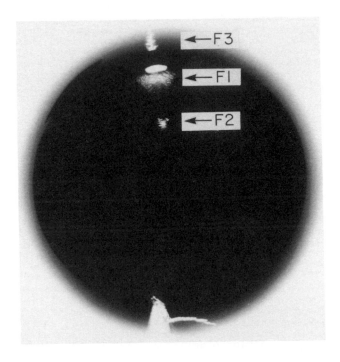

Fig. 2.   (a) Side view of the perspex sphere showing
internal fractures F-1, F-2 and F-3.

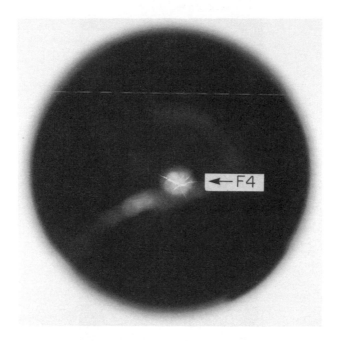

Fig. 2. (b) Plan view looking at the free antipole
showing surface fracture F-4.

The antipodal surface cracking F-4 is thought to be due to focussing
of the Rayleigh waves emanating from the loaded pole. Full details
of the theory of the Rayleigh waves behaviour over the surface of
the sphere in comparison with strain gauge measurements are given
in reference [17].

It is interesting to note that when a segment of the sphere is cut
and the load is applied to the centre of the flat surface of the
remaining hemisphere, the antipodal surface cracks disappear. Numerous photographs of specimens loaded in this way are given in refs.
[4] and [5]. The reason for the disappearance of surface cracks at
the antipole is thought to be mainly due to the effect of the sharp
corners between the flat and curved surfaces, that the Rayleigh waves
have to encounter. A large proportion of their energy will be reflected back towards the loading location.

Surface waves are almost eliminated in specimens having the geometry
of a rod with a hemispherical end. Ref. [11] reported the occurrence
of only fracture types F-1, F-2 and F-3 when such specimens were
loaded at the plane end. However, internal cracks were eliminated
whilst the surface antipodal crack was exaggerated when a flat circular disk was loaded at the edge in the same way. Ref. [3] shows how
a radial crack is generated at a point diametrically opposite the
point of loading. In a plate, the surface area of the flat edge is

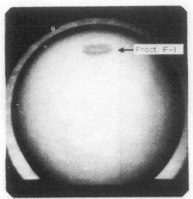

ENLARGEMENT OF FRAME 2

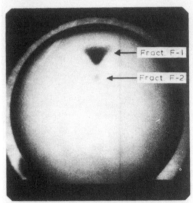

ENLARGEMENT OF FRAME 11

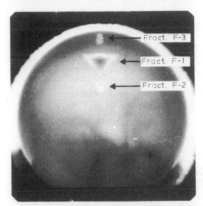

ENLARGEMENT OF FRAME 26

Fig. 3.   High speed photographs of 7.37 cm diameter
          sphere loaded at lower pole. Times are
          (top) 49 μsec (middle) 62.5 μsec and (lower)
          85 μsec after detonation.

too small to produce high amplitude reflected waves. Yet the circular edge allows the Rayleigh waves to travel around and meet again at the antipole.

Especially interesting features were found, ref. [16], in the neighbourhood of the loading point when hemispherically ended rods were loaded at their hemispherical end.   Figure 4 shows a typical set of fracture patterns found in such specimens. Of particular interest is an inward curving truncated cone fracture which develops towards the axis to produce a cardoid surface. This is explained in terms of propagating "Hertzian" cone cracks, normally found in rocks and glasses [23] and [24], which change direction due to reflected stress waves. The distant transverse crazings are thought to be incident, compressive P-waves reflected from the flat surface as a tensile P-wave (or S-waves reflected as S-waves). The next zone of crazing, the conical zone of crazing, is thought to correspond to the location of the caustic of the reflected S-wave derived from the incident P-wave. The other fracture features at the zone of impact are discussed in detail in ref. [16].

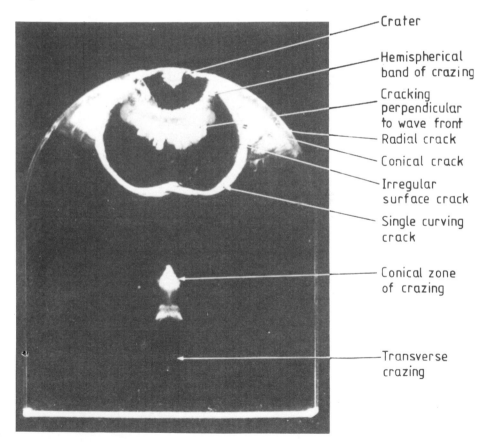

Crater

Hemispherical band of crazing

Cracking perpendicular to wave front

Radial crack

Conical crack

Irregular surface crack

Single curving crack

Conical zone of crazing

Transverse crazing

Fig. 4.   Section of a perspex specimen loaded at the hemispherical end and showing fracture features including cardoid fracture.   Ref. [16].

PARABOLOIDS AND HYPERBOLOIDS

The subject of fracture generation in paraboloids and hyperboloids
by explosive loading at the focus is fully discussed in reference
[13]. The main findings, however, may be summarised by referring
to Figs. 5-8. The manner in which P, S and H wavefronts propagate

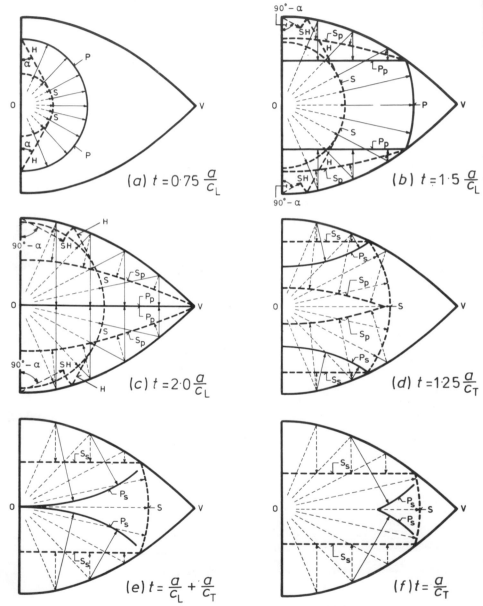

Fig. 5.   Wavefronts in a paraboloid loaded at centre
of base.

and reflect from the free boundary in paraboloids and hyperboloids loaded at the focus O, (i.e. centre of base whose radius is a), is indicated in Figures 5 and 6.   The reflected P_p wavefront in the paraboloid is a right cylindrical surface converging, and intensifying, on to the axis OV.   It arrives at OV at $t = 2a/C_L$ which is equal to 36.3 microseconds for perspex with 2a = 73.7 mm.   In the case of a hyperboloid, however, the reflected P_p wavefront has a curved cylindrical surface which reaches the axis first at the focal point O, the point of loading.   This focal point moves towards the apex along the axis with a finite speed governed by the meridional radius of curvature of the reflected wavefront and the dilatational wave speed of the material.

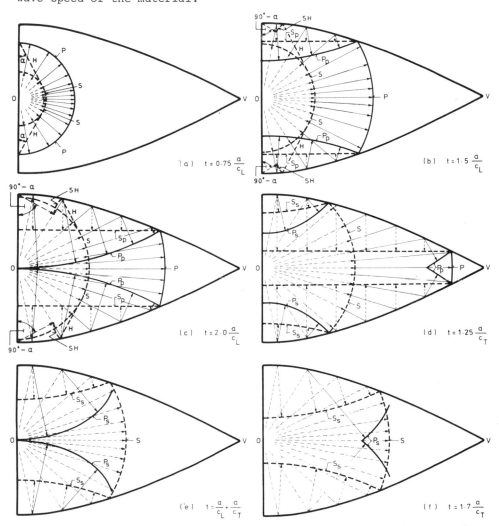

Fig. 6.   Wavefronts in a hyperboloid loaded at
           centre of base.

High-speed photographs for a paraboloid revealed that an axial fracture, F-1, appeared instantaneously at 39 microseconds after detonation. Whilst at the same framing speed, 10^6 frames/sec., the
hyperboloid, see Fig. 7, showed an axial fracture emanating from the
base centre, O, and propagating towards the apex with a high but
finite speed. Very close agreement was found between the instantaneous position of the head of fracture F-1 and the focal point of
the P_p wavefront along the axis.

Another type of fracture having the shape of a disc parallel to the
base was found to occur both in the hyperboloid and paraboloid.
Figure 8 shows this type of disc fracture, F-2, in a paraboloid.

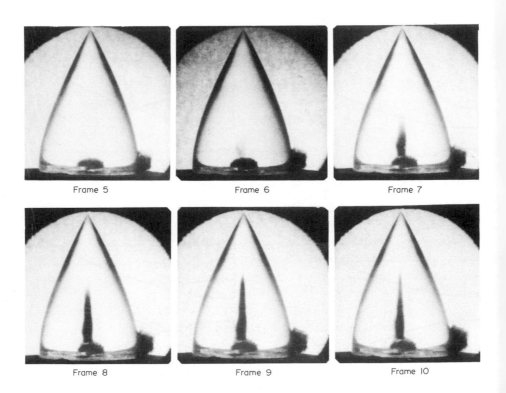

Frame 5   Frame 6   Frame 7

Frame 8   Frame 9   Frame 10

Fig. 7. Selected frames from a sequence of high-speed
photographs taken at 10^6 frames per second of
internal fractures in a 73.7 mm base diameter
Perspex hyperboloid of revolution due to explosion of an ICI 6 detonator at its base.

The high speed photographs revealed that F-2 appeared at t = 68 microseconds. This matches well with the instant of arrival at the axis of symmetry of P_s, the reflected dilatational wave from the incident shear S wavefront. On the other hand, its position corresponds to reflected rays resulting from waves incident at angles greater than 26° to the normal to the boundary. This, according to the ray theory [8], produces large-intensity reflected waves. A P_s ray reflected from a ray incident at 26° reaches the focal point on the axis of symmetry at a time approximately t = 65 microseconds after detonation. Therefore, it seems reasonable to associate this second type of fracture with the reflected P_s wavefront.

Another possible cause for F-2 was proposed in [13] to be associated with the focussing of the reflected head wave SH, see Fig. 5. The fact that SH originates from the H wave which has less curvature than P and S wavefronts, could result in a relatively larger amplitude SH wave. The focus of the SH wave, however, does coincide with the position of the upper disc fracture.

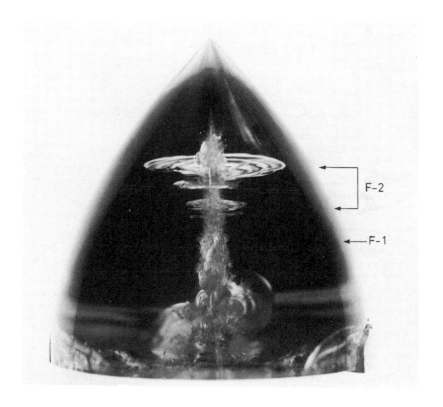

Fig. 8.   Internal fractures in a Perspex paraboloid
          loaded at its base (diameter = 73.7 mm).
          Axial fracture F-1 and disc fractures F-2.

ELLIPSOIDS OF REVOLUTION

For the case of an ellipsoid of revolution impulsively loaded at one
focus, F_1, two spherical stress waves emanate which upon reflection
at the boundary produce a total of four distinct wavefronts as shown
in Fig. 9.   The dilatational wave, P_p, resulting from the incident
P wave, is a spherical wavefront converging on to F_2, which is its
only focal point.   The reflected S-wavefront, S_p, originated by

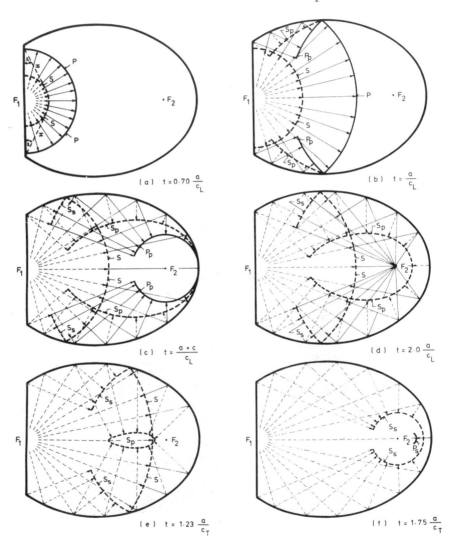

Fig. 9.   Wavefronts in an ellipsoid loaded at focus
F_1.   Major axis = 2a, F_1F_2 = 2c.

the incident P-wave, first reaches a focal point on the axis of symmetry and later a second focal point at the end of the major axis.
These two focal points move towards each other. There is a caustic
$C_s^{(P)}$ associated with the reflected wavefront S_p, Figure 10. It is
seen that each reflected ray touches the caustic on the other side
of the axis of symmetry after having passed through a first focal
point on the axis of symmetry. The S_s-wavefront, is geometrically
similar to the P_p-wave, converging on to F_2. The reflected P_s-wave-
front originates at the far end of the ellipsoid from shear wavefronts
incident at angles below 28.7°. This has a caustic $C_p^{(S)}$ as shown in
Fig. 10. For surface loading, Rayleigh waves as well as head waves
will be generated. These are excluded from the present discussion.

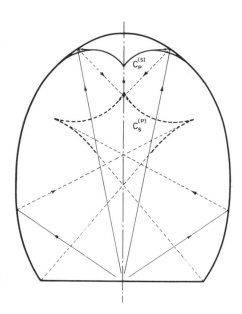

Fig. 10.   Caustic of reflected wavefronts in an
ellipsoid loaded at lower focus.

Experiments are reported in ref. [14] on perspex ellipsoid models
with major semi-axis equal to a = 50.8 mm and eccentricity e = 0.677.
The final observation of an exploded specimen, Fig. 11, shows that,
apart from the fracture in the immediate vicinity of the loaded region, three other distinct regions of internal fractures can be
identified. An upper fracture, F-1, concentrated around the axis
of symmetry, in the region near the focus. The fractured zone
extends to 15 mm. The lower fracture, F-2, is also of a narrow,
elongated type about 12 mm long and located in the central region
of the specimen. The middle fracture, F-3, is very small and appears
just below fracture F-1.

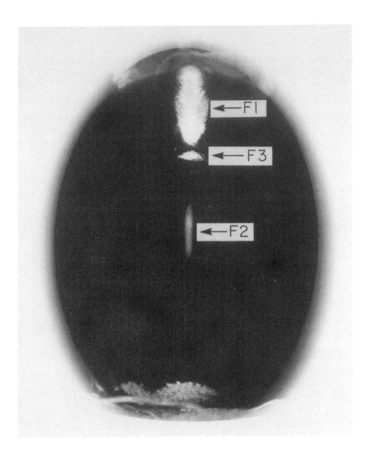

Fig. 11.   Internal fractures in ellipsoid loaded at
           lower focus. (Upper) fracture F-1 (lower)
           F-2 and (middle) F-3.

Figure 12 shows selected photographs taken at a rate of 500,000 frames
per second, (2.0 μsec between frames). The full film shows that the
fracture F-1 initiates in frame No. 17 (50 μsec after detonation)
soon stops growing but develops completely at a later time. Fracture
F-2 initiates in frame No. 22 (60 μsec after detonation) and by frame
No. 24 (64 μsec after detonation) it has developed completely. Then,
by frame No. 28, (72 μsec after detonation), it is observed that fracture F-1 extends further towards the centre of the ellipsoid. Finally,
in frame No. 29 (74 μsec after detonation), the third fracture F-3
occurs.

The formation of the above regions of internal fractures correspond
to the arrival of the different types of stress waves, which were
discussed above.

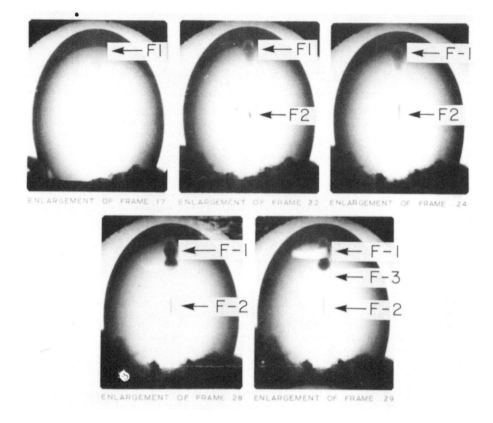

Fig. 12.   Selected photographs taken at 5×10^5 frames/
sec. of internal fractures in an ellipsoid
loaded at lower focus (a = 5 cm, b = 3.7 cm).

Fracture F-1 is thought to have been initiated by the reflected P_p-
wavefront. The period of t = 50 μsec corresponds approximately to
the time that a dilatational wavefront, (c_L = 2011 m/sec), takes to
travel from the centre of loading focus $\overline{F}_1$, to the boundary and then
to the other focus $\overline{F}_2$. If B is any point on the surface of the
boundary, we have the total length $\overline{F}_1 B + B \overline{F}_2$ is a constant equal to
the major axis of the ellipsoid i.e. equal to 2 a = 10 cm, thus
giving t = $2a/c_L$ = 49.7 μsecs.

The lower fracture, F-2 is thought to be associated with the S-wave-front, reflected from the incident spherical P-wave. Its location and time of initiation match well with the focussing of the S_p-wavefront, i.e. the point where it crosses the axis of symmetry. In fact, the period of t = 60 μsec corresponds approximately to the time at which the S_p-wavefront first intersects the axis of the ellipsoid, i.e. t = 56 μsec.

The time at which the second stage of development of fracture F-1 initiates, i.e. t = 72 μsec., suggests that this is due to the arrival of the S_p-wavefront in the region near the focus. As dis-cussed earlier, it has a second singular point running along the axis of symmetry. This singular point passes the focus at time t = 65 μsec. after detonation. Fracture F-3 is thought to be an ex-tension of the second stage of development of fracture F-1 and is thus also associated with the reflected S_p-wavefront.

Perhaps one of the most important applications of ellipsoidal stress wave guides is the development of an electrohydraulic discharge machine for destroying kidney stones. This was recently reported in ref. [22] where stress impulses are initiated at one focus of the ellipsoid and which are caused to be collected at the other focus, being the patient's kidney.

HOLLOW THICK CYLINDER

The case of a thick walled cylinder packed with an explosive charge along its axis is a very common situation in practise. The mechanics of fragmentation of ductile cylinders was fully discussed by the author in ref. [20]. When detonation is initiated at one end, a conical fracture normally occurs at the other end. This is es-pecially noticeable in brittle materials where a conical crack may be seen as sketched in Fig. 13 (a). Such a phenomenon was used by Nash and Cullis [21] to study the dynamic fracture of materials. In order to explain this fracture geometry it is envisaged that a detonation wave runs from the point of initiation end to the free end. Because of the movement of the point of loading along the axis of the cylinder with a speed faster than the P-wave of the material of the cylinder, the wavefront is oblique as shown in Fig. 13(b). In this case it is represented by lines parallel to AB. In a cylindrical situation, the wavefront is a conical surface moving along rays GB. As the P-wave is reflected back from the cylindrical surface it undergoes a change of sign. So that when the $P_p$wave (AF) reflected from the flat base meets with that reflected from the curved surface (BE), a region of high tensile stress is formed and a conical fracture occurs.

A most interesting situation occurs when the explosive is initiated simultaneously at both ends of the cylinder. The cylinder is found to be cut in two equal lengths; the fracture section, af, being perpendicular to the axis, as sketched in Fig. 14(a). This phenom-enon is finding a wide ranging application in the demolition industry and more recently the cutting of entrapped drill strings in the oil production industry. The particular advantage of such a feature is the ability to design a charge which cuts a pipe with no excessive diametral expansion. This reduces the pull-out forces on the portion of the pipe to be recovered.

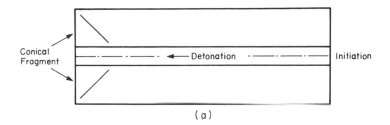

(a)

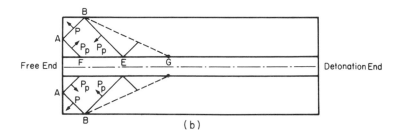

(b)

Fig. 13.   (a)   Conical fragment about to be formed
at the distal end.
(b)   Stress wave fronts just after complete
detonation of the explosive.

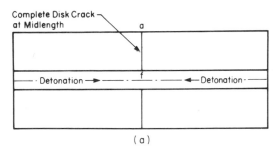

(a)

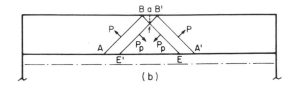

(b)

Fig. 14.   (a)   Disk crack completely separates the
two halves of the cylinder.
(b)   Stress wave fronts after arrival of
the two detonation fronts at the
middle of the cylinder.

The occurrence of the disk crack, af, is thought to be due to the interaction of the two reflected P_p waves BE and B'E' as the P wavefronts AB and A'B' pass each other at the middle of the cylinder, see Fig. 14 (b). The triangular region BB'f at this instant is subject to two tensile waves. The stresses can be resolved into two components, parallel and perpendicular to af. A charge may be designed with a suitable detonation speed such that the obliquity of the wavefronts can be optimised to maximize the tensile stress perpendicular to af and thus cause a brittle fracture along af. As the P_p waves BE and B'E' pass each other the point, f, propagates towards the inner surface of the cylinder until complete fracture is accomplished.

CONCLUSION

A variety of fracture features are observed inside Perspex (PMMA) solids of revolution loaded by an explosive charge. These features are mainly governed by the manner in which body waves interact with each other to form regions of tensile stress of high magnitude which cause failure. The location of the onset of fracture and the extent and direction of the fracture zone were found to be predictable using simple ray theory. Stress wave focusing seems to play a major role in determining the fracture behaviour of brittle solids with convex surfaces which are concave when viewed from inside the solid and which reflect the incident wavefronts back into the body of the solids with different radii of curvature and increasing amplitude. Such phenomena are finding increasing applications in the demolition industry and in the medical field.

REFERENCES

1. Johnson, W., Impact Strength of Materials, Edward Arnold, London (1973).
2. Nasim, M., S. T. S. Al-Hassani and W. Johnson, Stress wave propagation and fracture in thin curved bars, Int. J. Mech. Sci., 13, 599 (1971).
3. Lovell, E., S. T. S. Al-Hassani and W. Johnson, Fracture of spheres and circular discs due to explosive pressure, Int. J. Mech. Sci., 16, 193 (1974).
4. Silva-Gomes, J. F., S. T. S. Al-Hassani and W. Johnson, A note on times to fracture in solid perspex spheres due to point explosive loading, Int. J. Mech. Sci., 18, 543 (1976).
5. Johnson, W. and A. G. Mamalis, Fracture development in solid perspex spheres with short cylindrical projections (Bosses) due to point explosive loading, Int. J. Mech. Sci., 19, 309 (1977).
6. Rumpf, H. and K. Schonert, Die Brucherscheinungen in Kugeln bei elastischen sowie plastischen Verformungen durch Druckbeanspruchung. Drittes Europaisches Symposium Zerkleinern. Cannes (1971). Dechema-Monographien Band 69 (1972) 51.
7. Rumpf, H., Fracture Physics in Comminution, Dritte Internationale Tagnung uber den Bruch, Munchen, PL IX-142 (1973).
8. Silva-Gomes, J. F. and S. T. S. Al-Hassani, Internal fractures in spheres due to stress wave focusing, Int. J. Solids and Structures, 13, 1007-1017 (1977).
9. Stackwi, J. D. and O. H. Burnside, Acrylic plastic spherical shell windows under point impact load, Transactions ASME, Journal of Engineering for Industry B98, 563 (1976).

10. Johnson, W. and A. G. Mamalis, The fracture in some explosively end-loaded bars of plaster of paris and perspex containing transverse holes or changes in section, Int. J. Mech. Sci., 19, 169 (1977).

11. Al-Hassani, S. T. S., W. Johnson and M. Nasim, Fracture of triangular plates due to contact explosive pressure, J. Mech. Eng. Sci., 14, No.3, 173-183 (1972).

12. Kolsky, M. and A. C. Shearman, Investigation of fractures produced by transient stress waves, Research 2, 383 (1972).

13. Al-Hassani, S. T. S. and J. F. Silva-Gomes, Internal fracture paraboloids of revolution due to stress wave focusing, Conf. The Mechanical Properties of Materials at High Rates of Strain, Inst. Physo. 47, 187-196, (1979).

14. Al-Hassani, S. T. S. and J. F. Silva-Gomes, Internal fractures in solids of revolution due to stress wave focusing, shock waves and high strain rate phenomena in metals, Concepts and Applications ed: by Meyers and Murr, Plenum, Ch.10, 169-180 (1981).

15. Johnson, W. and A. G. Mamalis, High Velocity Deformation of Solids, IUTAM Symposium, Tokyo, Japan, ed. K. Kawata and J. Shioiri, 228-246 (1977).

16. Williams, D. J., B. J. Walters and W. Johnson, Crack patterns in cylinders explosively loaded at an hemispherical end, Int. J. Fracture, 23, 271-279 (1983).

17. Silva-Gomes, J. F., PhD. Thesis, University of Manchester, U.K. (1978).

18. Pao, Y. H. and C. C. Mow, Diffraction of elastic waves and dynamic stress concentrations, Adam Hilger, Bristol (1973).

19. Silva-Gomes, J. F., M.Sc. Dissertation, University of Manchester, (UMIST) (1976).

20. Al-Hassani, S. T. S. and W. Johnson, The dynamics of the fragmentation process for spherical shells containing explosives, Int. J. Mech. Sci., 11, 811 (1969).

21. Nash, M. A. and I. G. Cullis, Numerical modelling of fracture, Int. Phys. Conf. Ser. No. 70, 307-314 (1984).

22. Shock Treatment for Stones, News Week, June 25, 1984, p.54.

23. Swain, M. V., and B. R. Lawn, Indentation fracture in brittle rocks and glasses, Int. J. Rock Mech. Mining Science and Geomechanical Abstracts, 13, 311-319 (1976).

24. Knight, M. V., M. V. Swain and M. H. Chaudri, Impact of small steel spheres on glass surfaces, J. Mat. Sci., 12, pp. 1573-1586 (1977).

SUBJECT INDEX